JAMAICA

Primary Mathematics

Second Edition

NSC Edition

Grade 5

Steph King
Josh Lury

hachette LEARNING

The Publishers would like to thank the following for permission to reproduce copyright material.

Photo credits

t = top, *b* = below, *l* = left, *r* = right, *c* = centre, *b/g* = background

p.7: © Nadzin/stock.adobe.com; p.39: © VanderWolf Images/stock.adobe.com; p.72: (tl) © roberaten/stock.adobe.com, (tc) © tapui/stock.adobe.com, (t) © red2000/stock.adobe.com, (tr) © Pavel Parmenov/stock.adobe.com, (bl) © Winai Tepsuttinun/stock.adobe.com, (bc) © Primastock/stock.adobe.com, (br) © MigrenArt/stock.adobe.com; p.80: © Jazmine/stock.adobe.com; p.82: © 5second/stock.adobe.com; p.86: © dimazel/stock.adobe.com; p.89: © Minerva Studio/stock.adobe.com; p.91: © thayra83/stock.adobe.com; p.93: (t) © demarco/stock.adobe.com, (c) © WavebreakMediaMicro/stock.adobe.com; p.96: © edb3_16/stock.adobe.com; p.102: © mavoimages/stock.adobe.com; p.103: © pdesign/stock.adobe.com; p.106: © bakhtiarzein/stock.adobe.com; p.109: © Rafal Olechowski/stock.adobe.com; p.110: © Monkey Business/stock.adobe.com; p.133: © NattapanK/stock.adobe.com; p.145: © Good Studio/stock.adobe.com; p.171: (t) © Gabriel/stock.adobe.com, (c) © speed300/stock.adobe.com; p.191: © Yuriy Kulik/stock.adobe.com; p.201: © Africa Studio/stock.adobe.com; p.204: © psk55/stock.adobe.com; p.211: (c) © Valeri Luzina/stock.adobe.com, (b) © Art.Lantana/stock.adobe.com; p.228: © Ursula Page/stock.adobe.com; p.234: © Samuel B./stock.adobe.com; p.243: © Rido/stock.adobe.com; p.253: © stockshoppe/stock.adobe.com; p.260: © geargodz/stock.adobe.com; p.263: © prostooleh/stock.adobe.com; p.266: © Parilov/stock.adobe.com; p.276: © chungking/stock.adobe.com; p.293: © kolonko/stock.adobe.com; p.295: (c) © Flash concept/stock.adobe.com, (b) © Rawpixel.com/stock.adobe.com; p.301: (tl) © Stockbym/stock.adobe.com, (tr) © sculpies/stock.adobe.com, (c) © Pavelis/stock.adobe.com, (b) © jcg_oida/stock.adobe.com; p.325: © Nicola Katie/peopleimages.com/stock.adobe.com; p.329: © Faizzamal/stock.adobe.com; p.331: © Faizzamal/stock.adobe.com; p.333: © Faizzamal/stock.adobe.com.

To order, please visit www.hachettelearning.com or contact Customer Service at education@hachette.co.uk / +44 (0)1235 827827

ISBN: 9781036019976

First published in 2023

This edition published in 2026 by Hachette Learning (a trading division of Hodder & Stoughton Limited)

An Hachette UK Company

Carmelite House

50 Victoria Embankment

London EC4Y 0DZ

www.hachettelearning.com

Impression number 10 9 8 7 6 5 4 3 2 1

Year 2026

Cover illustration by Oxford Designers and Illustrators Ltd

Illustrations by Magriet Brink, Leo Daly, Amelia Rosato, Andy Cooke, Peter Lubach, Aptara Ltd., Oxford Designers and Illustrators Ltd. and Hyphen S.A.

Typeset by Hyphen S.A.

Printed and Bound in Great Britain by Bell & Bain Ltd, Glasgow

A catalogue record for this title is available from the British Library.

Contents

Term 2

Unit 1: Number

Unit 2: Measurement

Unit 3: Geometry

Unit 4: Algebra

Term 3

Unit 1: Number

TERM 1

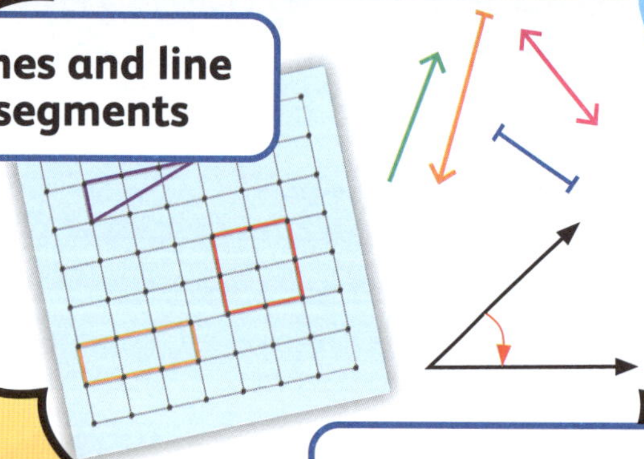

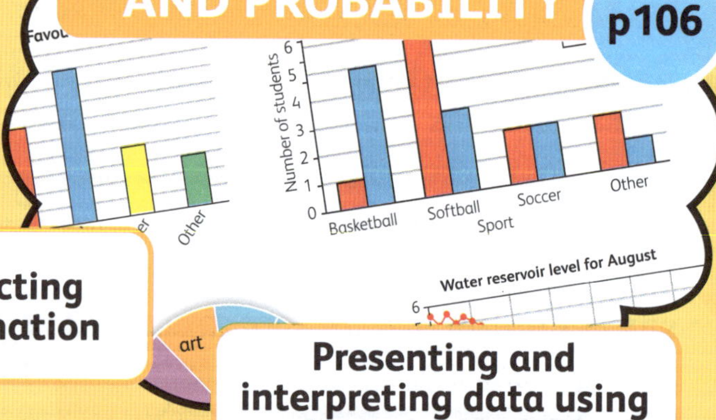

Number of students
Basketball
Softball
Soccer
Other
Sport
Water reservoir level for August
art

Unit 1: Number

Chapter 1 Sets

In this chapter you will explore this focus question:

To answer this question, you will:

- describe a set and differentiate between the set of real numbers
- describe a set as being finite, infinite or an empty set (null set)
- differentiate between sets of counting numbers, whole numbers, odd numbers, even numbers, prime numbers, composite numbers and fractional numbers.

Starting point

1 a Describe the objects in this group. What do they have in common? What is different about each object?

b Discuss with a partner how you could sort these objects into smaller groups. For example, you could sort them by colour. Think of two other ways you could sort them, and which would belong in each group.

2 2 | 9 | 15 | 30 | 12 | 10 | 17 | 27 | 5 | 7 | 22 | 25

Some children are grouping these numbers in different ways.

a Marva says: ‘The numbers that belong in my set are 2, 5, 7 and 9.’ What do the numbers in Marva’s group have in common?

b Amy says: ‘The numbers that belong in my set are 7, 17 and 27.’ Describe Amy’s set. List some other 2-digit numbers that could belong to this group.

c What other groups can you make from the set above? Describe them to a friend.

3 Tania also chooses a group of numbers. Each number can only be divided by 1 and the number itself. Explore Tania’s idea. Which numbers did she choose?

> You can use cubes or bottle tops to help you with dividing different numbers.

What is a set?

Key maths idea

A **set** is a collection or group of items that have a common attribute.

We can also say that the **members** (**elements**, or items) of a set share characteristics.

Example

There are seven members or elements in this set. All the members share the same attribute: they are living things.
We use a capital letter to name a set.
Let's use the letter L to name this set of living things.
We write the elements of the set in curly brackets (called braces), like this:
L = {monkey, frog, fish, owl, plant, person, butterfly}

An **empty set** has no members. It is also called the **null set**.
If set A is the set of living things that do not grow then it is a null set.
We write this null set as set A = { } or ∅

Key words

set
member
element
empty set
null set

1 Look at the objects that have been grouped in a set.

a Write a sentence to describe the set. Think about the attributes (or characteristics) all members of the set share.

b Use curly brackets (braces) to list the members of the set.

c How can you group the objects in the picture in a different way to make another set? Describe your new set and list the members.

2 Tamika groups some numbers in a set.
{0, 2, 4, 6, 8}

a How can you describe this set?

b Use curly brackets and create your own set of numbers. Challenge a friend to describe your set.

3 Look at these sets.

Set A	**Set B**	**Set C**	**Set D**
Days of the week	Five polygons	Continents of the world	Four fractions that are equivalent to one-half

a Name and list the members of each set. Remember to use curly brackets.

b Which sets have the same number of members?

4 Choose objects from the classroom to match each description of a set. Remember to name and list the elements of each set.

a colour is blue

b used for maths lessons

c used for writing

d longer than 15 cm

e have a curved surface

5 Here are some more sets. What attribute do the members of each set share?

V

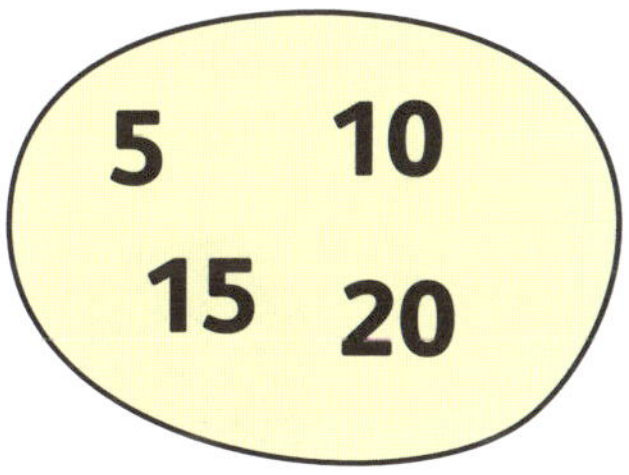

W

X

6 Think back to the objects you chose for the sets in question 4. Can you use some of the objects to create a new set, Z?

a Photograph, sketch or list the elements in your new set Z.

b Ask a friend to describe the set by identifying the common attribute.

Z

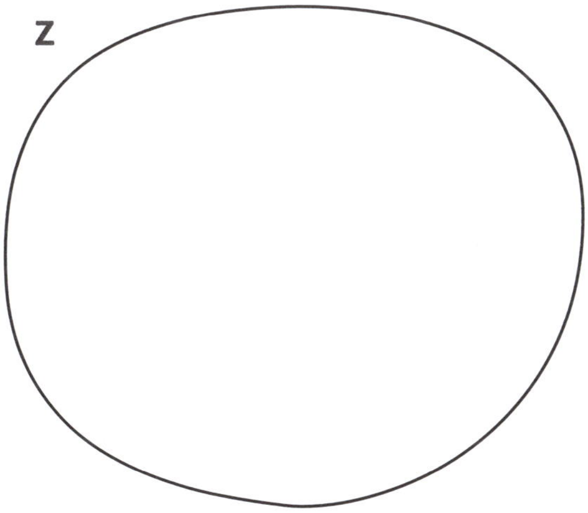

7 *I'm going to make a set of polygons with curved sides.*

Can you explain why this will be a null set?

Finite and infinite sets

Key maths idea

A set that has a limited or countable number of elements is called a **finite set**. For example, if set C is the set of colours in the Jamaican flag, then set C = {black, gold, green}.

This set is a finite set because it is possible to count how many elements or members it has.

When a finite set has a greater number of elements, we can use three dots (…) to show that the set continues in the same way, rather than listing them all.

For example, the set of counting numbers up to 50 = {1, 2, 3, …, 50}.

When a set has an endless number of elements, we call it an **infinite set**.

We can use the three dots to show that the set continues without end.

Examples

The set of **even** numbers = {2, 4, 6, 8, …}.

The set of **odd** numbers = {1, 3, 5, 7, …}.

Key words

finite set
infinite set
even
odd
whole numbers
counting numbers

The three dots are called an ellipsis.

1 Say whether these sets are infinite or finite. If they are finite, list the members:
   - a the set of **whole numbers**
   - b the set of even numbers between 60 and 70
   - c the set of **counting numbers**
   - d the set of 2-digit multiples of 5
   - e the set of objects on your table
   - f the set of stars in the universe
   - g the set of months in the year.

2 Describe each set and say whether it is finite or infinite.
   - a {3, 6, 9, 12, 15, 18, …}
   - b {1, 3, 5, 7, 9, …, 99}
   - c {0, −1, −2, −3, −4, −5, …}
   - d {Asia, Africa, North America, South America, Antarctica, Europe, Australia}

Remember to use an ellipsis to show when a set continues on.

3 Work with a partner to create three finite sets and three infinite sets. Name and list the elements of each set.

Subsets

Key maths idea

Sometimes we sort the elements of a set into further sets. These are **subsets** of the main set. The diagram shows the main set of animals inside the larger circle. Animals with a backbone (vertebrates) make a subset of the main set.

Let's think about a set of numbers.

The main set is all the counting numbers from 1 to 10.

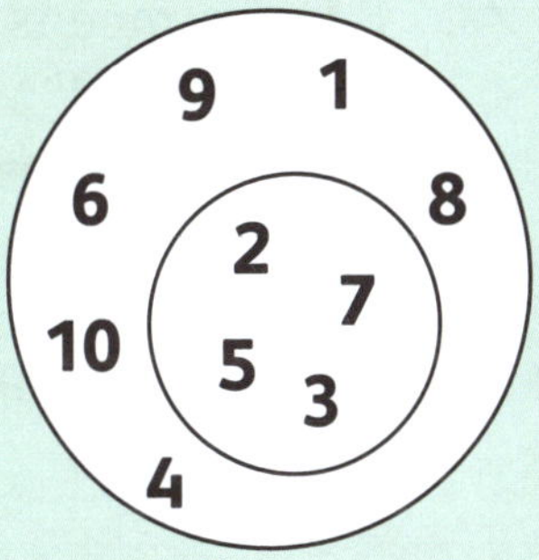

You will learn more about prime numbers in Chapter 3.

The set of **prime numbers** is a subset of the main set, so we can say that {2, 3, 5, 7} is a subset of {1, 2, 3, …, 10} because 2, 3, 5 and 7 are still part of the counting numbers from 1 to 10.

{11, 13, 17} is a set of prime numbers, but it is not a subset of {1, 2, 3, …, 10}.

Can you explain why?

2 is a prime number, because $2 \div 2 = 1$ and $2 \div 1 = 2$. It cannot be divided by any other number without leaving a remainder.

4, 6 and 8 are not prime numbers.

They are **composite numbers**, because they can be divided by more than two numbers.

Key words

subset

prime numbers

composite numbers

Why has an ellipsis been used for the main set? Is this set finite or infinite?

Remember that a prime number can only be divided by itself and 1.
So, only two division sentences can be made.

Can you explain why the number 1 is not a prime number or a composite number?

1 Use the main set of animals on the previous page and list the animals that belong to these subsets:

- **a** can live in water
- **b** can fly
- **c** have no backbone (are invertebrates).
- **d** Choose one of these three subsets and draw it as part of the main set.

2 Look at this diagram.

- **a** Describe and list the members of the main set.
- **b** Describe the subset. List the members as: { } is a subset of { }.

The symbol $\subset$ also means 'is a subset of'.

3 Work with a partner. Collect at least 6 different objects from the classroom for a main set. Some of the objects will need to make up a subset so think about common attributes. Represent your sets. For example, you can use a piece of paper for the main set and a loop of string or draw a hoop to show the subset. Invite other students from your class to describe your sets.

The symbol $\not\subset$ also means 'is not a subset of'.

4 True or false? Copy the statements that are true. Correct the statements that are false by writing 'is not a subset of'.

- **a**  $\subset$ {types of transport}
- **b**  $\subset$ {vertebrates}
- **c** {12, 15, 18} $\subset$ {0, 2, 4, 6, … 20}
- **d** {prime numbers} $\subset$ {composite numbers}
- **e** {multiples of 8} $\subset$ {multiples of 4}

5 Subsets can have 0 or more elements. How many different subsets of set F can you find? List them.

F = {apple, banana, cherry}

You may find it useful to organise your results in a table.

Venn diagrams

Key maths idea

We can use a **Venn diagram** to display different types of sets. The members of each set are chosen by **collecting** or **grouping** items according to their specific characteristics.

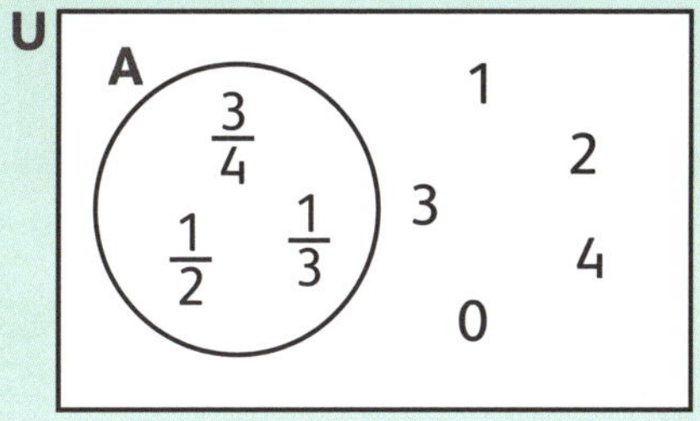

The **universal set** is the set of all the elements we are talking about in a specific case, so all the numbers inside the rectangle make up the universal set. We use the symbol U for the universal set.

The elements in set A are all fractional numbers!

The elements in the hoop make up a set within the universal set.

This Venn diagram has been drawn a little differently. It shows the numbers as two sets, A and B.

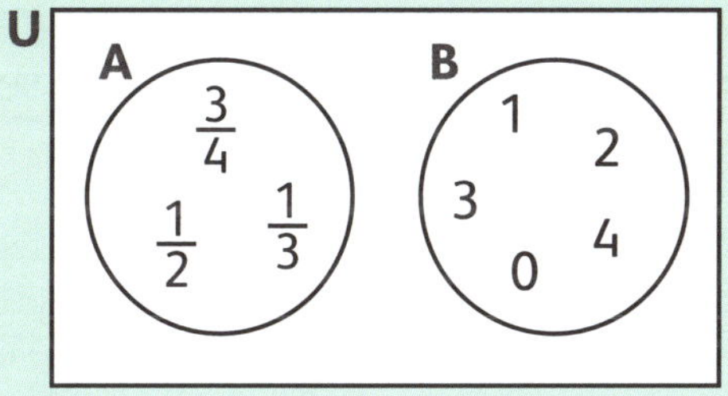

The elements in set B are all whole numbers!

We call these **disjoint sets** because they share no common elements.

Key words

Venn diagram
collecting
grouping
universal set
disjoint sets

1 Draw Venn diagrams to show the following:

	Universal set	Set
a	Days of the week	Weekend days
b	Months of the year	Months with 30 days
c	Numbers from 1 to 10	Composite numbers
d	Prime numbers from 1 to 10	Odd numbers
e	2-digit multiples of 5	Multiples of 10
f	$0, \frac{1}{5}, \frac{1}{4}, \frac{1}{3}, \frac{1}{2}, 1$	Fractional numbers

2 Look at this Venn diagram.

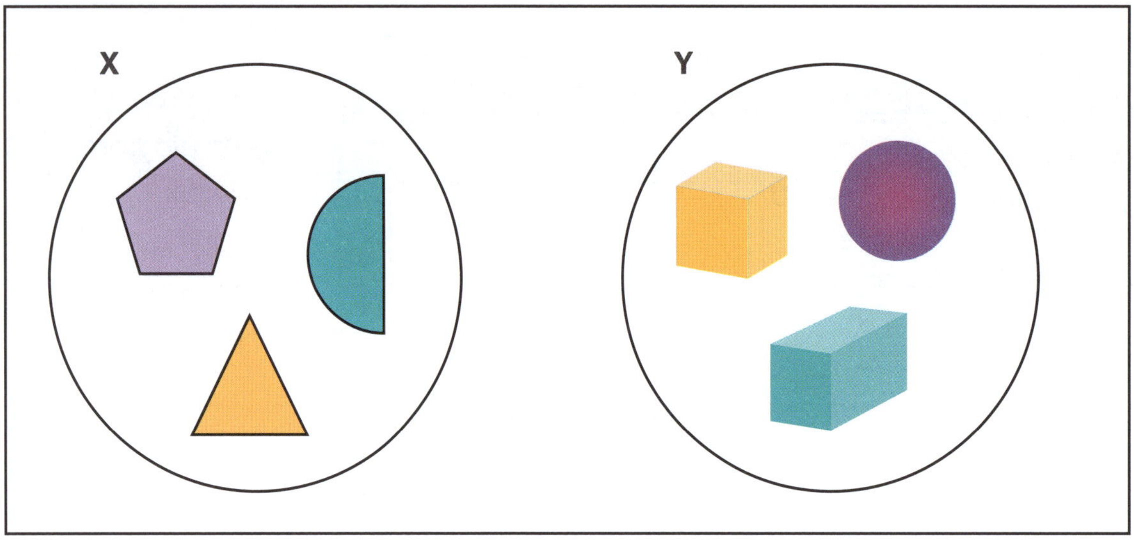

a Explain why sets X and Y are disjoint sets.

b List the members of each set.
X = Y =

c How many members are in the universal set?

d Choose another member of your own for each set.

3 Look at this universal set of living things:

U = {beetle, elephant, bird, crocodile, bee, snail, shark}
Draw a Venn diagram to show these disjoint sets: V, the set of vertebrates and W, the set of invertebrates.

Intersection of sets

Key maths idea

When some members of one set are also members of another, we can overlap the sets and show the **intersection** of the sets.

Example
The members of the intersecting set have both characteristics of set A and set B as they are single-digit numbers and multiples of four. We say these are **jointed sets** because they do share common elements.

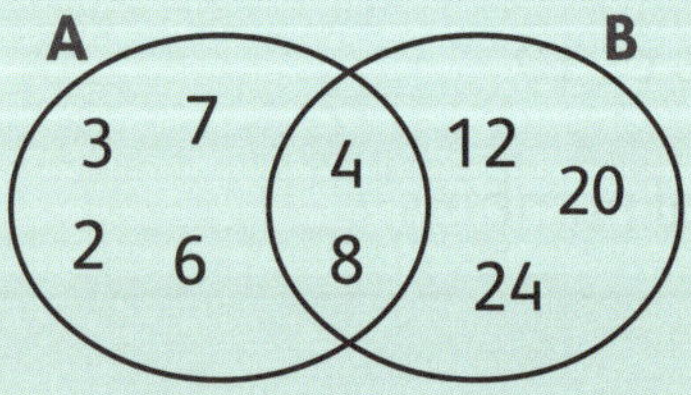

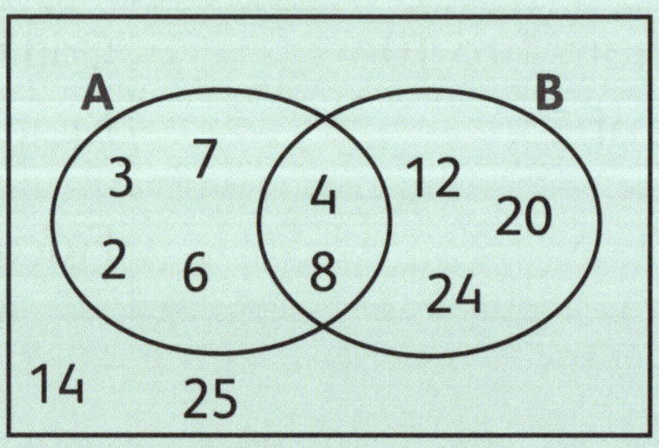

Key words
intersection
jointed sets

Can you explain why 14 and 25 are members of the universal set but not members of set A or set B?

1 Look at this Venn diagram.

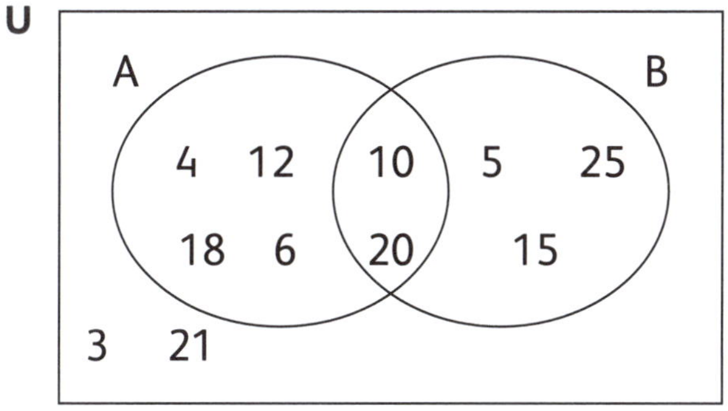

a List the members of set A. Remember to use curly brackets. A =

b List the members of set B. B =

c List the members of the intersecting set.

d Explain why these are members of both sets A and B.

2 Here is another Venn diagram.

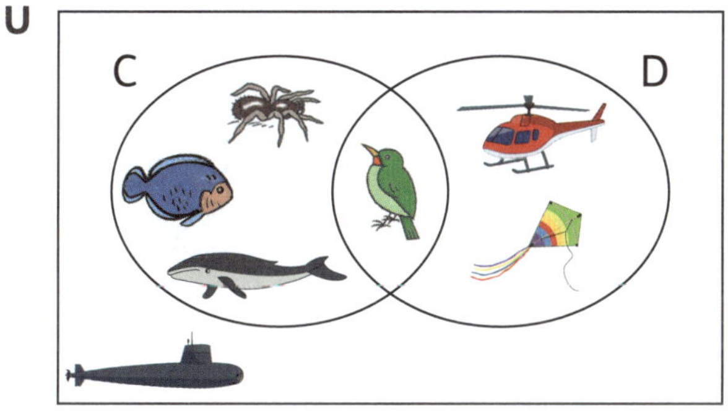

a List the members of the universal set. Remember to use curly brackets. U =

b What are the characteristics of the intersecting set?

c Why is the submarine outside the sets C and D?

3 Dana and Akiel have been thinking about this Venn diagram.
They write E ∩ F = {equilateral triangle}

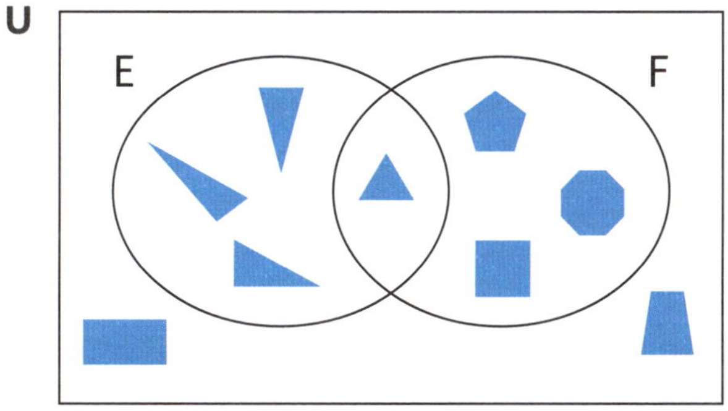

a What do you think the symbol ∩ means?

b What are the two characteristics that the equilateral triangle has?

c Why are the rectangle and the trapezium outside the sets E and F?

4 Work with a partner. Use a set of small objects to create your own Venn diagram to show jointed sets. List the members of each set.

Challenge:

5 Dana and Akiel write down what they notice about another Venn diagram.

U = {months of the year}

A = {months with 31 days}

B = {months that include the letter r in their names}

$A \cap B$ = {January, March, October, December}

Use the given information to draw the Venn diagram.

Real-world maths: Classifying animals

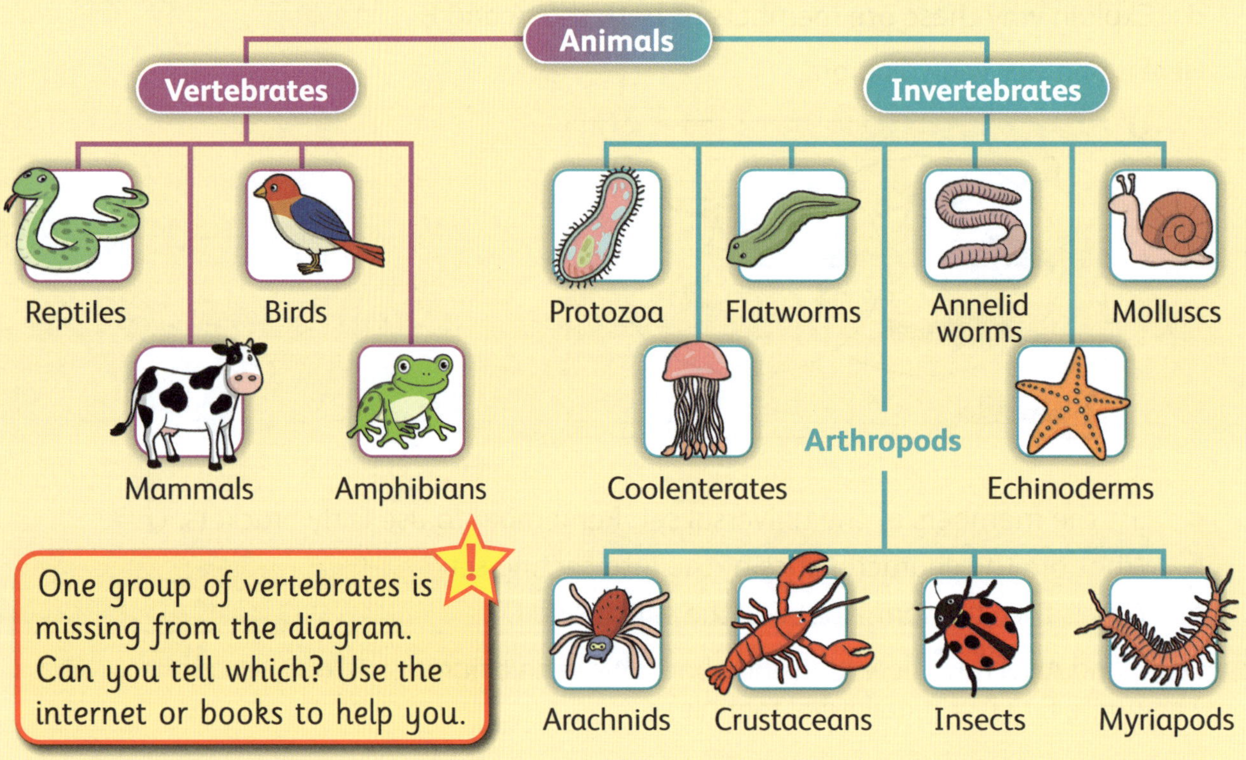

One group of vertebrates is missing from the diagram. Can you tell which? Use the internet or books to help you.

Animals can be grouped together by similar characteristics or attributes.

Discuss the classification of animals with a partner and talk about the following types of sets. You can research ideas in more depth using the internet:

a universal set
b set
c subsets
d disjoint sets
e jointed sets.

Sketch Venn diagrams to help you explain.

Taxonomists use classification for plants and animals to help them classify newly discovered plants or animals.

Find out more about John Venn, the creator of Venn diagrams.

Maths detective: Attributes of sets

Can you spot which members from the previous set need to be deleted each time to suit the attribute of the new set?

Set A = {0, 1, 2, 3, 4, 5, 6 } - whole numbers

Set B = { ____ , ____ , ____ , ____ , ____ , ____ } - counting numbers

Set C = { ____ , ____ , ____ } - even numbers

Set D = { ____ , ____ } - composite numbers

Set E = { ____ } - common factor of 16 and 24

What I have learned

Copy and complete these sentences and definitions.

1 The members or elements of a set all ____ .

2 A set with no members is called a null set or an ____ set.

3 A set that has a limited or countable number of elements is called a ____ set.

4 A set with an endless number of elements is called an ____ set.

5 A subset is ____ .

6 The universal set is ____ .

7 Disjoint sets are sets that ____ .

8 Jointed sets are sets that ____ .

9 The intersecting set is ____ .

Practice questions

1 Use what you know about sets to describe what you see. Remember to think about the types of sets and the characteristics of the members.

a

b

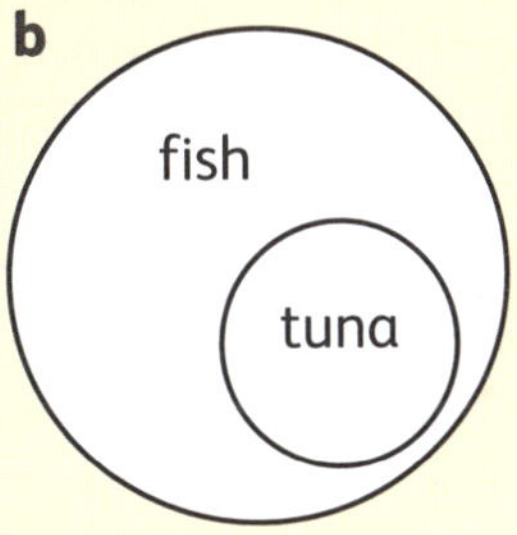

c

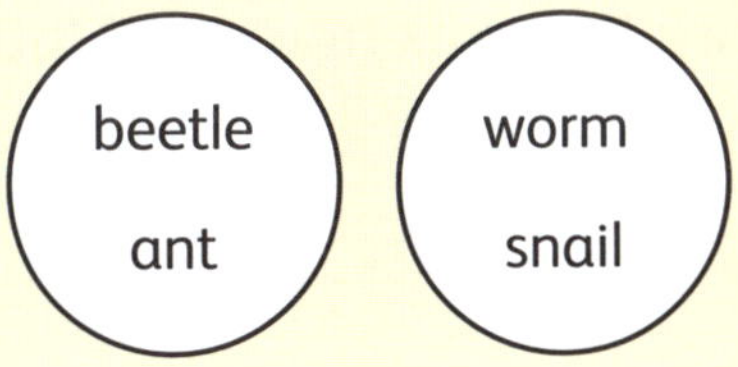

d

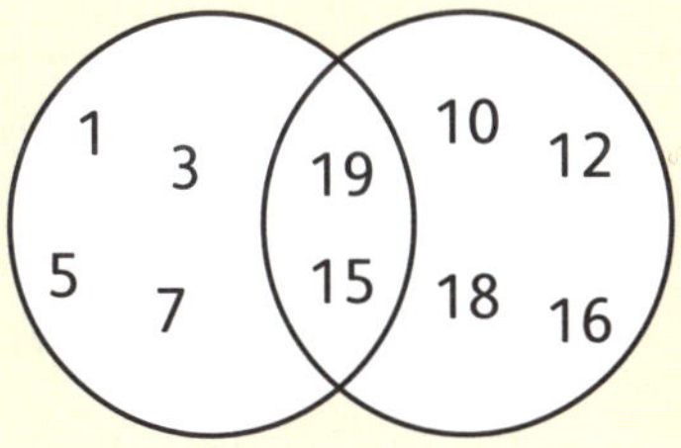

2 Look at this Venn diagram about pets.

a How many members are in the universal set?

b Use the Venn diagram to help write a sentence to describe what you know about:

- Sue
- Terrence
- Harry
- Kristen

c Write the members of each set.

CATS =

DOGS =

CATS ∩ DOGS =

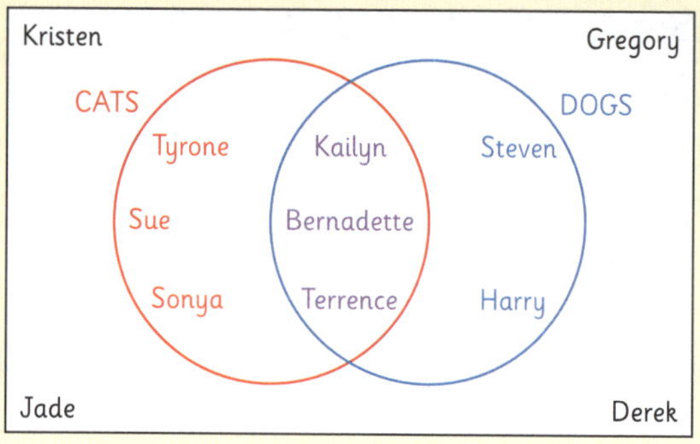

1 Self check

	Check you can do this:	If you have difficulty, try this:
1	Identify and list members of a set, and describe common attributes.	Go back to pages 8–9. Look at some of the sets you described and how you decided what belonged in the set.
2	Create your own sets and describe the common attributes.	Make groups of objects or numbers and say what they have in common.
3	Describe a set as being finite, infinite or a null set.	Go back to pages page 10. Remember that finite numbers can be counted. If something is infinite it goes on forever so you cannot count it.
4	Talk about and represent different types of sets, for example, universal, subset and disjoint sets.	Go back to pages 13–14 and go over the definitions of the different types of sets.
5	Work with Venn diagrams, explain what each region represents and use symbols to describe sets, for example, U for the universal set.	Revise the activities from the section on pages 13–16.
6	Identify and compare sets of counting numbers, whole numbers, odd numbers, even numbers, prime numbers, composite numbers and fractional numbers.	Revise the work from the Maths detective section on page 17.

Chapter 2 Number value

In this chapter you will explore these focus questions:

To answer these questions, you will:

- read and write number names, and numerals using the Hindu–Arabic place value system up to eight digits
- read and write whole numbers up to eight digits
- identify the place values of thousand, hundred, ten, tenths, hundredths or thousandths in numbers.

Starting point

1 Let's think about the number 3 425. Read the number and say it aloud. We write the number in words as 'three thousand, four hundred and twenty-five'. The digits that make up the number have the face values 3, 4, 2 and 5, but it is important to know the **place value** of each digit to get a real sense of the size of the number.

2 Look at the Gattegno chart below. Talk about any **patterns** you see.

1 000 000	2 000 000	3 000 000	4 000 000	5 000 000	6 000 000	7 000 000	8 000 000	9 000 000
100 000	200 000	300 000	400 000	500 000	600 000	700 000	800 000	900 000
10 000	20 000	30 000	40 000	50 000	60 000	70 000	80 000	90 000
1 000	2 000	3 000	4 000	5 000	6 000	7 000	8 000	9 000
100	200	300	400	500	600	700	800	900
10	20	30	40	50	60	70	80	90
1	2	3	4	5	6	7	8	9

3 **a** What place value parts do we need to make the number 3 425?

$3\,425 = \square + \square + \square + \square$

b What is the same and what is different about the place value parts we need to make the number 34 251?

4 Work with a partner. Read each number and write it in words. List the place value parts you need each time.

a 327 **b** 732 **c** 7 132
d 71 329 **e** 571 329 **f** 2 938 571

Key words
place value
patterns

Our number system

Key maths idea

We use the numerals 0 to 9 to write numbers in the Hindu–Arabic number system. The position of each **digit** in a number tells us its place value. Look at this place value chart. It is arranged in columns.

Ten millions	Millions	Hundred thousands	Ten thousands	Thousands	Hundreds	Tens	Ones
					3	5	5

The table shows the number 355. In words we say: 'three hundred and fifty-five'.

The digit 5 in the tens position has a value that is 10 times as large as the digit 5 in the ones position. In fact, the value of a digit in the column immediately to the left is always 10 times as large as the same digit in the column to the right.

Look at the number represented on this place value chart. Can you write it in words?

What can you say about the value of a digit in the column immediately to the right compared to the value of the same digit in the column to the left?

Ten millions	Millions	Hundred thousands	Ten thousands	Thousands	Hundreds	Tens	Ones
			9	9	9	9	9

Talk about

Talk about the value of the digit 9 in each position.
What happens when we count on one more from 99 999?
What happens when we count on one more from 999 999?

Remember that our number system is a base ten system, so we regroup 10 ones to make a ten, then 10 tens to make a hundred, and so on.

Key words
digit
million

1 What number does each group of counters represent? Write each number in words.

a

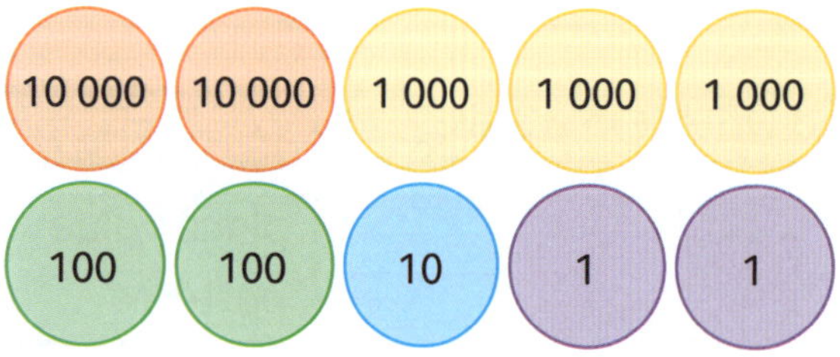

b

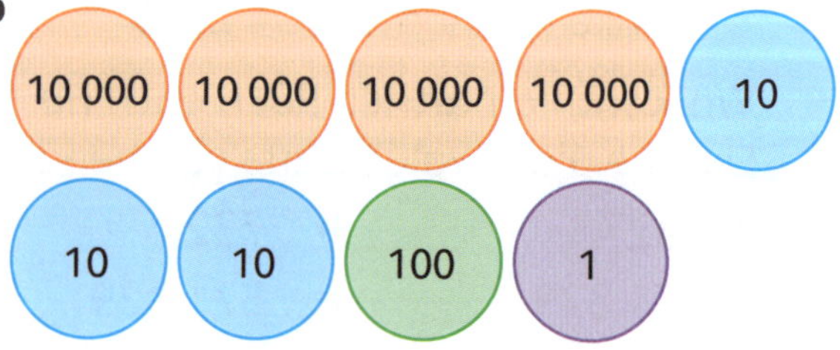

2 Work with a partner. Put a set of digits cards from 0 to 9 face down on the table.

a Pick four cards to make a number. Write the number in numerals and in words. Rearrange the four cards to make another number. Copy and complete this table to show the numbers you make.

My number	My number in words

b Now repeat with 5 cards, and then 6 cards, 7 cards, and 8 cards.

3 **a** Represent each of these numbers on a place value chart. Read the numbers to a partner. Talk about the face value and the place value of each digit.

243 405 | 408 499 | 4 314 212 | 44 513 867

b The digit 4 is repeated in each number. Dana says:

In the number 243 405 the first digit 4 has a value that is 20 times as large as the second digit 4.

Dana has made a mistake. Explain why.

c Now write the value of the digit 4 in each of the numbers. Can you work out how many times as large the first digit 4 is each time?

4 Play this game. You will need a ten-sided die, some counters and a place value grid. The aim is to finish with a number greater than 3 000 but less than 5 000 after eight rolls of the die.

- Roll the die to find how many counters to use to make your 2-digit starting number. For example, if you roll a 5, you could use the five counters to make 5, 14, 50, or 23, etc.
  Put the counters on the place value chart.
- Roll the die a second time to find how many more counters you can use. You may not put counters in an empty column unless it is the result of regrouping. Position your second set of counters and regroup 10 counters if needed. Write down the new number on your place value grid after each turn.
- Continue until you have made eight rolls of the die.

5 When do we use or hear about large numbers such as **millions**? List some ideas. Then work out whether you have been alive for more or less than a million minutes.

Try this millions challenge!

Number value

Key maths idea

Three numbers are represented on the place value chart. What are they?
What is the same and what is different about them?

Ten millions	Millions	Hundred thousands	Ten thousands	Thousands	Hundreds	Tens	Ones
●●	●●●●●		●●●	●●●●	●	●●●	
	●●	●●●●●		●●●	●●●●	●	●●●
		●●	●●●●●		●●●	●●●●	●

Each number contains the digits 2, 5, 0, 3, 4 and 1 but the position of each digit is different.
Say which number each time has a digit with the place value:

- 50 000
- 5 000 000
- 20 000 000
- 200 000
- 3 000

Which number is 100 less than 2 503 514?
Talk about what might cost these amounts in Jamaican dollars?

Read the numbers aloud to help you.

1 What are the values of the underlined digits?

a 45<u>6</u> 782 **b** 4 567 8<u>2</u>5 **c** <u>2</u> 370 304 **d** 2<u>3</u> 703 047

2 Here is a place value chart.

10 000	20 000	30 000	40 0000	50 000	60 000	70 000	80 000	90 000
1 000	2 000	3 000	4 000	5 000	6 000	7 000	8 000	9 000
100	200	300	400	500	600	700	800	900
10	20	30	40	50	60	70	80	90
1	2	3	4	5	6	7	8	9

What number is represented by:

a the rectangular shapes

b the oval shapes?

If we add another row at the top of the table, what will the numbers be? And if we add a row at the bottom of the table? Make up some other numbers and ask a partner to find them.

3 **a** Write the number each abacus represents.

i ______

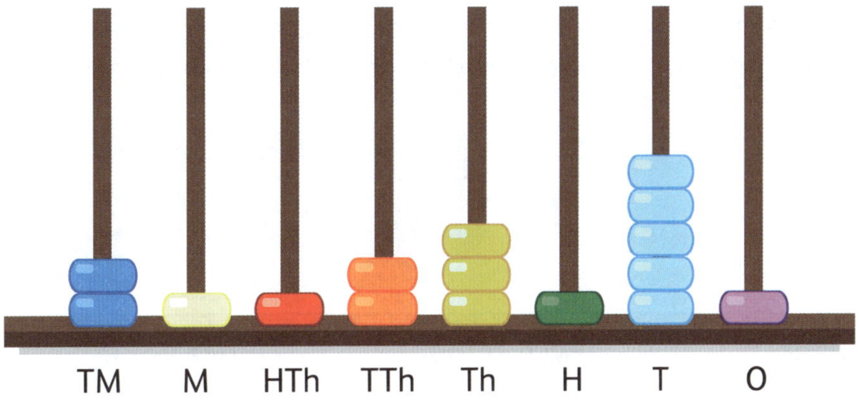

ii ______

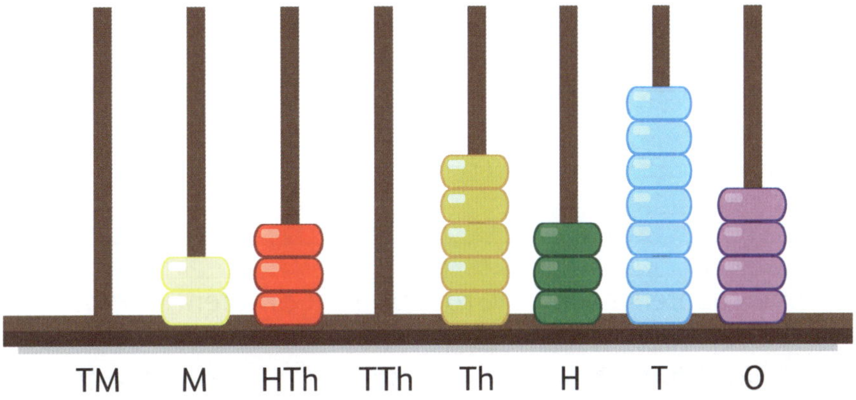

iii ______

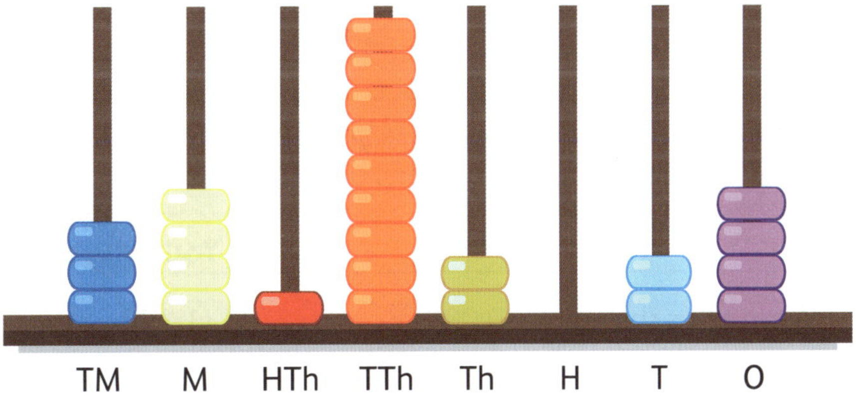

b Increase each number on the abacus by 10. Write the new numbers.

c Now increase each number on the abacus by 10 000. Write the new numbers.

d Decrease each number on the abacus by 100. Write the new numbers.

4 Work in groups of four. You need five dice – four white and one of any other colour. Roll the dice and form a number with the digits shown by the dice. Record the numbers in a table like the one shown here. Circle the digit from the coloured die. Fill in the face value, place value and value of the circled digit of each number. One number has been done for you.

Number	Face value	Place value	Value
1(6)432	6	1 000	6 000

5 Use nine counters each time to represent different numbers on this place value chart.

Ten millions	Millions	Hundred thousands	Ten thousands	Thousands	Hundreds	Tens	Ones
			●●●	●	●	●●	●●

Make a table to record the numbers you make that fit these descriptions. You can use only nine counters. The first one has been done for you.

		Your number
a	A digit has the value 30 000	31 122
b	The largest possible odd number	
c	The smallest possible multiple of 5	
d	A 7-digit number less than 5 million	
e	The largest possible number	
f	The value of the first digit is 100 times as large as the value of the third digit	

6 Malcolm is describing a number. What could it be? Find all possibilities.

> My number has 7 digits.
>
> The digit 8 appears three times in the number.
>
> The value of the second digit in the number is 10 times as large as the value of the third digit.
>
> One of the digits has the value 5 000.
>
> The number is a multiple of 100.
>
> The digit in the millions position is odd.

> There are five possibilities for question 6. How many can you find?

Expanding numbers

Key maths idea

We can **expand** numbers to show their place value parts. The **face value** of the 5 digit is 5. The **place value** of the 5 is millions. When we multiply the face value by the place value, we get the **true value** of this digit, which is 5 million.

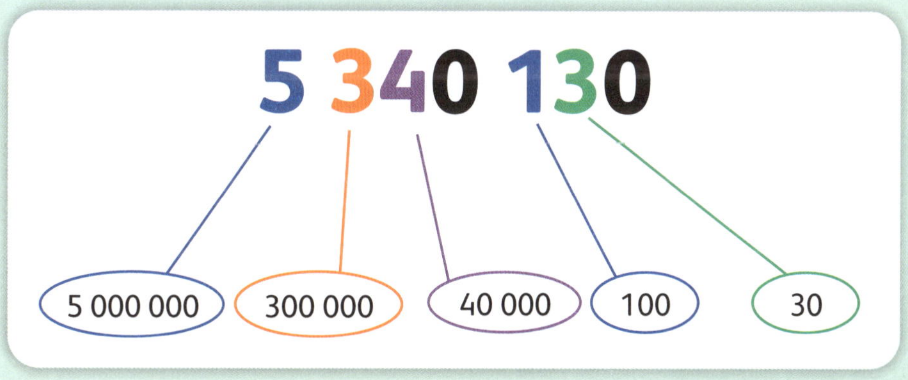

We can show how the number has been expanded as:

5 340 130 = 5 000 000 + 300 000 + 40 000 + 100 + 30

Try adding each of the place value parts on a calculator. What is the total? Why?

The expanded form must be **equal to** the number, because each **addend** in the sum shows the true value of the digit in each place of the number.

Key words

- expand
- face value
- place value
- true value
- equal to
- addend

> In this number, we can see there are no extra thousands in the thousands position and no extra ones in the ones position.

1 Try this activity. You will need digit cards 0–9.

a Make a 4-digit number and then expand it. Use a calculator to check you are correct. Record it as □□□□ = □ + □ + …

b Now add in a fifth digit. Expand your new number.

c Continue adding digits until you make an 8-digit number.

d Talk to a partner about how the value of each digit changes each time another digit is added.

2 True or false? Correct all the statements that are false.

a 45 321 = 40 000 + 5 000 + 300 + 20 + 1

b 256 822 = 20 000 + 50 000 + 6 000 + 80 + 2 + 2

c 3 702 145 = 3 000 000 + 700 000 + 20 000 + 100 + 40 + 5

d 37 021 459 = 30 000 000 + 7 000 000 + 20 000 + 1 000 + 500 + 40 + 9

e 909 357 = 900 000 + 9 000 + 300 + 50 + 7

3 Look at the costs of these flight tickets.

a Expand each amount to show the place value parts.

b The cost of a six-week cruise is ten times as expensive as the $3 475 flight ticket. What is the cost of the 6-week cruise?

Expand the cost to show the place value parts.

4 Look on the internet to find the prices of some houses in Jamaica.

a Draw a table to record the different costs and to show each one in expanded form.

b What is the cost of the most expensive house you can find?

Representing numbers in different ways

Key maths idea

Look at these numbers, which are represented with base ten blocks. The longs each represent ten. The units represent one.

We can expand 65 to show the place value parts: 65 = 60 + 5

We can **partition** a number in many different ways as long as the sum of the partitioned parts is still equal in value to the whole.

Example

65 = 50 + 15 or 65 = 60 + 3 + 2

Key word

partition

Use blocks to make the number 65. How many other ways can you represent it? Record what you find.

1 Use blocks to make the number 342. Copy and complete these partitions. Use the blocks to help you.

- **a** 342 = ☐ + ☐ + 2
- **b** 342 = ☐ + 42
- **c** 342 = 200 + ☐
- **d** 342 = 200 + ☐ + ☐
- **e** 342 = ☐ + ☐ + ☐ + 2
- **f** 342 = ☐ + 230 + 12

There is more than one possible answer for some of these partitions!

2 Tamila has 845 cents. She spends more than 400 cents on Monday. She spends 200 cents on Tuesday and the rest of her money on Wednesday. Find at least five different ways to make this true.

3 True or false? Write a correct answer for any answer you think is false.

a	48 017 = 40 000 + 8 000 + 100 + 7
b	248 017 = 200 000 + 40 000 + 10 + 7
c	208 107 = 200 000 + 80 000 + 107
d	308 117 = 308 000 + 117
e	23 250 = 23 000 + 250
f	23 250 = 23 000 + 200 + 50
g	23 250 = 20 000 + 3 200 + 5
h	423 502 = 40 000 + 23 000 + 500 + 2

4 Work with a partner. Copy the spinner shown on the right. Put the point of a pencil through one end of a paper clip and then put it on the centre of the spinner. You can then spin the paper clip. You also need a calculator.

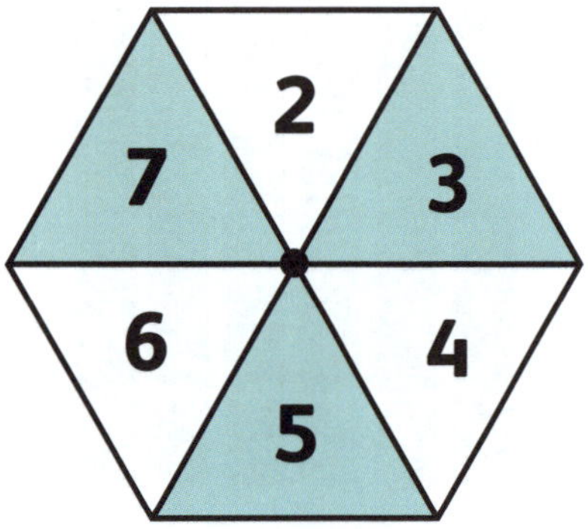

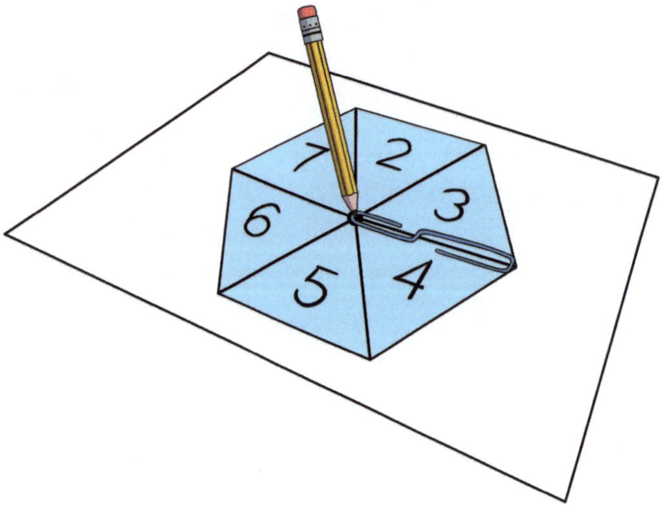

Use a place value chart and counters to help with any difficult partitions.

- You each start with the number 523 486.
- Take it in turns to spin to find the number of parts you must partition the number into.
- Partition the number and write the parts out using addition.
- Your partner uses a calculator to check that your partitions are correct.
- Score the matching number of points. For example, partition correctly into 4 parts and score 4.
- Who has the highest score after five turns each?

Comparing numbers

Key maths idea

Look at the two numbers on the place value chart.

Ten millions	Millions	Hundred thousands	Ten thousands	Thousands	Hundreds	Tens	Ones
	5	2	3	4	9	1	5
5	4	2	3	4	7	9	4

Read each of the numbers. Which is larger? Which is smaller?

The numbers share the same first digit, but we can see that 54 234 794 is **greater than** 5 234 915 because the digit 5 is in the ten millions position and has the value 50 000 000.

We can use the 'greater than' symbol to show that 54 234 794 < 5 234 915.

Now let's **compare** the number 5 234 915 with another number.

Millions	Hundred thousands	Ten thousands	Thousands	Hundreds	Tens	Ones
5	2	3	4	9	1	5
5	2	3	4	7	9	4

The first four digits in both numbers are the same and they have the same place value positions.

The hundreds position is important, because it shows us that the top number has two more hundreds than the bottom number, so 5 234 915 > 5 234 794.

We can also write 5 234 794 < 5 234 915 to show that 5 234 794 is **less than** 5 234 915.

When you order and compare numbers, it is important to check if they have the same number of digits and what position they have. The number 9 874 may look bigger than 10 404 because it starts with the digit 9.

What about decimal numbers? The number 12 645 only has five digits, but is bigger than the **decimal number** 7 345.23 even though it has fewer digits.

Key words
greater than
compare
less than
decimal number
greater than or equal to
less than or equal to

1 Represent each pair of numbers on a place value grid and **compare** them. Which is larger? Use the symbols $<$ and $>$ to show what you found out.

a 560 123 and 65 184

b 146 235 and 142 235

c 25 644 and 256 042

d 10 702 222 and 10 699 999

2

85 324 must be larger than 211 473 because the first digit is 8 and this is larger than 2.

Do you agree with Akiel? Explain or show your thinking.

3 Look at the prices of these cars. Which car is more expensive in each case? Copy the prices and write the symbol $<$ or $>$ between them.

a $14 995 ☐ $14 999

b $19 500 ☐ $18 995

c $99 500 ☐ $112 250

d $55 400 ☐ $54 500

4 Copy and complete the table to show where you will position each of these numbers.

5 999 999 85 745 60 000 605 378

600 000 599 999 1 325 327

Less than or equal to 599 999	Greater than or equal to 600 000

We can use the symbol $\geq$ to mean **'greater than or equal to'**. What do you think the symbol $\leq$ means?

Write a number of your own that will fit in each column of the table.

5 The area of each continent is given below in square kilometres.

Continent	Area (square km)
Africa	30 065 000
Antarctica	13 209 000
Asia	44 579 000
Australia	7 687 000
Europe	9 938 000
North America	24 256 000
South America	17 817 000

a Which continents have an area of less than 15 million square kilometres?

b Which continent has the greatest area?

c Which is larger, North America or South America?

d Is Africa bigger or smaller than North America?

Ordering numbers

Key maths idea

It is often useful to arrange numbers or quantities in order of size.

We can order numbers in **ascending** order (from smallest to largest).

23 456 23 564 32 465

smallest ⟷ largest

We can also order numbers in **descending** order (from largest to smallest).

32 456 23 564 23 465

largest ⟷ smallest

Key words

ascending

descending

Think of a number that is greater than 23 564 but smaller than 32 465.

Explain where you will put the number on each of the number lines.

1 Write each set of numbers in size order, smallest first.

a 315 634 32 584 35 824 308 492

b 920 413 921 437 921 493 43 219

c 78 106 45 000 71 992 700 021

2 Look at the numbers in this table.

Country	Area (in km²)	Population (approx. in 2017)
Antigua	442	89 000
Barbados	430	283 000
Dominican Republic	48 442	10 776 000
Grenada	344	104 000
Jamaica	11 425	2 729 000
St Lucia	617	172 000
St Vincent and the Grenadines	389	110 000
Trinidad and Tobago	5 124	1 357 000

a Write the names of the countries in order of area, from largest to smallest.

b Write the names of the countries in order of population size, from most people to fewest people.

c Which three countries are largest in area?

d Which three countries have the fewest people?

e Which of these countries have more than a million people living in them?

3 Numbers A, B, C, D and E have been positioned on a number line. Write a set of possible values in order from smallest to largest.

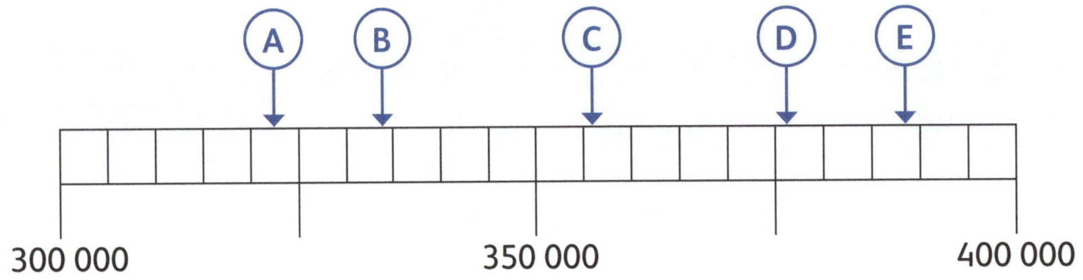

4 Use each of the digits cards once to make this ordering true. Is there more than one possible solution?

Decimal fractions

Key maths idea

Decimal fractions are positioned between 0 and 1 on the number line. When we write them as proper fractions, they have denominators 10, 100, 1 000, and so on.

This number line shows **tenths**. Read the proper fractions and decimal fractions aloud. Remember to say the decimal fractions as 'zero point one'.

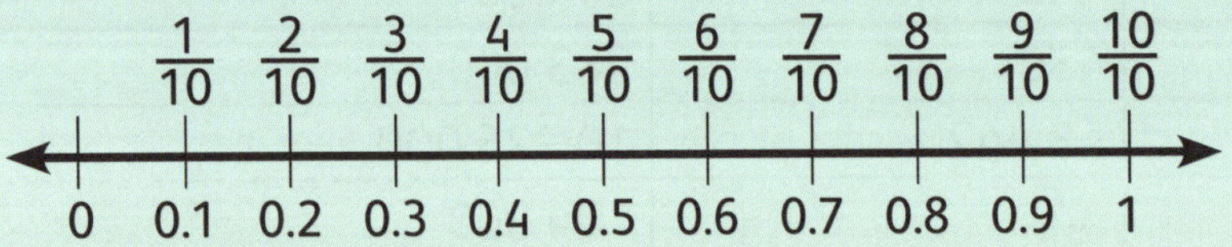

$\frac{5}{10}$ is the same as the decimal fraction 0.5. What other proper fractions do you know that are equivalent to 0.5?

We can see tenths on a place value chart. A decimal point separates the fraction from whole part of the number.

Ones	Tenths	Hundredths	Thousandths
0	5		
0	9		

Key words

decimal fraction
tenth
increase
decrease
hundredth
thousandth

Talk about

What happens when we **increase** the decimal fraction 0.9 by 1 tenth? What happens if we **decrease** the decimal fraction 0.9 by 1 tenth?

Look at these diagrams.

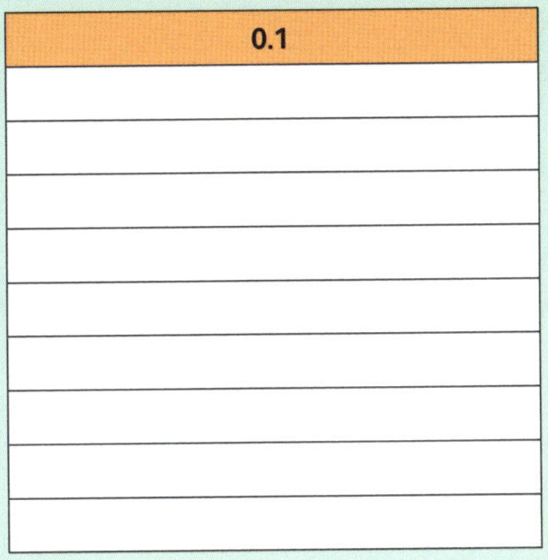

The whole is divided into 10 equal parts: $1 \div 10 = \frac{1}{10}$
So each of the 10 equal parts is one-tenth ($\frac{1}{10}$) or 0.1.

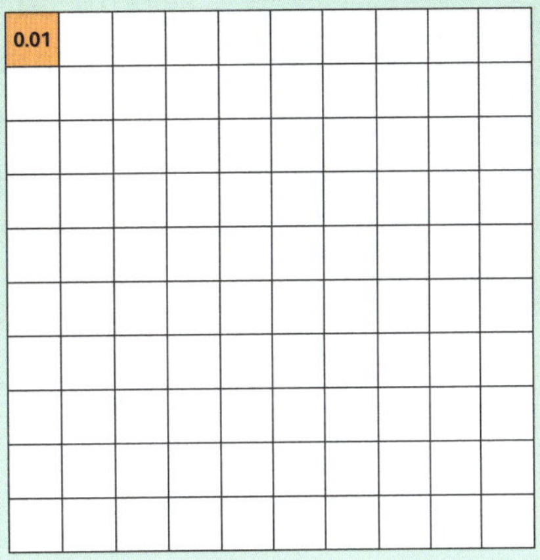

The whole is divided into 100 equal parts: $1 \div 100 = \frac{1}{100}$
So each of the 100 equal parts is one-**hundredth** ($\frac{1}{100}$) or 0.01.

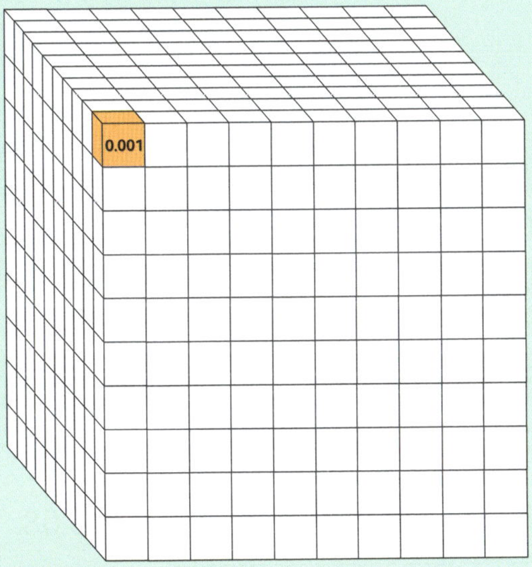

The whole is divided into 1 000 equal parts: $1 \div 1\,000 = \frac{1}{1\,000}$
So each of the 1 000 equal parts is one-**thousandth** ($\frac{1}{1\,000}$) or 0.001.

1 Write a proper fraction and decimal fraction to match each diagram.

a

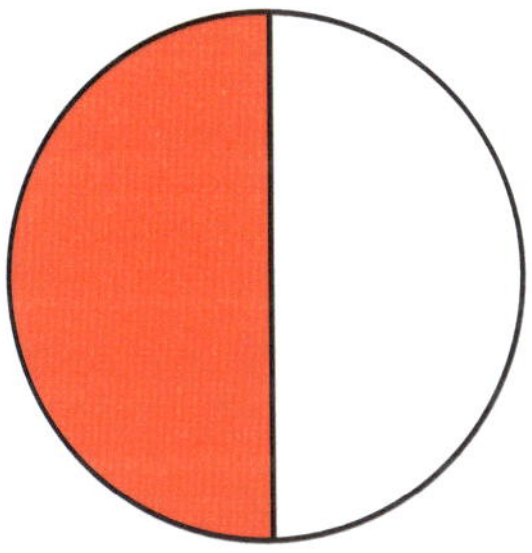

b

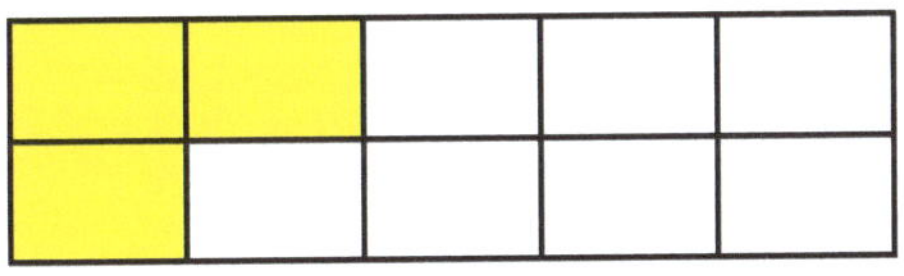

c

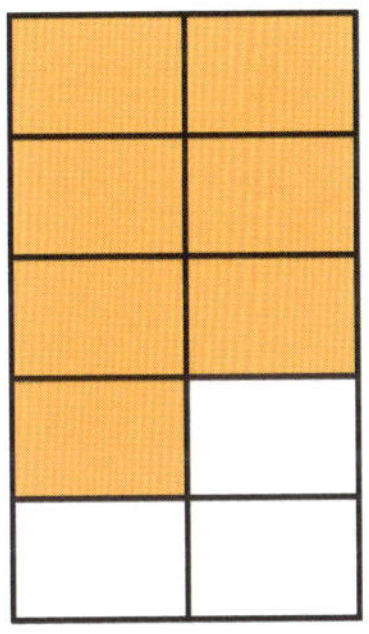

d

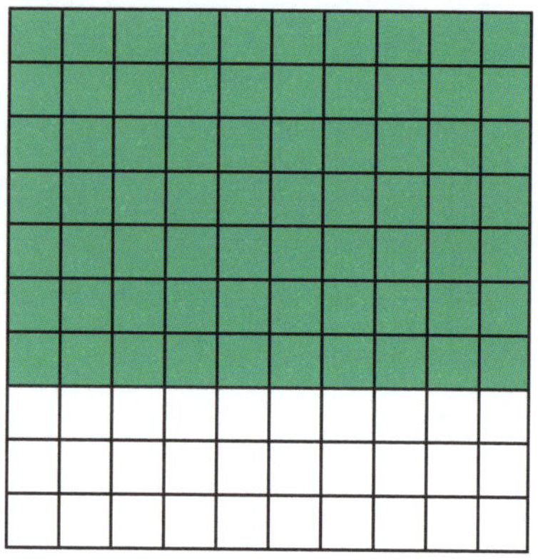

Why can we write part d as a fraction with denominator 10 or denominator 100?

e

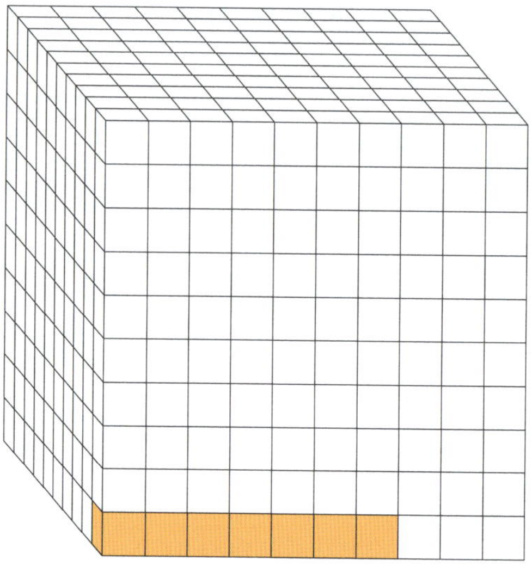

2 Write these decimal fractions in words.

a 0.5 b 0.05 c 0.1

d 0.001 e 0.06 f 0.006

3

Do you agree with Amy's statement? Explain why or why not.

4 Look at the decimal fractions on this place value chart.

Ones •	Tenths	Hundredths	Thousandths
0 •	8		
0 •	0	8	
0 •	8	8	

a Use grid paper to make two squares like this. Shade squares to show 0.8 and 0.08.

b Talk to a partner about how you can use what you know about 0.8 and 0.08 to represent the decimal fraction 0.88 on a third square.

Decimal number values

Key maths idea

The decimal fraction 0.42 is made up of 4 tenths and 2 hundredths.

We read it as 'zero point four two'.

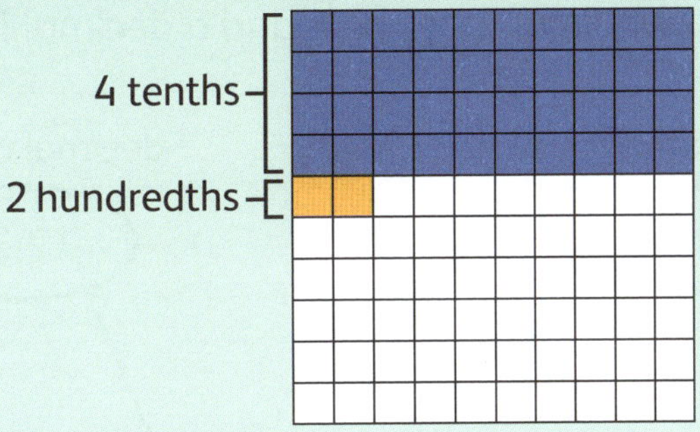

Look at the two decimal fractions on this place value chart.

Ones	Tenths	Hundredths	Thousandths
0 •	4	2	
0 •	4	2	5

The digit 4 has the value 4 tenths or 0.4.

The digit 2 has the value 2 hundredths or 0.02.

What is the same and what is different about the numbers 0.42 and 0.425?

1 What decimal fractions are represented here?

a

b

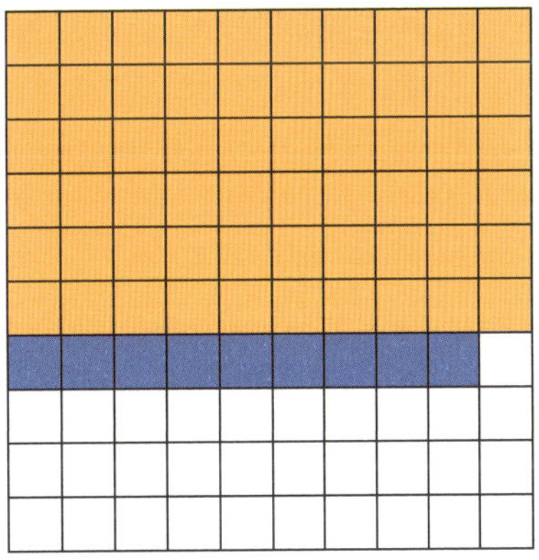

2 Represent these decimal fractions on a number line, place value chart or on grid paper.

a 0.75 **b** 0.83 **c** 0.25 **d** 0.99

e Which decimal fraction is equivalent to $\frac{1}{4}$? Which is equivalent to $\frac{3}{4}$? How do you know?

3 Write the value of the underlined digit 3 in each number.

a $0.\underline{3}$ **b** $0.6\underline{3}$ **c** $0.\underline{3}6\underline{3}$ **d** $0.0\underline{3}6$ **e** $0.\underline{3}6$

Decimal numbers

Key maths idea

A decimal number has a whole number part and a decimal fraction part.

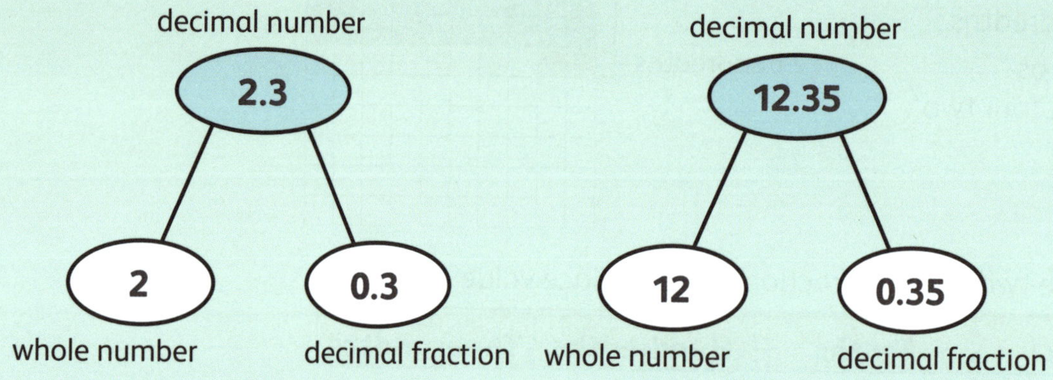

We read 2.3 as 'two point three'.

How do we read 12.35?

What is the value of the digit 5?

The decimal number 2.3 is less than half-way between 2 and 3 on the number line.

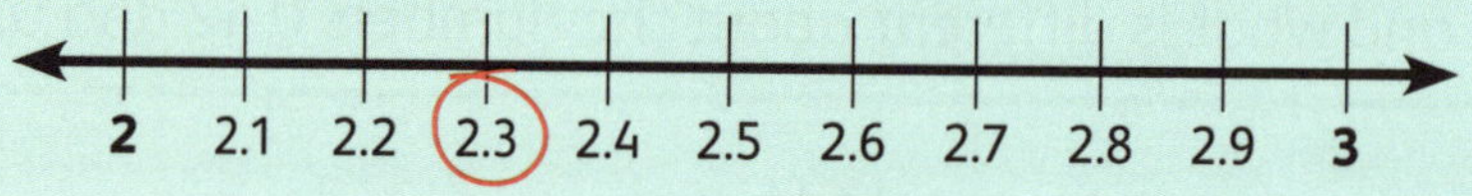

The decimal number 12.35 is exactly half-way between 12.3 and 12.4 on the number line.

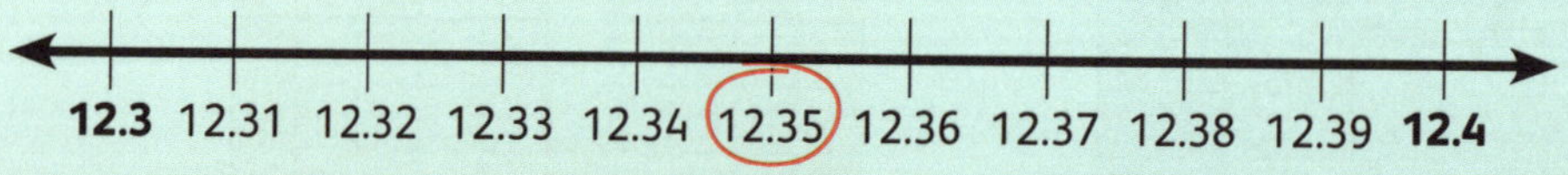

What is the whole number part and the decimal fraction part of the number 3.275? What is the value of the digit 5?

1 Copy and complete these diagrams.

a

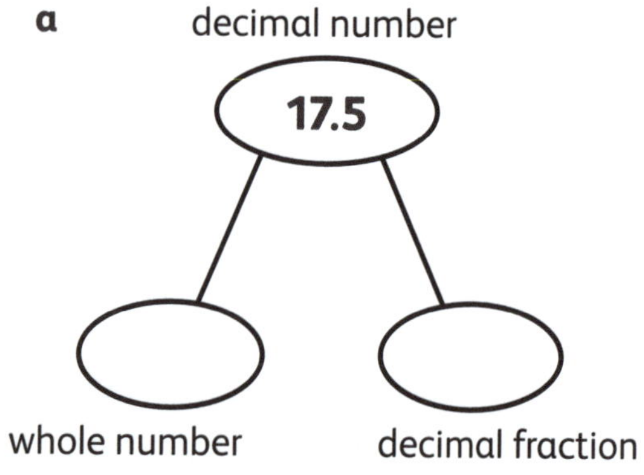

b

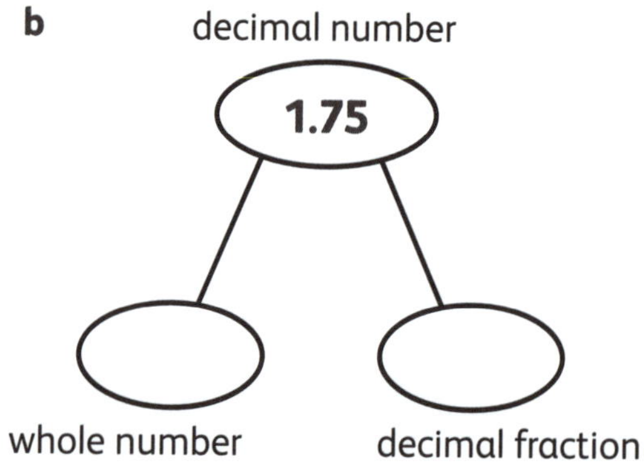

2 Look at this number line. Then answer the questions below.

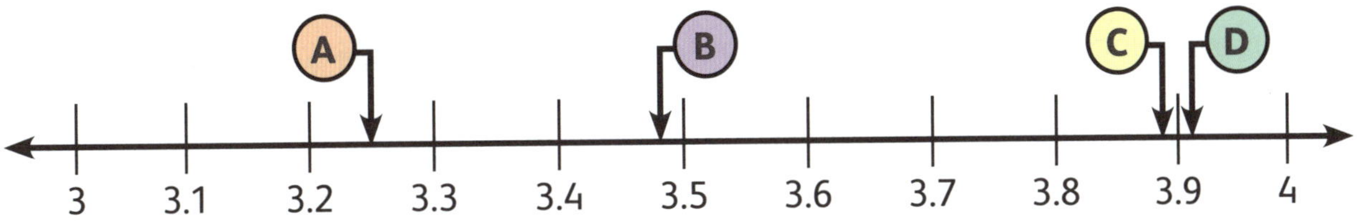

a Which letter represents 3.89?

b Which letter represents a decimal number with 4 tenths?

c Which letter represents a decimal number with 5 hundredths?

d What decimal number do you think the remaining letter represents? What is the value of each digit in the number?

3 Romano is sorting these decimal numbers.

62.5 | 4.16 | 3.16 | 10.66 | 5.62 | 1.546 | 3.45 | 0.66 | 4.016

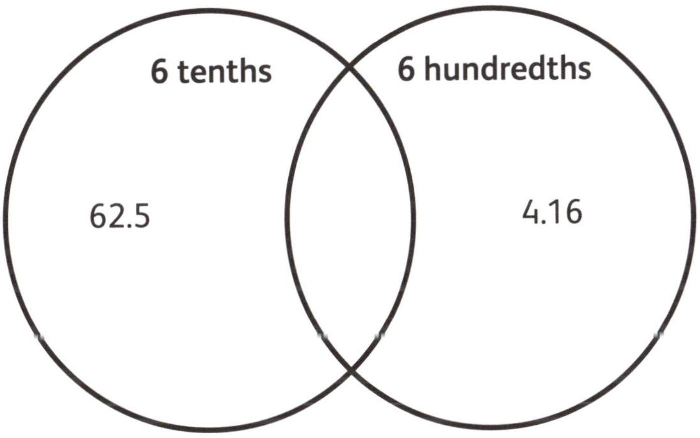

a Do you agree with what Romano has done so far? Explain why.

b Copy and correctly complete the Venn diagram.

c Write your own number in each of the three sections.

Maths detective: Exploring bases

Use your maths detective skills to explore working with different bases. Our decimal system works in base 10.

Case 1

True or false? The same number is represented on both charts.

Explain how you know.

Case 2

Draw your own place value charts, one for base ten and one for base five. Use coloured cubes or counters to match the numbers shown in the previous task.

> Always start each task below with these starting numbers.

1 Put four more cubes in the ones position (1st column) of each number. What will you need to do now? Are the new numbers still the same?

2 Now put four more cubes in the 2nd column of each number. What happens now? Are the new numbers still the same? Why or why not?

3 Finally put two more cubes in the 3rd column of each number. Can you predict which number will be larger?

Case 3

This place value chart uses base six.

216s	36s	6s	1s

> You can also try some numbers of your own.

Copy and complete the table below to show how you can represent the following base ten numbers in base six. The first one has been done for you.

Base 10 number	216s	36s	6s	1s
10			1	4
21				
36				
45				
100				
250				

> Is it easier to work in base ten, base five or base six?
> Explain your thinking.

> What do you notice about the relationship between the base you are working in and the largest digit that is written in each place value position?

Real-world maths: Megaship loads and computer calculations

Case 1

Goods are distributed all over the world on large container ships. Trading companies work with huge numbers when importing or exporting goods.

- Each container shown has a mass of 2 170 kg.
- Each container can be used to transport around 27 300 kg.
- Megaships can transport up to 19 700 containers.
- They can load and transport up to 173 500 tonnes of goods.

1 Which number has the greatest value?

2 Which number is the least value?

3 Say each number.

4 Which number has the 7 in the thousands place?

5 What is the value of the 7 in the other three numbers?

Case 2

The number pi is the ratio of the circumference to the diameter of a circle. As a proper fraction, pi is $\frac{22}{7}$.

1 Express pi as a decimal fraction rounded off to the nearest hundredth.

2 Computers have calculated pi to many millions of decimal places. Discuss with a partner how you could work out how much printer paper you would need to print out the first million digits of pi.

3 In 1874, William Shanks calculated the first 707 decimal places of pi. He worked it out by hand (without a calculator). How long do you think this would take you to do? Discuss how you could work it out.

What I have learned

Copy and complete the diagram to show what you know about the number 23 505 675. What else can you add to the diagram?

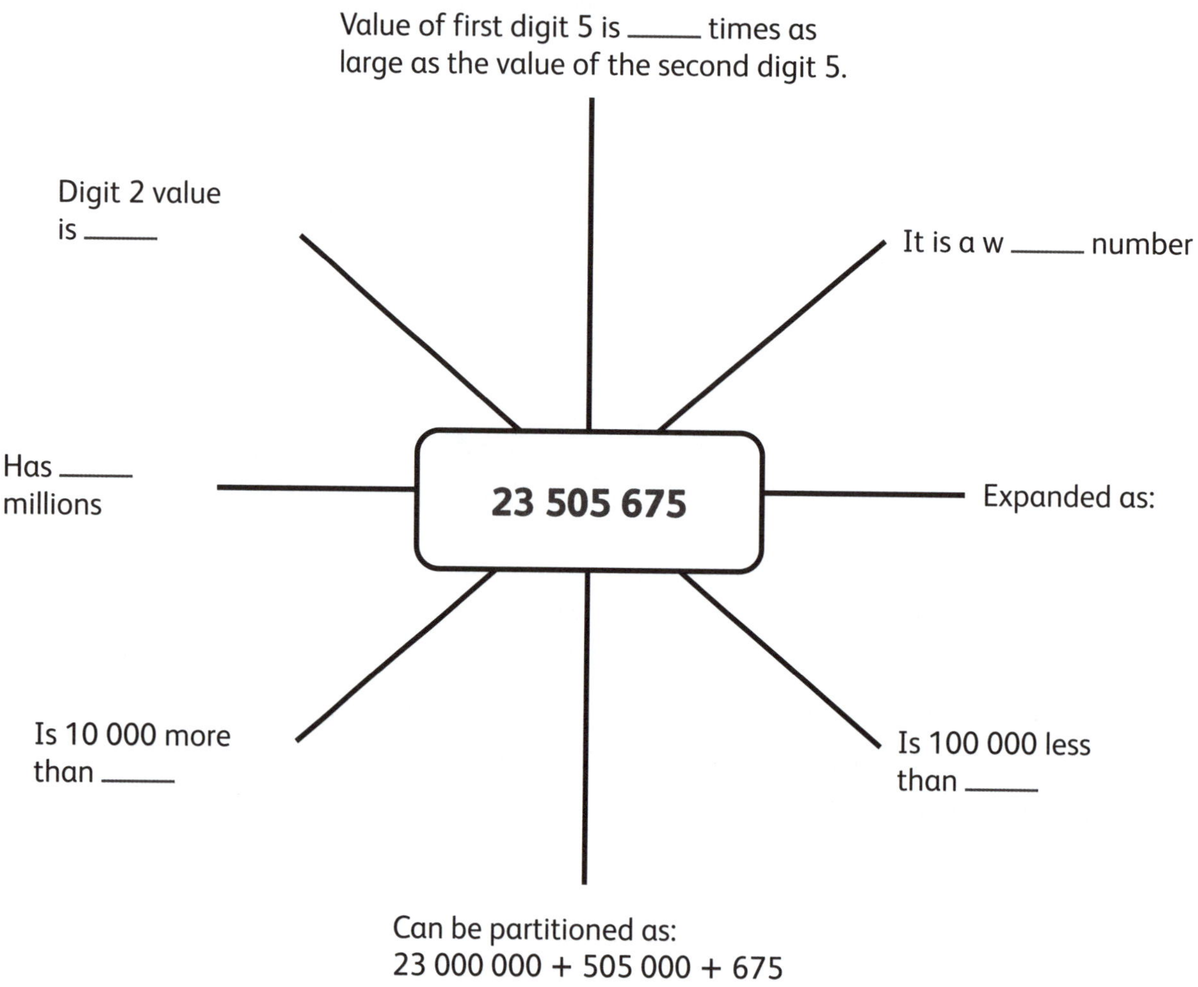

Now create a similar diagram for 16 635.

Practice questions

1 What are the values of the underlined digits?

a 67$\underline{2}$ 198 **b** 6$\underline{2}$7 918 **c** 5 37$\underline{5}$ 763 **d** $\underline{5}$4 757 630

Now expand the largest number to show the place value parts.

2 What number is:

a 1 more than 9 999 999

b 1 000 less than 240 000

c 10 000 more than 12 395 000?

3 Compare these numbers. Use < or > to show which is larger.

a 34 567 250 □ 3 456 725

b 276 402 □ 276 042

c 1 250 175 □ 1 250 275

4 Order these numbers from largest to smallest.

a 45 650 219 45 560 219 54 650 219 54 650 129

b 1 252 000 12 520 000 125 200 21 250 000

5 Which decimal fractions are shaded in each diagram?

a

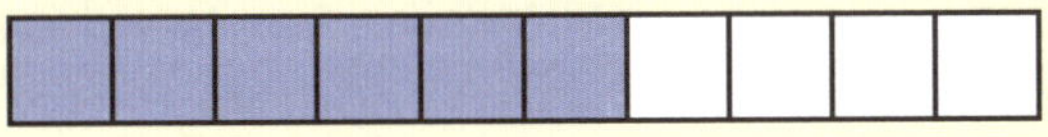

b

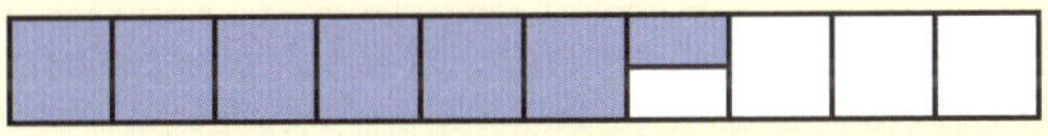

c

2 Self check

	Check you can do this:	If you have difficulty, try this:
1	Read and write numbers with up to eight digits.	Go back to pages 19–20. Start with three-digit numbers, then make longer numbers (thousands, then ten thousands, then hundred thousands, and so on).
2	Give the face value and place value of a digit in a whole number, and expand whole numbers to show their place value parts.	Revise the work on pages 25–28. Practise writing a number with between five and eight digits, then expressing it as a sum of each digit's true value.
3	Partition numbers in different ways, e.g. 34 545 = 34 000 + 540 + 5.	Go back to pages 27–28.
4	Compare whole numbers and use the symbols $<$ or $>$.	Go over pages 29–30. Remind yourself how we show that one value is less than or greater than another.
5	Order whole numbers.	Revise the work on pages 31–32. Remember to use what you have learned about place value. You can use number lines or place value charts to help you.
6	Identify and represent decimal fractions, e.g. with blocks, on a place value chart or as pictures.	Revise the diagrams and place value charts on pages 32–34.
7	Read and write decimal fractions and decimal numbers.	Revise the work on decimal numbers on pages 36–37.
8	Give the face value and place value of a digit in decimal numbers.	Revise the definitions of face value, place value and true value on page 25.
9	Explain the function of the decimal point in a number.	Go back to the section on decimal fractions (page 32) and remind yourself of how we write decimals and what the decimal point shows.

Chapter 3 Types of numbers

In this chapter you will explore this focus question:

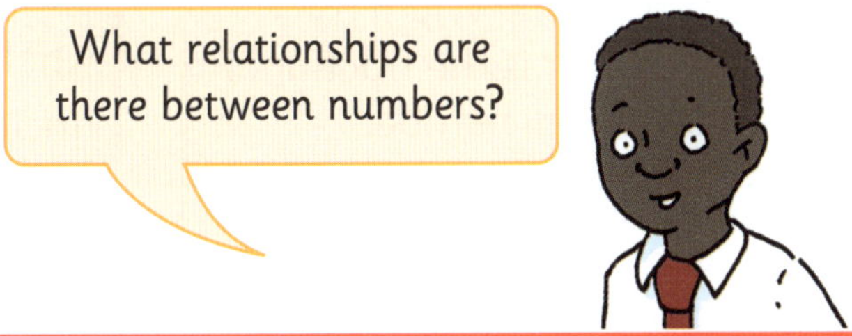

To answer this question, you will:

- understand and use prime, composite and fractional numbers
- distinguish between counting, whole, odd, even, prime, composite and fractional numbers
- identify the factors of a number
- identify common factors of two given numbers
- identify multiples of a number and the common multiples of two or more numbers.

Starting point

Key word
counting numbers

These children are talking about different types of numbers they can find from 0 to 10.

Which of the numbers from 0 to 10 can each child find?
Explain how you know this.
Why are **counting numbers** also part of the set of whole numbers?

Fractional numbers

Key maths idea

A **fractional number** represents part of a **whole**. When a whole is divided into five equal parts, each part is the fractional number $\frac{1}{5}$. Three of these parts is the fractional number $\frac{3}{5}$. Fractional numbers written in this way are called **fractions**.

1 whole				
$\frac{1}{5}$	$\frac{1}{5}$	$\frac{1}{5}$	$\frac{1}{5}$	$\frac{1}{5}$

A decimal fraction is also part of a whole. It uses place value to show the size of the fraction. We can use what we know about tenths to help us:

$\frac{3}{5}$ is equivalent to $\frac{6}{10}$ or 0.6

Key words

fractional number
whole
fractions

Talk about

Talk about where you use fractional numbers and whole numbers in real life. Think about the coins that make up the Jamaican currency: why do we often round amounts of money involving decimal numbers to the nearest whole number or to the nearest tenth or hundredth?

1 **a** Sort these numbers. Copy the table and write the numbers in the correct columns.

8 0.8 $\frac{8}{10}$ $\frac{3}{4}$ 0.75 75 34 $\frac{2}{5}$

Whole numbers	Fractional numbers

b Add some more numbers of your own in each group.

2 Which fractional numbers are represented here? Record your answers as fractions and as decimal fractions.

a
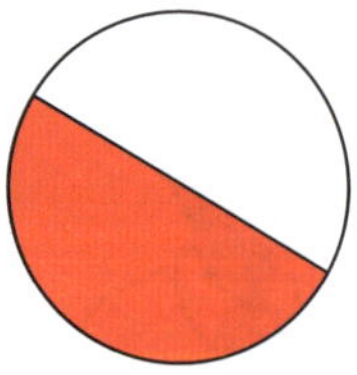

b
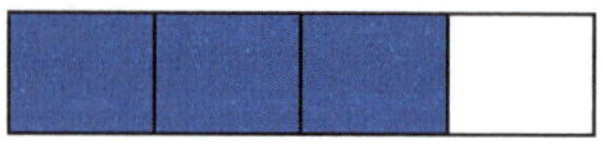

c
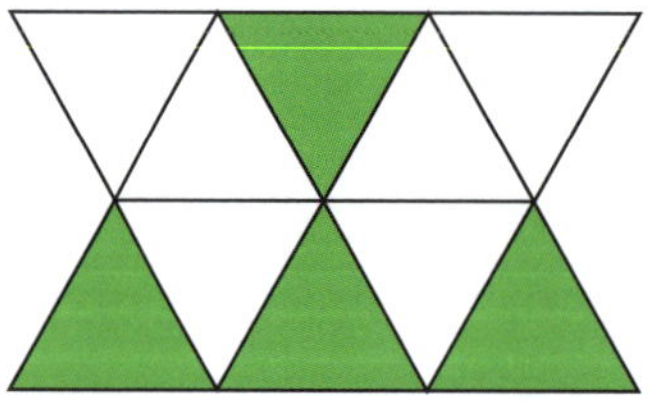

3 Make or draw your own representations for these fractional numbers.

a $\frac{4}{10}$ **b** 0.25 **c** $\frac{2}{5}$ **d** 0.9 **e** $\frac{9}{100}$

Odd and even numbers

Key maths idea

Even numbers are whole numbers that can be divided by 2 without a remainder. They always end in 0, 2, 4, 6 or 8.

Odd numbers have a remainder of 1 when we divide them by 2. They always end in a 1, 3, 5, 7 or 9.

Remember that zero (0) is neither odd nor even.

Key words

even numbers
odd numbers

Can you see a pattern in the diagram?

1 Use the pattern above to complete these sentences using the words 'odd' or 'even'.

a The difference between any pair of even numbers is always ______.

b The difference between any pair of odd numbers is always ______.

c The difference between any odd and even number is always ______.

2 Use curly brackets to show the following sets:

a even numbers between 50 and 80

b odd numbers between 30 and 50

c the first ten 3-digit even numbers

d the last ten 2-digit odd numbers.

3 

Rearrange the digits to make:

a the smallest possible odd number

b the smallest possible even number

c the largest possible odd number

d the largest possible even number

e the odd number closest to 7 000

f the even number closest to 9 000.

Factors

Key maths idea

A **factor** divides into a given number exactly, without leaving a **remainder**.

In the multiplication 2 × 3 = 6, both 2 and 3 are factors of the product 6.

6 ÷ 2 = 3 and 6 ÷ 3 = 2

There are no remainders.

We can use **arrays** to help find all the factors of 6.

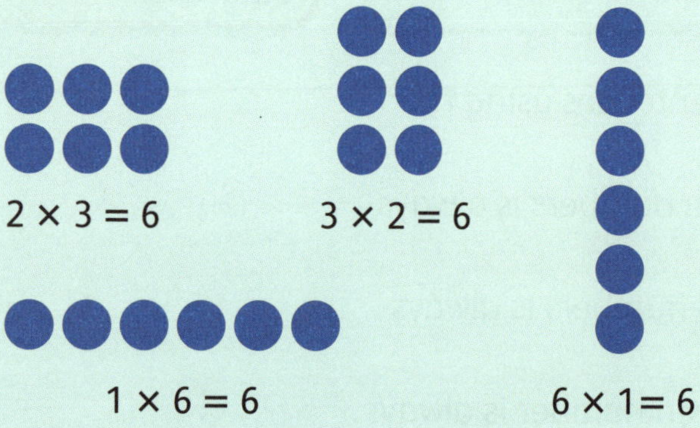

So 1, 2, 3 and 6 are the factors of 6.

Some numbers share the same factors.
We call these common factors.

Look at the different arrays for the number 15.

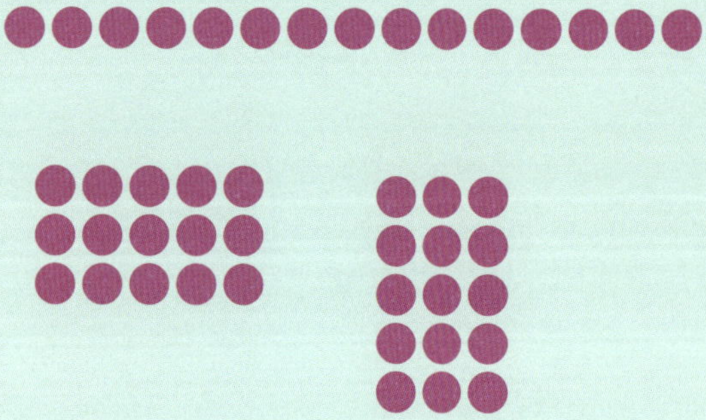

Now compare these arrays with the ones for 6. What do you notice?

Both 6 and 15 can be represented as arrays with rows of 3, and as arrays with rows of 1.

So 1 and 3 are the common factors of 6 and 15.

The **highest common factor** (HCF) is 3 as it is the largest factor shared by both numbers.

Key words

factor
remainder
array
highest common factor (HCF)

1 Work with a partner. Use counters to make the different arrays for each number below. Copy and complete the table.

Number	**Factors**	**How many factors?**
9		
11		
14		
28		
36		

2 Malcolm uses bars to show the factors of 18. What is the value of each of the bars from a to f?

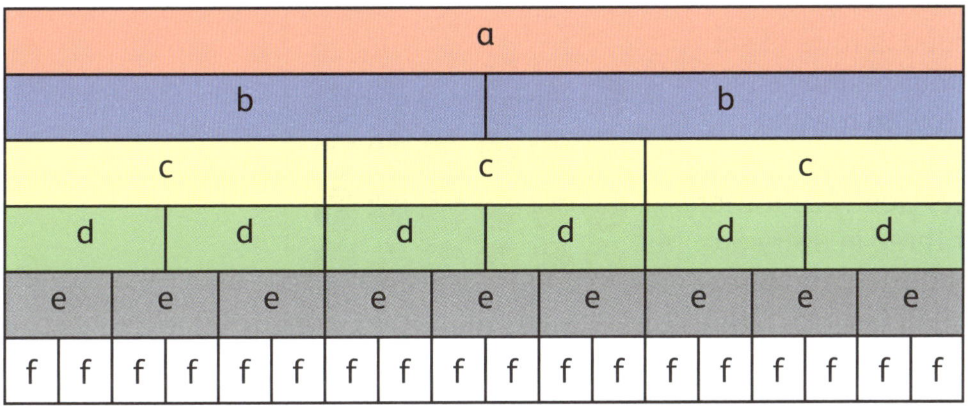

3 List the factors for each pair of numbers.
Circle the common factors.

a 8 and 12

b 12 and 21

c 10 and 20

d 16 and 24

Try to work out the highest common factor (HCF) each time too!

4 **a** Sort these numbers. Copy and complete the Venn diagram.

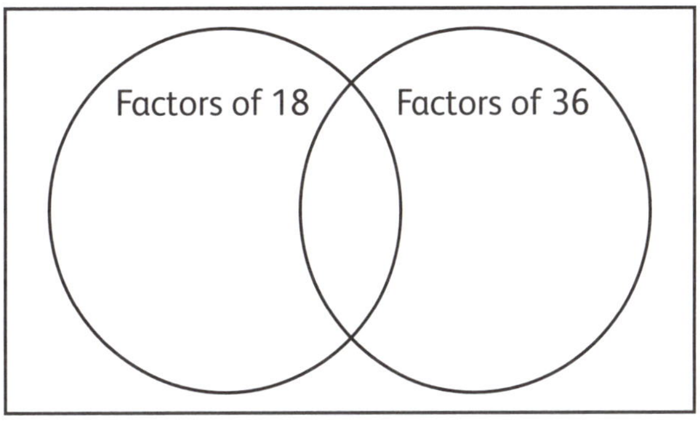

b Why are all the numbers in the intersection of the sets the common factors of 18 and 36?

c What else do you notice about factors of 18?

d What is the HCF of 18 and 36?

Multiples

Key maths idea

A whole number is a **multiple** of each of its factors.

The number 6 has the factors 1, 2, 3 and 6 so 6 is a multiple of each of these:

Multiples of 1: 1, 2, 3, 4, 5, **6**, …
Multiples of 2: 2, 4, **6**, 8, 10, 12, …
Multiples of 3: 3, **6**, 9, 12, 15, 18, …
Multiples of 6: **6**, 12, 18, 24, 30, 36, …

The array is arranged in rows of six. It shows the first four multiples of 6 as we move from row to row. The array also shows the multiples of 3. The diagram shows clearly that there are two groups of three in every group of six. Multiples of both 3 and 6 are called common multiples.

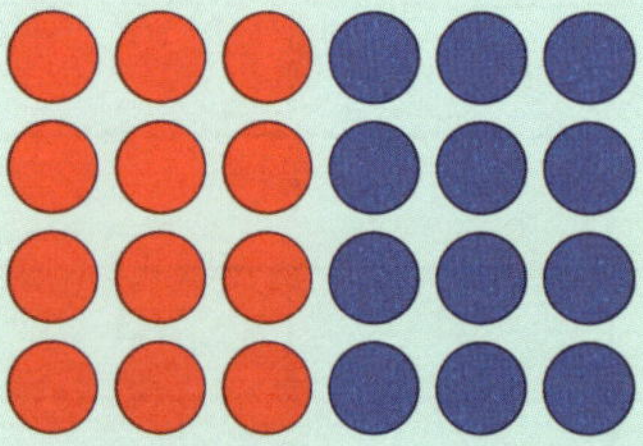

So, the common multiples of 3 and 6 are: 6, 12, 18, 24, …
What are the first five common multiples of 2 and 3?
And the first five common multiples of 2, 3 and 6?

If we look at the common multiples of 3 and 6, we see that **6** is the **lowest common multiple (LCM)**. The LCM is the first multiple that is found in both sets.

Key words

multiple
lowest common multiple (LCM)
sequence
pattern

1 Complete these **sequences** to show the first ten multiples of each number.

- **a** Multiples of 4: 4, 8, ____, ____, ____, … , 40
- **b** Multiples of 5: 5, 10, ____, ____, ____, … , 50
- **c** Multiples of 8: 8, 16, ____, ____, ____, … , 80
- **d** Multiples of 10: 10, 20, ____, ____, ____, … , 100

2 Use the multiples from question 1 to find:

- **a** common multiples of 4 and 8
- **b** common multiples of 5 and 10
- **c** common multiples of 4 and 5
- **d** common multiples of 5 and 8
- **e** the lowest common multiple (LCM) of 4 and 8
- **f** the lowest common multiple (LCM) of 4 and 5.

3 **a** Sort these numbers. Copy and complete the Venn diagram.

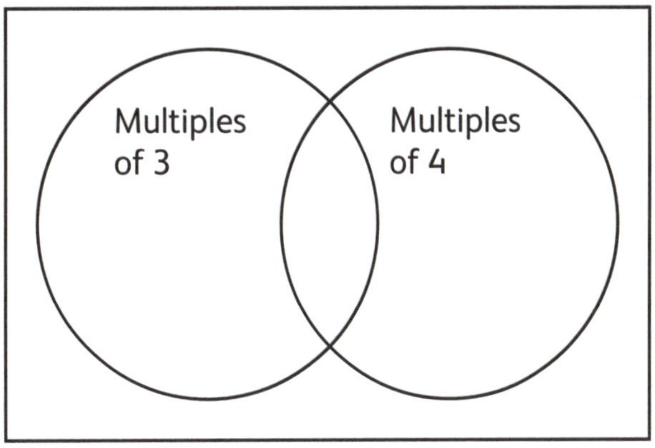

b How can you describe the numbers in the intersection of the two sets?

c What is the LCM of 3 and 4?

d Add a number of your own to each set.

4 Use what you know to find the first three common multiples of:

a 2, 3 and 5 **b** 3, 4 and 5 **c** 3, 6 and 9.

5

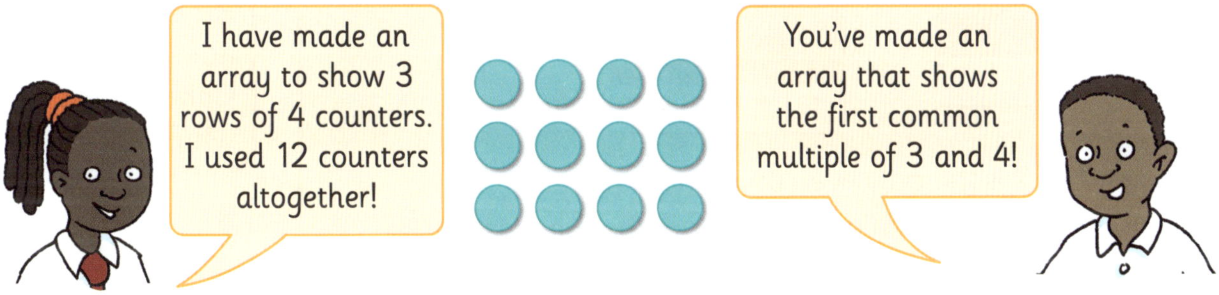

a What happens when you make arrays for other common multiples of 3 and 4?

b What **patterns** do you notice?

c Can you use what you have found out to explain why 48 and 60 are common multiples of 3 and 4, but why 40 is not?

6 Amari and Romano are arranging baskets of balls. They want each basket to have the right number of balls so that equal groups of four or equal groups of five children can have a ball each.

a How many balls could be in a basket? Find at least four solutions.

b What do you notice?

Prime and composite numbers

Key maths idea

We say that the number 3 is a **prime** because it can only be divided exactly by itself (3) or by 1 without leaving any remainders:

$3 \div 3 = 1$ and $3 \div 1 = 3$

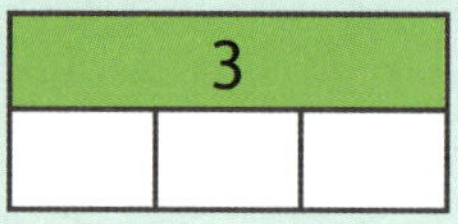

So prime numbers have only two factors, the number itself and 1.

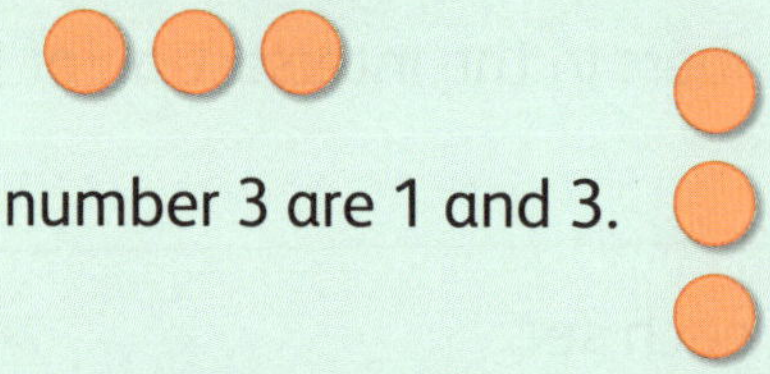

The factors of the prime number 3 are 1 and 3.

Prime numbers can be represented by only two different arrays.

Look at these diagrams.

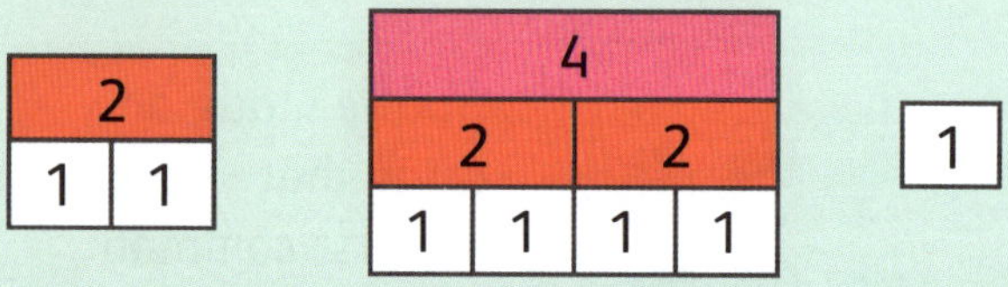

The number 2 is the only even prime number.

The number 4 is called a **composite** number because it has more than two factors.

Key words

prime

composite

The number 1 is neither prime nor composite. Can you explain why?

1 **a** Make arrays to help find all the 1-digit prime numbers.

b Use arrays to help identify all the 1-digit composite numbers.

2 Copy and complete the table.

Number	Factors	Prime or composite?
24		
38		
83		
15		
21		
17		
64		
91		
82		

3 Akiel wants to place these numbers on a Venn diagram.

15 9 5 2 53

29 1 36 49 81

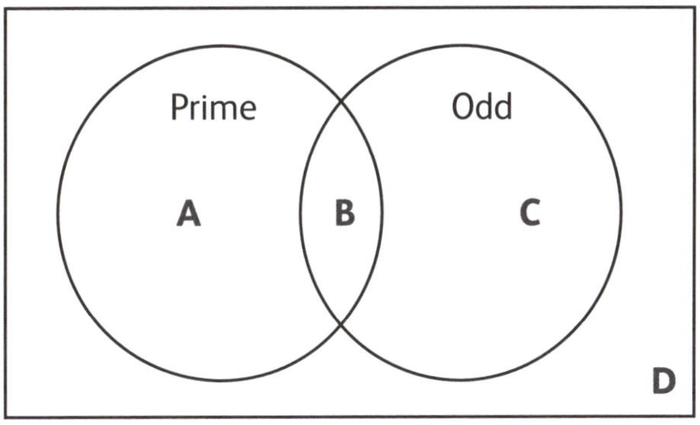

Should he place each number in section A, B, C or D?

4 Jalisia is thinking of a 2-digit prime number.
Malcolm can ask only three questions to find her number.
He asks: 'Is it an odd number?'
Is this a good question to ask? Explain your reasoning.

5 True or false?

a A multiple of 5 can never be prime.

b A multiple of 10 is always composite.

c The number 3 is prime, so the numbers 23, 33, 43, 53 and 63 must also be prime.

Maths detective: Prime factors

Look at this factor tree for the composite number 12. Talk about how it works. What do you notice about the numbers that are circled?

Case 1

You can start with any factor pair.

Here we chose 3 and 4.

3 is prime. Circle it.

4 is not prime, so we break 4 into a factor pair.

2 is prime, so we circle both 2s.

All the circled numbers are the prime factors of 12.

Now we write them as a multiplication:

$2 \times 2 \times 3 = 12$

Always write the prime factors in ascending order.

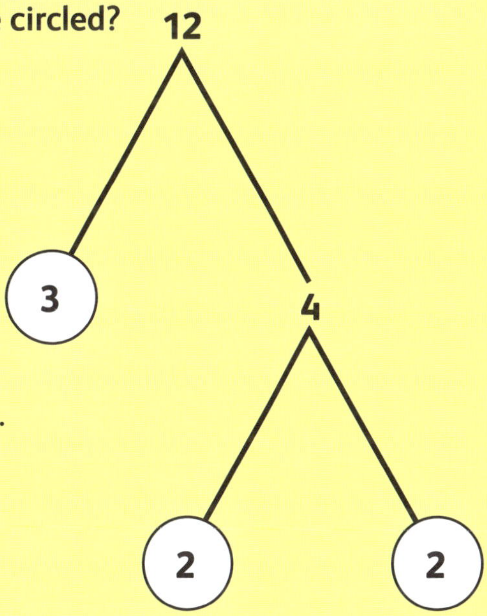

Try to find out if it is true that every composite number up to 20 can be written as a product of its prime factors.

Case 2

The number 24 is double 12.

1 Can you think how you can use what you know to help write 24 as a product of its prime factors without using a factor tree?

2 What about 36? What relationship is 36 to 12? How will this help you?

Real-world maths: Design an egg carton

Eggs come in their own special packaging, but their shells are still very vulnerable. So eggs are packed in **cartons** or trays to keep them safe.

- **Design** your own packaging to hold 6 eggs.
- How many different rectangular cartons can be designed?
- What if the carton needs to hold 12 eggs? Can you use what you know to help with your designs?
- Now try 24 eggs. What do you notice this time?
- Make a display of your different cartons. Talk about what you noticed about factors and multiples of 6, 12 and 24.

Key words
carton
design

Practice questions

1 Write out these numbers. Circle the whole numbers and put a square around the fractional numbers. An example of each has been done for you.

(5)	[$\frac{1}{3}$]	0.5	25	$\frac{2}{5}$	0.75	75	14	$\frac{1}{4}$

2 a Which of these numbers can be divided exactly by 2 with no remainders?
13 34 75 48 120 99 56

b What do we call these types of numbers?

c What do we call numbers that cannot be divided in this way?

3 a List the factors of 12.

b List the factors of 15.

c Now write the common factors of 12 and 15.

d What is the highest common factor (HCF) of 12 and 15?

4 a List the first ten multiples of 3.
b List the first ten multiples of 5.
c Now use your list to find common multiples of 3 and 5.
d Which of these numbers will also be common multiples of 3 and 5? Why?
35 40 45 50 55 60 72
e What is the lowest common multiple (LCM) of 3 and 5?

5 a Sort the numbers under the following headings in the table.
9 19 11 5 35 27 2 82 1 7 17 21 31

Prime numbers	Composite numbers

b One number is left over. Which one is it?
c Why does the number not belong under either heading?

What I have learned

1 Write down the terms from a to h. Match each term to the definitions from A to H. Write each definition next to the matching term.

a Product
b Fractional number
c Factor
d Multiple
e Prime number
f Even number
g Composite number
h Odd number

A a number that can be divided by 2 with no remainder
B a number into which the given factor divides exactly
C a number that is part of a whole
D a number with only two factors, itself and 1
E the result of multiplying two numbers
F a number that leaves a remainder when divided by 2
G a number that divides exactly into another number
H a number with more than two factors

2 Check with a partner to see if your definitions are correct.
3 Give an example of each type of number in your list.
4 Work together to write a definition for a 'common factor' and a 'common multiple'.

3 Self check

Check you can do this:	If you have difficulty, try this:
1 Explain the difference between whole numbers and counting numbers.	Go back to the starting point on page 43, and remind yourself what counting and whole numbers are.
2 Identify odd and even numbers and explain why a number is odd or even.	Go back to page 45. Remember, even numbers have 2 as a factor, and they always end with a 0, 2, 4, 6 or 8.
3 Identify fractional numbers as parts of a whole.	Go back to page 44, and remind yourself how we write fractions as a part of a whole.
4 Identify factors of a number and common factors of two numbers.	Go back to pages 46–47 and remind yourself about dividing a number by another number. A factor goes into another number without a remainder.
5 Identify multiples of a number and common multiples of two or more numbers.	Go back to pages 48–50 and remind yourself about skip counting by a given number, or multiplying factors by each other. This creates multiples.
6 Explain the difference between prime numbers and composite numbers, and explain why the number 1 is neither prime nor composite.	Go back to pages 50–51. Remind yourself that a prime number only has itself and 1 as a factor. Composites have other factors. Think about why 1 does not fall into either group.

Unit 2: Measurement

Chapter 4 Time and temperature

In this chapter you will explore this focus question:

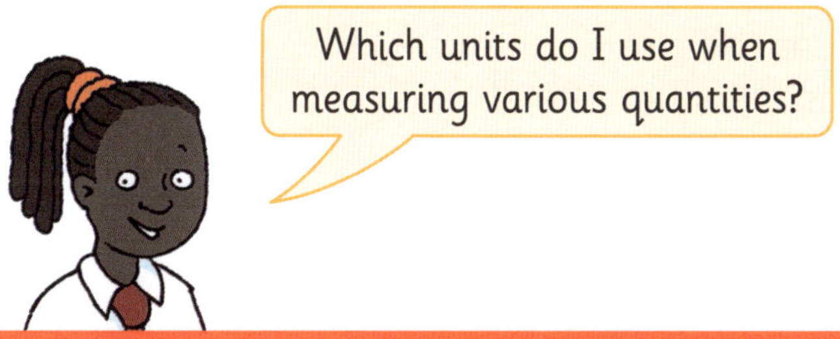

To answer this question, you will:

- read and write time, and work with units of time
- identify a decade or a century, and use AD and BC
- calculate time intervals between events
- perform the four operations on units of time
- estimate and measure temperature in given situations
- record temperatures above and below zero
- tell the difference between two temperatures when one or both is below zero.

Starting point

3 July

11:35 a.m. dentist
4:00 p.m. swimming
7:30 p.m. dinner with Grandma

1 This is Malcolm's schedule on 3 July. He looks at the clock and it shows this time:

a Malcolm has already been to the dentist but not to swimming. Does the clock show a.m. or p.m. time?

b How many minutes until swimming begins?

c Malcolm plans to walk to Gran's from swimming. The walk takes 25 minutes. He needs 20 minutes to get changed. What is the latest time he can leave the pool?

d Gran's birthday was one week earlier. What was the date of her birthday?

2 Look at a clock or a smartphone that shows the time. Write the time:

a right now **b** one hour from now **c** two and a half hours ago.

3 Compare answers from Exercise 2 with others in your group. Try to find at least three different ways of writing each time.

> Remember: you can write the time in words, using a.m. and p.m. time, or using the 24-hour clock.

4 Write sentences to explain the relationships between:

a seconds, minutes and hours **b** hours, days and weeks **c** weeks, months and years.

The 24-hour clock

Key maths idea

The hours on the **24-hour clock** are numbered from midnight on one day to midnight on the next. This is a period of 24 hours. Look at the times on the clock in the picture: the p.m. times (the times after noon) are on the outer scale.

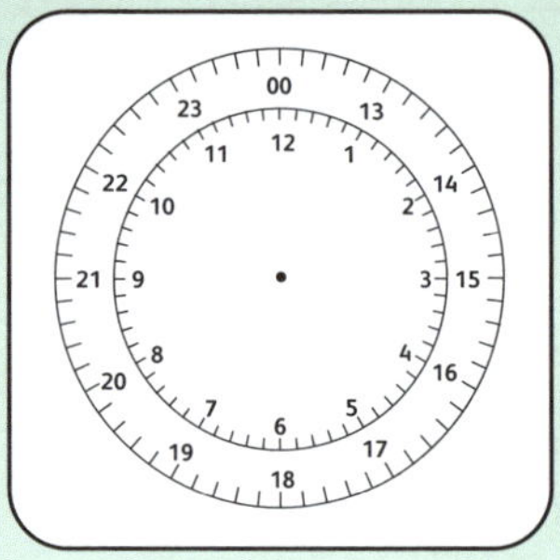

Clock faces usually do not represent all 24 hours, but it is useful to imagine that each hour on the clock face represents two different times.

You can use **partitioning** to convert between the **12-hour clock** and 24-hour clock.

Example

16:00 tells us that the time is 16 hours after midnight. 16 hours can be partitioned into 12 + 4: the 12 hours up to noon (midday) and the additional 4 hours past noon.

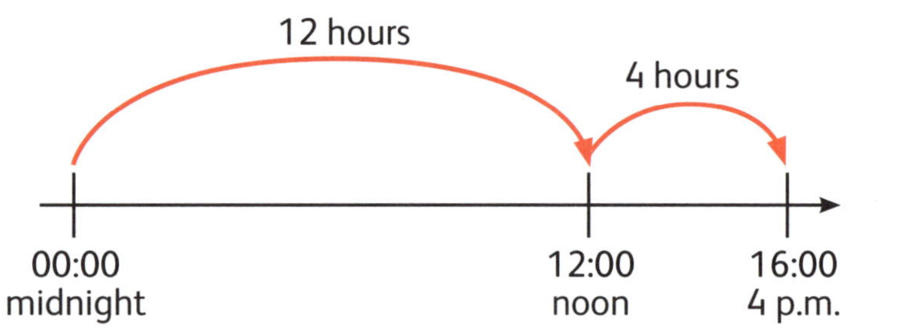

So 16:00 is 4 p.m., which is 4 hours past noon.

Key words

24-hour clock

partitioning

12-hour clock

1 Use the 24-hour clock face above to help you write the times a–j.

12-hour time	24-hour time
12:00 a.m.	a
5:45 a.m.	b
8:55 a.m.	c
10:30 a.m.	d
11:15 a.m.	e

12-hour time	24-hour time
12:00 p.m.	f
2:45 p.m.	g
4:20 p.m.	h
7:55 p.m.	i
10:30 p.m.	j

2 Use a 12-hour clock to write these times as a.m. or p.m.

a 19:35	**b** 07:35	**c** 03:05
d 15:05	**e** 10:10	**f** 22:10

3 Look at these watches.

a Write the times in order from earliest to latest.

b Write the time that is half an hour earlier than each of the four watches in the left column of the picture.

c Write the time that is quarter of an hour later than each of the four watches in the right column of the picture.

4 Mr Harris has booked a seat on the 08:50 flight to Boston. He needs to arrive at the airport two hours before this time. Mr Harris arrives at 6:50 p.m. Explain the mistake that Mr Harris has made.

Schedules and timetables

Key maths idea

We use **schedules** and timetables to show when an event takes place – its start time and its end time. This table shows the times that Malaki spent on some of his activities.

Activity	Start	Finish
Jogging	6:00 a.m.	6:45 a.m.
Breakfast	8:15 a.m.	8:30 a.m.
Library	9:35 a.m.	11:30 a.m.
Lunch	12:10 p.m.	1:00 p.m.

Look at the same information on a **time line**. A time line is a number line on which the numbers indicate time.

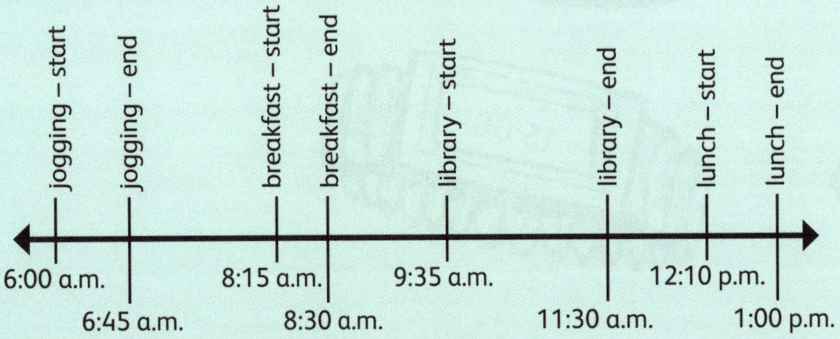

We can use the information to help calculate the length of each activity.

We call this a **time interval**, a time **lapse** (lapse in time) or **elapsed** time.

Let's work out how long Malaki spent at the library.

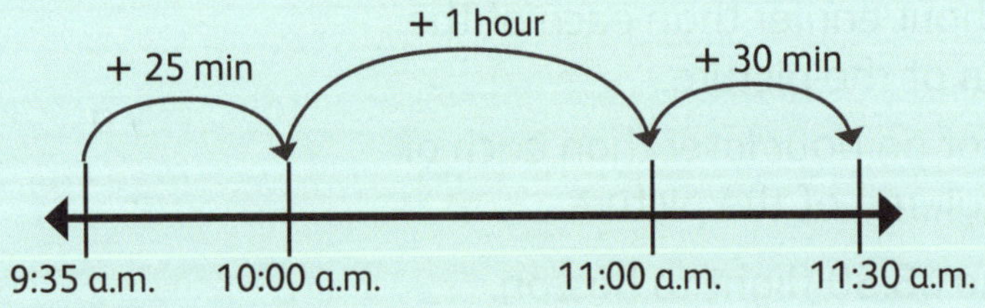

25 minutes + 1 hour + 30 minutes = 1 hour 55 minutes

Malaki spent 1 hour 55 minutes at the library. We can also convert mixed units of time to minutes, because we know there are 60 minutes in an hour.

60 min + 55 min = 115 min

Key words

- schedule
- time line
- time interval
- lapse
- elapsed

This table shows the times that Malaki spent on some of his activities.

Why can we also calculate the time interval by counting on two hours from 9:35 a.m. and then subtracting 5 minutes?

Malaki has football practice in the afternoon. It starts 2 hours 45 minutes (2 h 45 min) after the end of lunch. We can calculate the start time by counting on 2 hours and 45 minutes from 1:00 p.m. to reach 3:45 p.m.

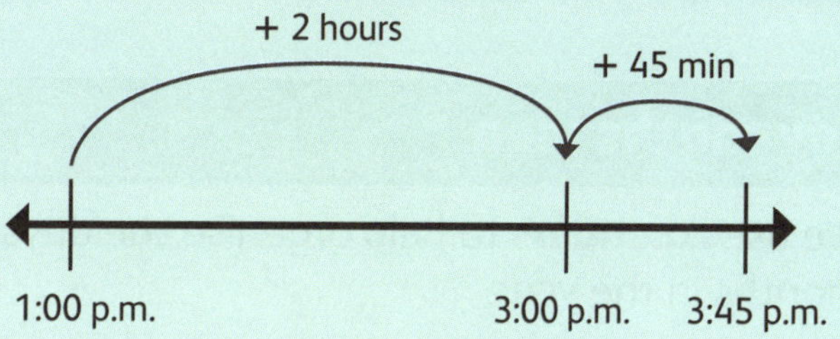

Football practice starts at 3:45 p.m.

1 Calculate how long each of Malaki's other activities last.

2 Use Malaki's time line to answer these questions.

- **a** What is the time interval between the end of library and the start of lunch?
- **b** How much time elapsed between the start of breakfast and the start of lunch?
- **c** How much longer did Malaki spend at the library than at lunch?

3 Here are some time intervals. Write the missing information for a–f.

Time interval in mixed units	1 h 12 min	b	2 h 45 min	d	3 h 29 min	f
Time interval in minutes	a	75 min	c	112 min	e	179 min

4 Look at this flight timetable, which shows departure and arrival times to three destinations.

Flight number	**Destination**	**Departure time**	**Arrival time**
705	Barbados	15:25	17:10
543	Port of Spain	15:45	18:15
432	New York	16:30	20:20

- **a** How long is the flight to Barbados?
- **b** How much longer is the flight to Port of Spain than to Barbados?
- **c** The flight to New York departs 23 minutes late. What time should it now arrive?
- **d** A passenger arrives at the airport at 13:18. How long must they wait until the flight to Port of Spain?
- **e** Explain why you think that times for air, ferry and train travel and other modes of transport are usually shown using 24-hour clock notation.

5 Tamila finishes school at 3:30 p.m. The last lesson of the day is art. This lasts for 55 minutes. Before the art lesson, the school afternoon starts with a 1 hour 15 minute history lesson and then a 20-minute maths test. What time does the school afternoon begin?

6 Use a table or a time line to create a schedule for six activities in your own day. Remember to show the start and finish times. Use 24-hour clock notation. Then make up some questions to ask others about your schedule.

Working with calendars

Key maths idea

We use **calendars** to help organise the days, weeks and months in the year.

What information might people put on a calendar? When have you used a calendar?

Look at the month shown on the calendar here. Which month do you think it is? Why? What clues can you use to help you?

SUN	MON	TUE	WED	THU	FRI	SAT
29	30	1	2	3	4	5
6	7	8	9	10	11	12
13	14	15	16	17	18	19
20	21	22	23	24	25	26
27	28	29	30	31	1	2

Today is the 7th day of the month.

Amy will celebrate her birthday on the 30th day of the month.

We can count on from the 7th to the 30th.

Let's start by counting the weeks and then the days.

We can use a time line or the calendar to help us.

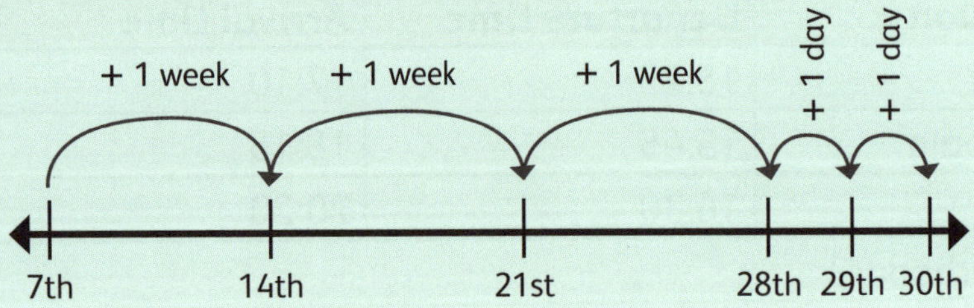

SUN	MON	TUE	WED	THU	FRI	SAT
29	30	1	2	3	4	5
6	7	8	9	10	11	12
13	14	15	16	17	18	19
20	21	22	23	24	25	26
27	28	29	30	31	1	2

Amy's birthday is in 3 weeks and 2 days' time.

We can use multiplication to help show this in days.

7 days × 3 = 21 days

21 days + 2 days = 23 days

Key word

calendar

We can also find 7 + ☐ = 30 or 30 − 7 = ☐ to solve this problem. Explain why.

1 Look at the calendar shown here.

2021 FEBRUARY						
Sun	Mon	Tue	Wed	Thu	Fri	Sat
31	1	2	3	4	5	6
7	8	9	10	11	12	13
14	~~15~~	16	17	18	19	20
21	22	23	24	25	26	27
28	1	2	3	4	5	6

a How do you know that 2021 was not a leap year?

b How many days have passed in 2021 so far, not including 15 February?

c What will the date be in 3 weeks' time?

d Mr Baker ordered a new oven on 15 February. He expects it to arrive on 1 April. How many days later is this?

e Mrs Baker started her new job exactly 5 weeks before 15 February. On what date did she start the new job?

2 Draw your own time lines to answer these questions.

a What date is 2 weeks and 6 days later than 23 November?

b What date is 28 days before 14 June?

c How long is it in weeks and days from 1 May to 1 July?

d How long is it in weeks and days from 16 October to 31 December?

3 Use these calendars to answer the question.

MAY						
Su	Mo	Tu	We	Th	Fr	Sa
			1	2	3	4
5	6	7	8	9	10	11
12	13	14	15	16	17	18
19	20	21	22	23	24	25
26	27	28	29	30	31	

JUNE						
Su	Mo	Tu	We	Th	Fr	Sa
						1
2	3	4	5	6	7	8
9	10	11	12	13	14	15
16	17	18	19	20	21	22
23	24	25	26	27	28	29
30						

JULY						
Su	Mo	Tu	We	Th	Fr	Sa
	1	2	3	4	5	6
7	8	9	10	11	12	13
14	15	16	17	18	19	20
21	22	23	24	25	26	27
28	29	30	31			

The West Indies cricket team is scheduled to play six one-day matches with 5 days' rest in between. Their first match is on 5 May. What are the dates of their other matches?

4

Mr Price received five letters on 24 February with these postmarks:

a Write down the place where each letter was posted. Next to each, write down how long it took for the letter to arrive.

b Which letter took the longest?

5 Christie is born at 20:41 on 28 April. When will she be exactly 60 hours old?

Years, decades and centuries

Key maths idea

Every day we tell the time in seconds, minutes and hours. We tell longer **periods** of time in days, weeks, months and years. We can group years into even longer periods of time.

10 years = 1 **decade**	100 years = 1 **century**	1 000 years = 1 **millennium**

The millennium in the year 2000 marked the start of the 21st century.
This time line shows some important dates in Jamaican history.

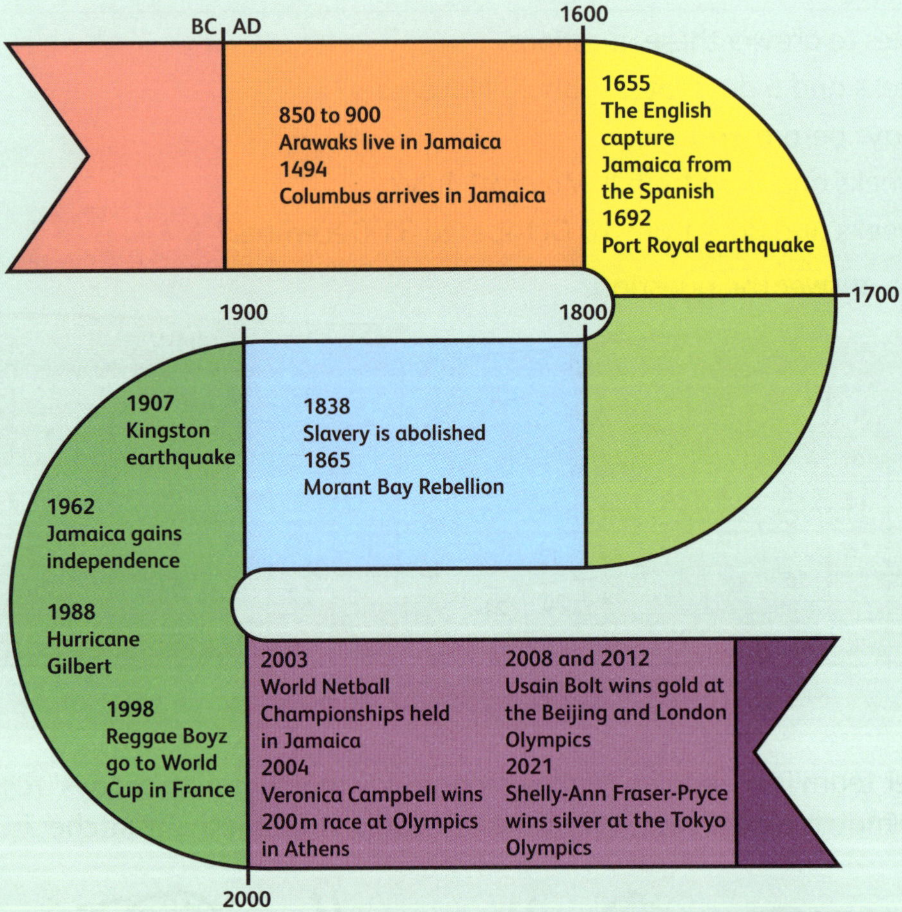

Notice the two letters **AD** at the start of the time line. This tells us that the dates are '**after Christ**' ('anno domini'). Dates before the birth of Jesus Christ use the letters **BC** for '**before Christ**'. For example, King Tutankhamun of Egypt died around 1324 BC. This date is earlier than the start of the Roman Empire in 27 BC.

Look at the events on the time line.
Which events took place in the 21st century? How do you know?

Key words
period
decade
century
millennium
AD (anno domini, or after Christ)
BC (before Christ)

Hurricane Gilbert happened in 1988. This was in the 20th century. We can also say that Hurricane Gilbert happened in the decade of the 1980s or the 9th decade of the 20th century.

Talk about the other events on the time line in a similar way.

1 Copy and complete.

a _____ years = 1 decade

b 30 years = _____ decades

c _____ years = 5 decades = $\frac{1}{2}$ _____

d 100 years = _____ century

e _____ decades = 1 century

f _____ years = 1 millennium

g _____ centuries = 1 millennium

2 Compare these sets of dates. Which date is earlier in each pair?

a 1998 AD or 1989 AD

b 1066 AD or 1066 BC

c 100 BC or 27 BC

d 2021 AD or 2012 AD

3 **a** Sort these dates for the summer Olympic Games into the correct decades. One date will be left over.

Athens 2004	Los Angeles 1984	Rio de Janeiro 2016
Atlanta 1996	Melbourne 1956	Rome 1960
Barcelona 1992	Mexico City 1968	Sydney 2000
London 2012	Munich 1972	Tokyo 1964

1960s	1970s	1980s	1990s	2000s	2010s

b Which Olympic Games took place in the 1st decade of the 21st century?

4 Compare these sets of dates. Which date is earlier in each set? Sort these dates into the correct centuries.

1939 Start of World War 2	2011 Earthquake and tsunami in Japan
1850 Human population reached 1 billion	1502 Columbus made his last voyage to the Caribbean
1961 First humans in space	1788 European settlers arrive in Sydney, Australia
1653 Taj Mahal completed	1876 Alexander Graham Bell invents the telephone

16th century	17th century	18th century	19th century	20th century	21st century

Years, events and time intervals

Key maths idea

This time line shows important dates in the life of Jamaica's national heroes.

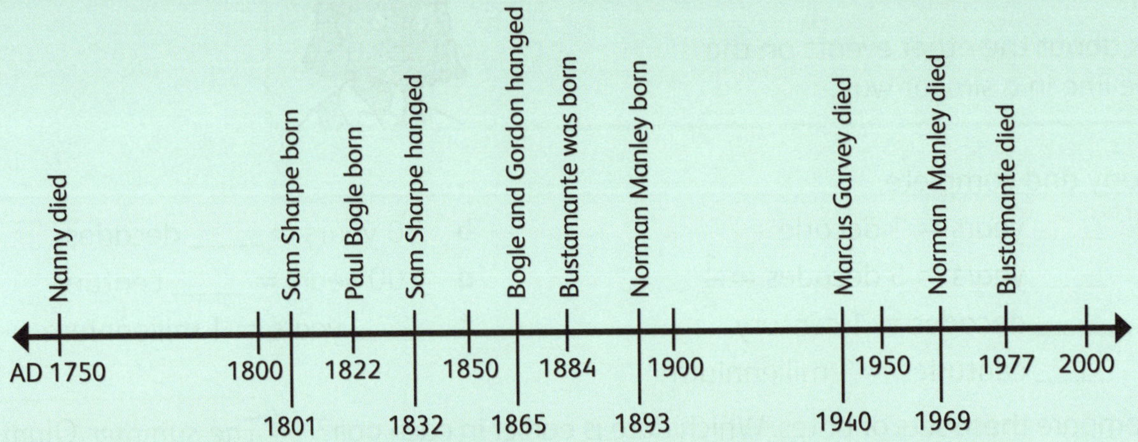

The time line shows us that Sam Sharpe was born in 1801 and died in 1832. We can calculate the time interval between these two events.

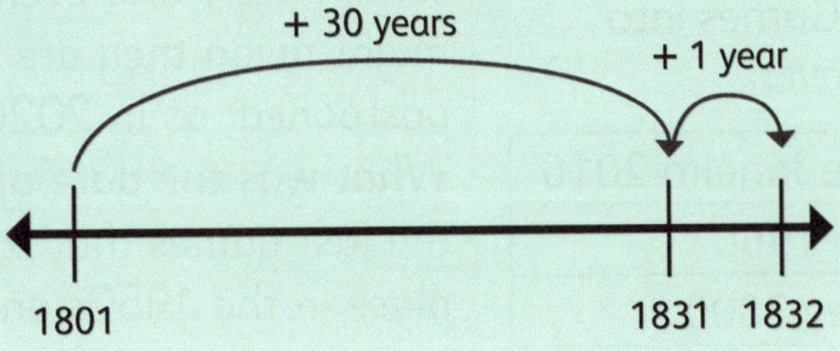

The time interval between 1801 and 1832 is 30 years + 1 year = 31 years.

1 Use the time line about Jamaican heroes to answer these questions:

a In which century did Nanny die?

b For approximately how many decades did Sam Sharpe live?

c Which hero was born exactly one decade before Sam Sharpe's death?

d How many years later did Marcus Garvey die than Paul Bogle?

e Which hero was born in the last decade of the 19th century?

f Which hero lived for almost a century? Calculate the number of years.

2 Calculate the time intervals between these events.

a	1914 AD Start of World War 1	1939 AD Start of World War 2
b	1850 AD Human population reached 1 billion	1975 AD Human population reached 4 billion
c	10 AD Population of Rome reaches 1 million	1850 AD Population of London reaches 1 million
d	776 BC First Ancient Olympic Games in Greece	535 BC Construction on Isle of Samos of oldest known tunnel still in use

3 **a** Make a time line to show your own life and main milestones in years (your birth, when you started school, and so on). You can expand the time line to include your parents, grandparents and other interesting people, both from and outside of Jamaica.

b Create some questions for others to answer about your time line.

Temperature

Key maths idea

We use **degrees Celsius** (°C) or **degrees Fahrenheit** (°F) to measure **temperature**.

This **thermometer** shows a temperature of −4 °C.

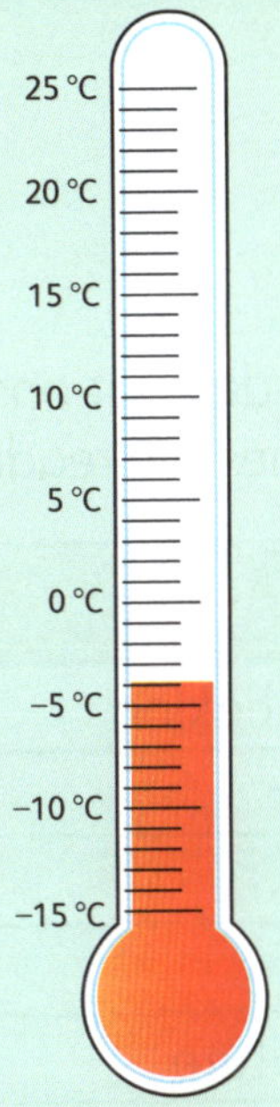

As the temperature falls further below zero, it gets colder. So −10 °C is colder than −4 °C.

As the temperature rises above zero, it gets warmer. So 10 °C is warmer than 4 °C.

This table shows the average temperatures for December in four cities.

	Beijing	Moscow	Anchorage	Munich
High	3 °C	−4 °C	−4 °C	4 °C
Low	−7 °C	−9 °C	−10 °C	−2 °C

The highest average temperature in Moscow is 7° lower than the highest average temperature in Beijing. We can check this on a number line.

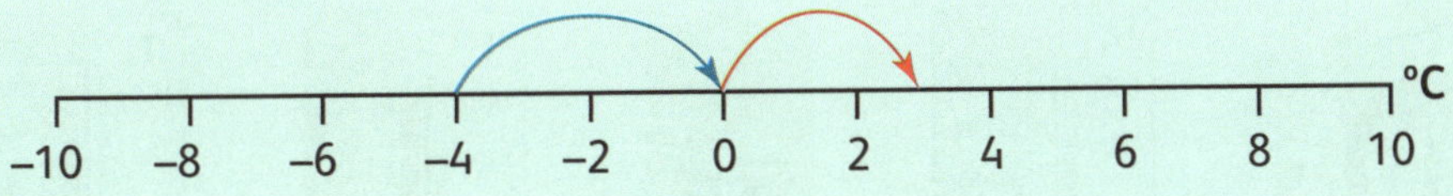

Key words

degrees Celsius
degrees Fahrenheit
temperature
thermometer

Remember that when a calculation crosses zero, it is useful to stop at zero first before counting on or back.

1 Answer these questions.

a How much warmer is the average low temperature for Munich than the average low temperature for Anchorage?

b What is the difference between the average low and the average high temperature for Beijing?

c The average low temperature in Moscow is 7 degrees colder than for _____ .

d What is the midway temperature between the high and low values for Anchorage?

e Which city's average temperatures have a difference of 5 degrees?

2 The thermometer shows the starting temperature each time. For each temperature, write the new temperature after:

- **a** a rise of 12 °C
- **b** a fall of 8 °C
- **c** a rise of 10 °C
- **d** a fall of 12 °C
- **e** a fall of 7 °C followed by a fall of 5 °C
- **f** a fall of 11 °C followed by a rise of 10 °C.

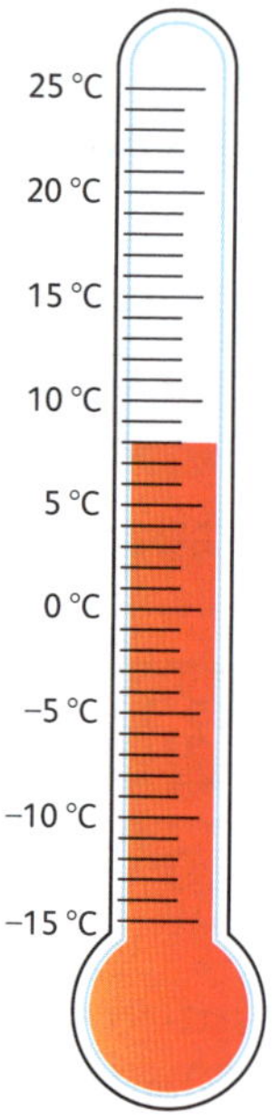

3 Copy the table. Use the descriptions and then write the letters for the thermometer column that best matches the readings on each thermometer.

Description	°C	°F	Thermometer
Oven temperature for baking	180	356	
Boiling point for water	100	212	
Cup of tea	60	140	
Hot bath	40	104	
Healthy body temperature	37	98.6	
Warm summer day in the Caribbean	28	82.4	
Freezing point of water	0	32	

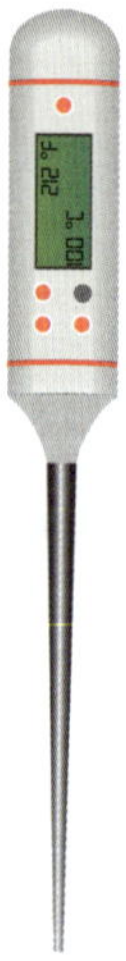
a

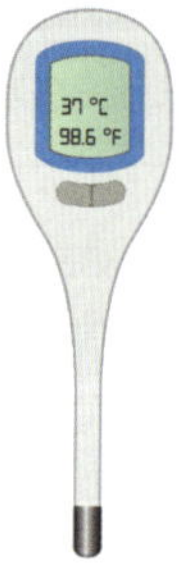
b

c

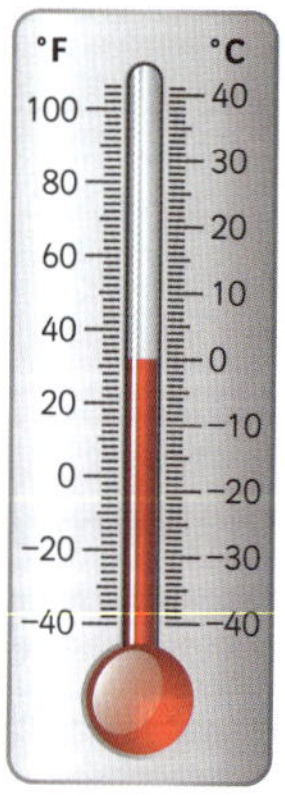
d

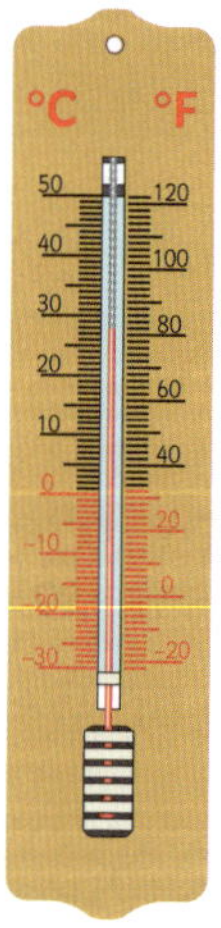
e

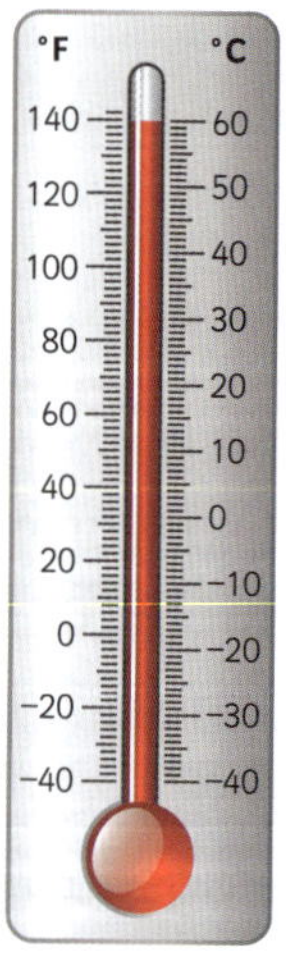
f

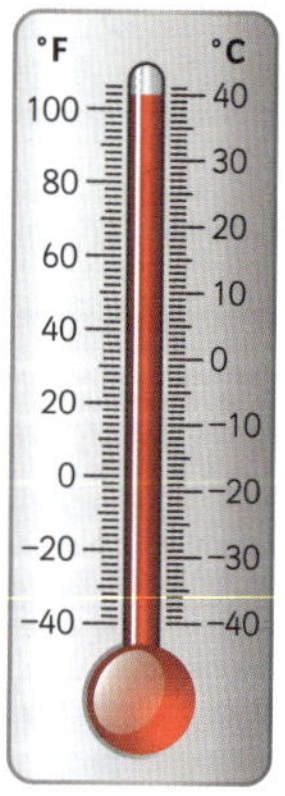
g

4 Write the temperatures for each pair of thermometers. Then calculate the difference in temperatures.

a

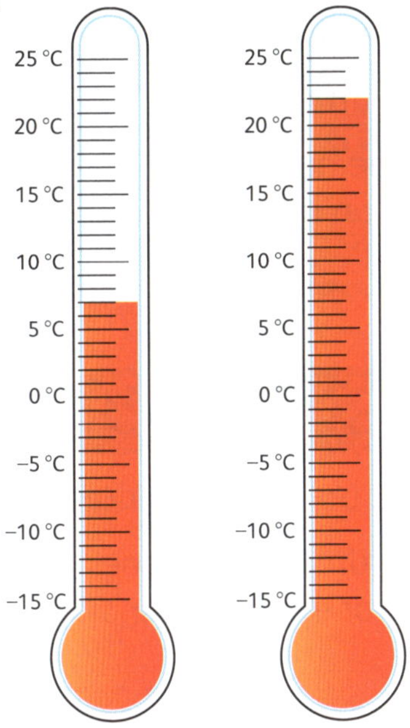

b

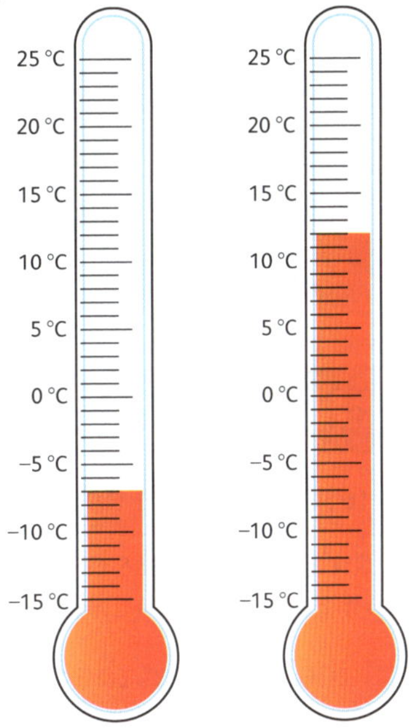

5 A temperature given in degrees Celsius can be converted to an approximate temperature in degrees Fahrenheit by doubling and adding 30. Copy and complete this table of temperature conversions.

> 10 °C is approximately 50 °F because
> 10 × 2 + 30 = 50

Degrees Celsius (°C)	14 °C	18 °C	20 °C	25 °C	33 °C	42 °C	100 °C
Degrees Fahrenheit (°F)							

Maths detective: Work it out

Use your maths detective skills to solve these cases.

Case 1

The table shows the cost for parking a car in a car park. David parks his car at 13:40 and leaves at 16:25. Cora parks her car at 13:55 but pays $70 less than David. What is the latest time that Cora could have left the car park?

Parking time	Amount
Up to 1 hour	$150
1–2 hours	$210
2–3 hours	$280
3–4 hours	$310
Over 4 hours	$380

Case 2

Julie saves $90 each day in June and $45 each day in July. How much money does she need to save each day in August so she has enough money to buy a computer game costing $4 025?

Case 3

Mr Jackson switches on his new freezer. The temperature inside the freezer is currently 12 °C. On average, the temperature in the freezer drops 6° every two hours.

1 How many hours should it take to reach the target temperature of −18 °C?

2 If the freezer reaches −18°C at 19:35 and takes exactly the number of hours you calculated in question 1, what time did Mr Jackson switch on the freezer?

Real-world maths: History of clocks

Work with a partner to research more about time. You can use books or the internet.

1 Long ago, people didn't have clocks and watches to tell the time like the ones we use today. Find out about the different instruments they used, for example, sundials.

2 How did the days of the week get their names?

3 When was the calendar we use today introduced? What is it called?

4 Nowadays we also use lots of expressions to talk about time. What do you think they mean? Write your own explanations.

There is no time like the present.

I am just killing time.

We should call it a day.

It is crunch time.

This can help you to save time.

What I have learned

1 Choose two islands in the Caribbean that you would like to visit.

2 Look at the school calendar and find out the dates of the school holidays.

3 Check the temperatures of the two destinations for the dates of the school holidays.

4 Write a paragraph explaining which would be a better choice of holiday destination. Take into account the travel time and the weather. Add any other reasons of your choice.

5 Work out a travel itinerary for a one-week holiday. This is a time line or schedule that gives details of departure and arrival times, and what you would do on each day of your holiday.

You will need to use the internet to do research for this question.

Practice questions

1 Write all of these times using 24-hour time notation.

a 7:15 a.m. b 10:40 a.m. c 12:00 noon

d 3:48 p.m. e 11:30 p.m. f 12:00 midnight

2 Work out the end time for each activity.

a I go for a run at 11:25 a.m. I run for an hour and ten minutes.

b A TV show starts at 20:05. It lasts for 45 minutes.

c Sports day starts at 8:15 a.m. It finishes five and a half hours later.

3 Work out the start times.

a I get home at 5:10 p.m. after a 55-minute car journey. What time did I start the journey?

b A movie finishes at 7:55 p.m. It was 2 hours 35 minutes long. What time did it start?

4 How long did it take? Work out the elapsed times.

a Devon put a cake in the oven at 11:07 a.m. and took it out at 11:42 a.m. How long did it bake for?

b Melisha started a car journey at 10:32 a.m. and arrived at her destination at 12:05 p.m. How long was the journey?

c Candace started her run at 18:20 and finished at 19:07. How long did she run for?

5 Look at these dates.

14 April 1985	30 September 2003	23 June 1861	5 March 1901
11 December 1875	2 August 1949	1 October 2003	19 April 1967

a List the dates that are in the 20th century.

b Which date is in the 7th decade of the 19th century?

c Which date is 3 weeks and 3 days before 24 October 2003?

6 This chart shows the minimum and maximum temperatures for one day, as recorded at different locations.

a Calculate the difference in the maximum and minimum temperatures at each location.

b The minimum temperature at location E was 9 °C lower than the maximum temperature that day. Calculate the minimum temperature.

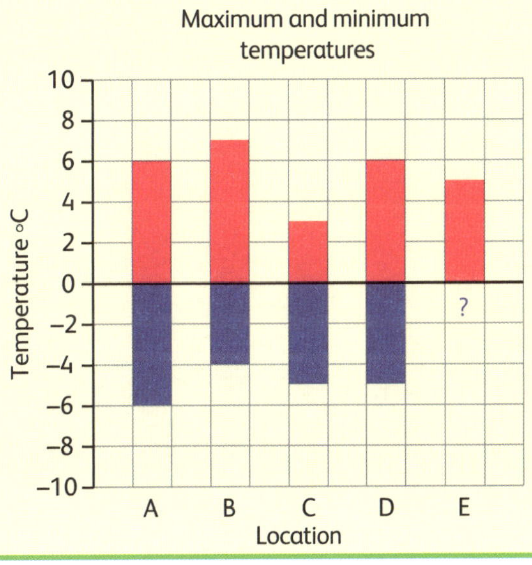

4 Self check

	Check you can do this:	If you have difficulty, try this:
1	Create and read schedules, and calculate time intervals involving shorter or longer periods of time.	Go back to pages 58–59, and remind yourself how to read the start and finish time from the chart. When you subtract start times from finish times, don't forget that there are 60 minutes in one hour. You can also look back at pages 61–62.
2	Say the decade and century for a given year; interpret and use the letters AD and BC after a year.	Remember that a decade is a period of ten years and a century is a period of 100 years, so look at the tens or hundreds place in the year. Also go back to pages 62–63 and remind yourself of the work you did on longer periods of time.
3	Create and read time lines.	Go back to pages 64–65, and remind yourself how a time line works a bit like a number line.
4	Interpret and record temperatures, and calculate the differences between temperatures.	Revise the work on pages 65–67 and remind yourself what degrees are and how to read a thermometer. Make sure you can calculate the difference between positive temperatures, as well as the difference between positive and negative temperatures. You can use a number line to help you.
5	Use calculating skills to solve problems about time and temperature.	Revise the problems you solved on page 67.

Chapter 5 Length, mass and capacity

In this chapter you will explore this focus question:

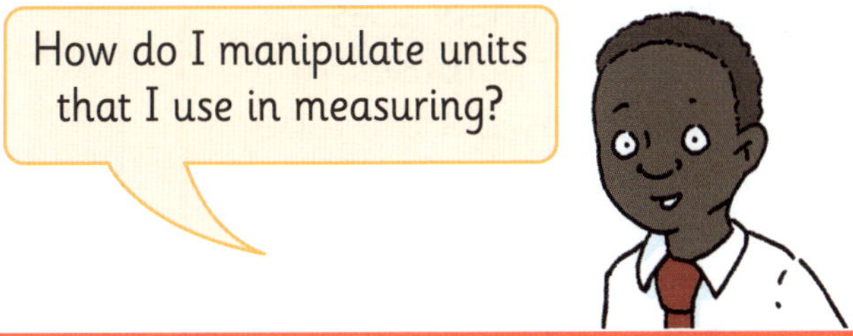

To answer this question, you will:

- use the relationships between units of measurement of length, mass and capacity
- estimate and measure distances
- solve problems involving millimetres, centimetres, metres and kilometres
- work with grams, kilograms and tonnes
- choose the most appropriate metric units to solve real-life problems.

Starting point

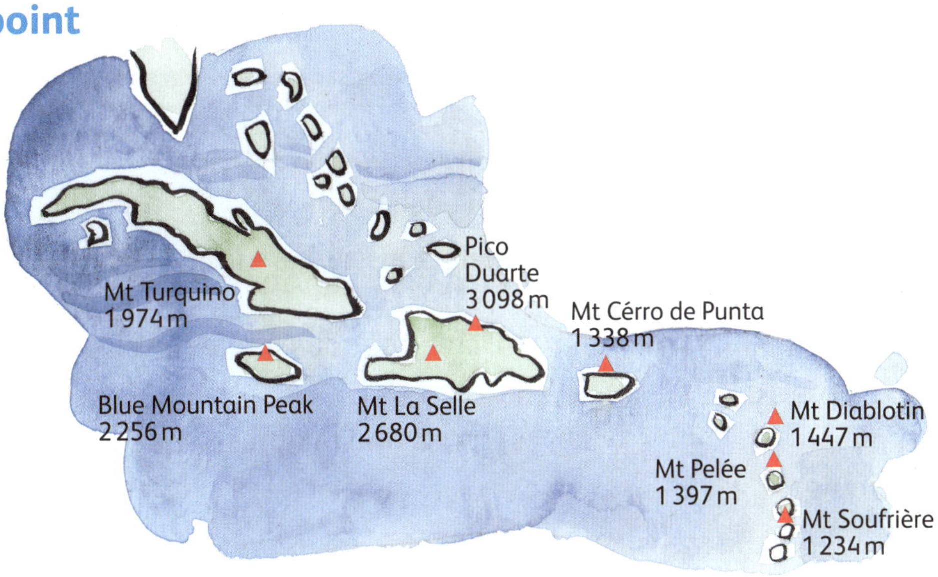

1 The map shows the eight highest mountains in the Caribbean.

a Can you name the mountains in order of height, from lowest to highest?

b How much lower is Mount Turquino than Mount La Selle?

2 A travel company wants to offer a 2-week trip for tourists to visit the three highest mountains in the Caribbean. All trips will depart from Miami and return there.

Work with a partner to plan the trip. Remember to think about distances, transport to and from the islands, and timings. Consider which months of the year may be better for the trip and give advice about average temperatures so visitors know what to bring with them.

Present your trip to the rest of the class.

Revising what you know about measurement

Key maths idea

These are **measuring instruments**. Each instrument uses **units of measurement** that are suitable for its purpose. For example, bathroom scales measure in kilograms, but the tape measure is marked in centimetres and metres.

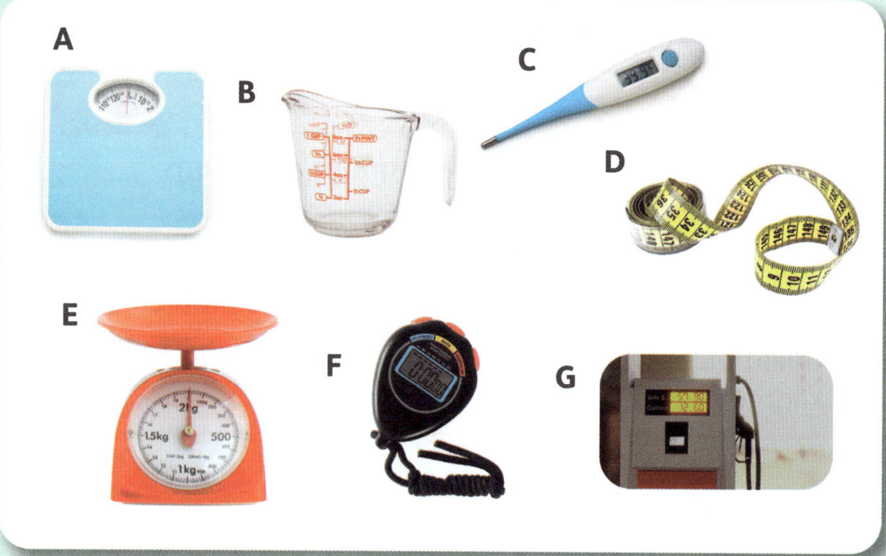

Key words

measuring instrument

unit of measurement

1 Work with a partner. Talk about each measuring instrument. Discuss what you would measure with each one.

2 Match the units of measure to the correct instrument in the picture.

a millilitres
b millimetres and centimetres
c degrees Celsius
d litres
e grams
f seconds and minutes
g kilograms

3 Choose three of the measuring instruments. Discuss what you would use if you needed this instrument but you did not have one available. What other way could you use to measure?

Remember there are 1 000 grams in 1 kilogram, and 1 000 kilograms in 1 tonne.

4 Convert from kg to tonnes. The first one is done as an example.

a 1 455 kg = 1.455 t
b 1 500 kg = _____ t
c 485 kg = _____ t
d 17 kg = _____ t
e 2 209 kg = _____ t
f 2 kg = _____ t

5 Convert from tonnes (t) to kilograms (kg). The first one is done as an example.

a 1.5 t = 1 500 kg
b 0.5 t = _____ kg
c 0.25 t = _____ kg
d 0.001 t = _____ kg
e 1.005 t = _____ kg
f 65.081 t = _____ kg
g 100.018 t = _____ kg
h 19.084 t = _____ kg

Units of measurement

Key maths idea

Look at the meaning of the prefixes '**centi**', '**milli**' and '**kilo**' and how they relate to the metre.

Term	'Centi' (hundredth)	'Milli' (thousandth)	'Kilo' (thousand)
Length	A **centimetre** (cm) is $\frac{1}{1\ 00}$ of a metre (m)	A **millimetre** (mm) is $\frac{1}{1\ 000}$ of a metre (m)	A **kilometre** (km) is 1 000 metres

The words help us to remember the **relationships** between the units of measurement. This is useful when we need to convert from one to the other.

Centimetres, metres and millimetre are examples of **units of measurement**. You will notice the same prefixes in the units for different measures. For example, when we measure how heavy an object is, we may measure in kilograms, and when we measure how much a container holds, we may measure in kilolitres.

3 kilometres		
1 000 m	1 000 m	1 000 m

We can use a part-whole diagram to help convert 3 kilometres to metres.

We can also use a place value chart to convert 3 kilometres to metres.

Th	H	T	O	
			3	
3	0	0	0	3 × 1 000

Can you see how each example works?
How many metres are there in 3 kilometres?
What should we do differently to convert 3 metres to centimetres?
We can use multiplication to convert a larger unit of measure to a smaller unit.

We can use division to help convert from a smaller unit of measure to a larger unit.

Example 1

There are 1 000 grams in a kilogram, so 5 kg is equal to 5 000 g because 5 × 1 000 = 5 000

Example 2

5 000 g in kilograms is 5 kg because 5 000 ÷ 1 000 = 5

Key words

centi-
milli-
kilo-
centimetre
millimetre
kilometre
relationship
unit of measurement
centilitre
decimetre

1 How many times as long is:

- **a** 2 kilometres than 2 metres
- **b** 10 kilometres than 10 metres
- **c** 2 kilometres than 1 metre
- **d** 5 metres than 100 centimetres?

2 Complete the sentences:

- **a** A _____ is a thousandth ($\frac{1}{1\,000}$) of a litre.
- **b** A **centilitre** is a _____ ($\frac{1}{100}$) of a litre.
- **c** A _____ is a thousand (1 000) grams
- **d** A milligram is a _____ of a gram.

3 Complete these conversions. You can use part-whole diagrams or place value charts to help you.

- **a** 5 kilograms = _____ grams
- **b** _____ kilograms = 10 000 grams
- **c** 12 litres = _____ millilitres
- **d** 2.5 metres = _____ centimetres
- **e** 4 500 millilitres = _____ litres
- **f** 40 millimetres = _____ centimetres

4 'Deci-' means one-tenth ($\frac{1}{10}$).

- **a** How many centimetres do you think are equal to one **decimetre**?
- **b** Explain or show how you know this.

5 Look at these buckets. Three amounts of water are poured into each one.

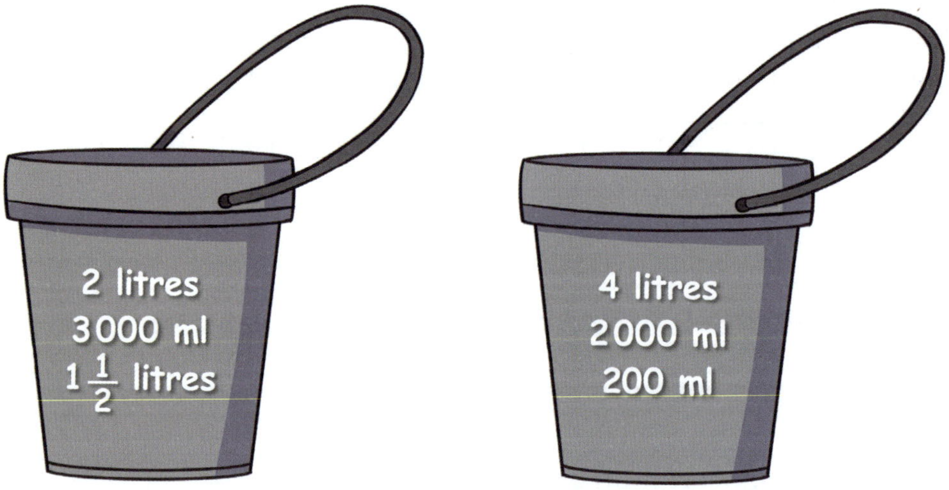

Bucket A

Bucket B

- **a** Which bucket has the most water in total? Explain your thinking.
- **b** Make up a similar problem of your own about mass or length for others to solve.

Working with length and height

Key maths idea

You can use rulers and tape measures to accurately measure the **length** and **height** of objects or spaces, for example, the dimensions of a room.

We measure smaller objects in millimetres and centimetres.

The length of the straw shown here is 12 cm 5 mm or $12\frac{1}{2}$ cm or 12.5 cm or 125 mm.

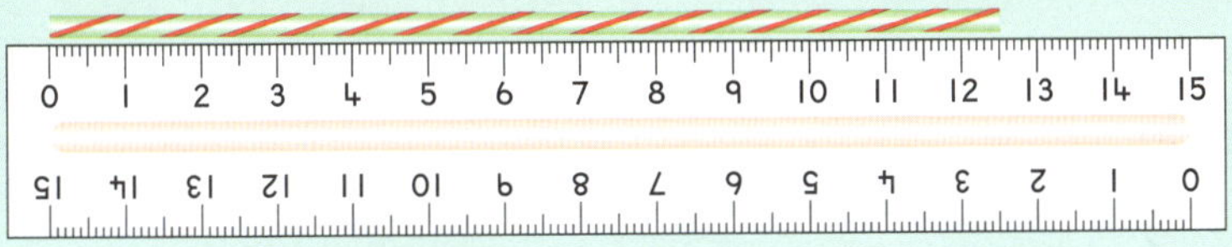

There are 100 centimetres in a metre (1 m = 100 cm), so it will take 8 of these straws placed end to end to equal 1 metre.

Metres, centimetres and kilometres are **metric** measures of length. They use a base ten system. In some places, people still also use **imperial** measures such as **inches** and **feet** to measure length.

12 inches (12") = 1 foot (1 ft.)   3 ft. = 1 yard (1 yd.)

1 760 yards = 1 mile (1 mi.)

An inch is approximately 2.5 cm. Approximately how many inches long is the straw?

How can you work out the approximate number of **centimetres** in a foot?

Key words

length

height

metric

imperial

inch

foot

centimetre

Measurement

1 Look at items A to D.

a Estimate the length of each item shown here. Write down your estimates.

A

B

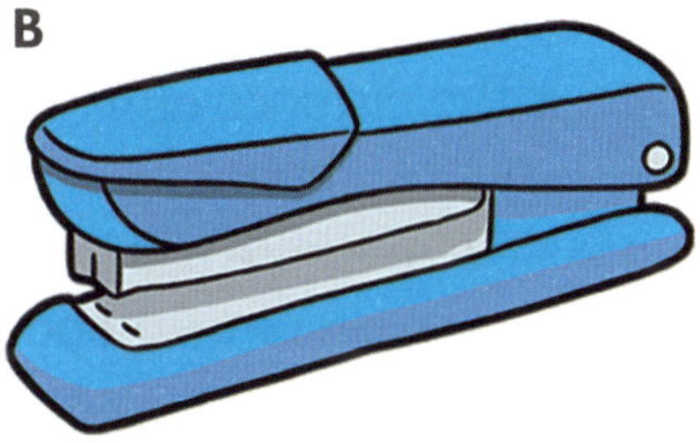

C

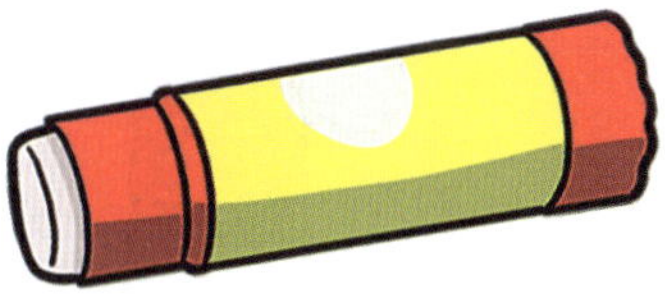

D

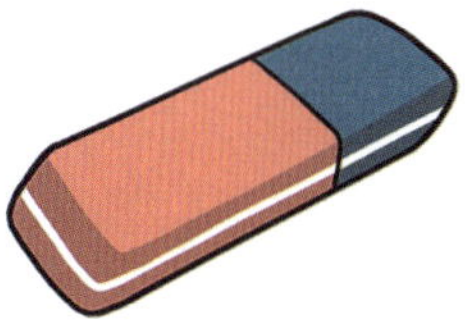

b Now use your own ruler to measure each item. How accurate were your estimates? Copy and complete the table to show the actual measurements in centimetres and also in millimetres.

	Centimetres (cm)	**Millimetres (mm)**
Pencil		
Stapler		
Rubber		
Glue stick		

c Which of the objects is approximately 2 inches long?

2 A crayon is $9\frac{1}{2}$ cm long. Accurately draw straight lines that are:

a 6 cm longer than the crayon

b $3\frac{1}{2}$ cm shorter than the crayon

c half the length of the crayon

d double the length of the crayon

e a decimetre longer than the crayon.

3 Work in groups of five and measure the height of the students in your group. Complete a table like the one below.

Name	**Height**

4 Use your table to find this information:

a the tallest student in your group

b the shortest student in your group

c a student with a height that rounds to 2 metres when rounded to the nearest metre

d the combined height of the five students in your group

e the difference in height between the tallest and shortest students in the group.

5 **a** Work with a partner. Use a metre ruler to estimate the length, width and perimeter of your classroom. Remember that the perimeter is the distance around a 2-D shape. Record your estimates.

b Now measure the actual dimensions of the classroom. Compare these with your estimates. How close were you? How can you improve your estimating skills?

More about perimeter

Key maths idea

Look at these regular **polygons**. What makes a shape regular?

3 cm

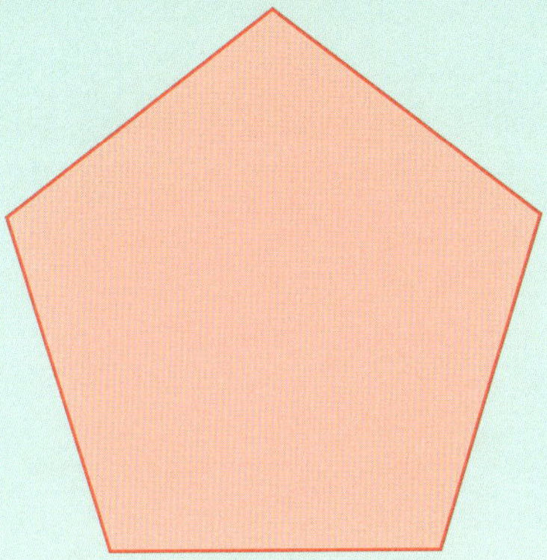

3 cm

Let's try out Junelle's idea:

The length of one equal side is 3 cm, so the **perimeter** can be calculated as

$$3\text{ cm} \times 4 = 12\text{ cm}$$

We can check this with addition:

$3\text{ cm} + 3\text{ cm} + 3\text{ cm} + 3\text{ cm} = 12\text{ cm}$

The perimeter of the square is 12 cm.

The length of one side of the regular pentagon is also 3 cm. This time we calculate the perimeter as $3\text{ cm} \times 5 = 15\text{ cm}$ because there are 5 equal sides.

Key words

polygon

perimeter

I think that the perimeter of a square can be calculated by multiplying the length of one side by 4.

1 Name, measure and then calculate the perimeter of the regular 2-D shapes. Think carefully about what you need to do differently each time.

a

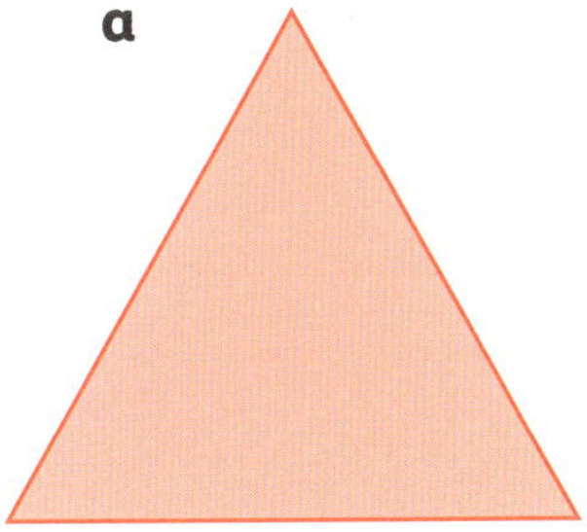

b

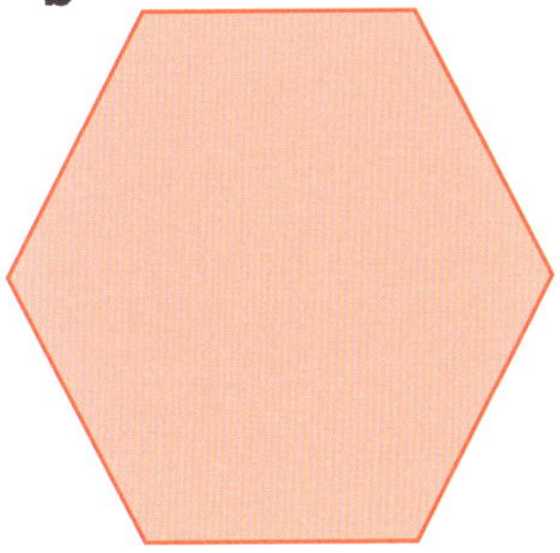

c

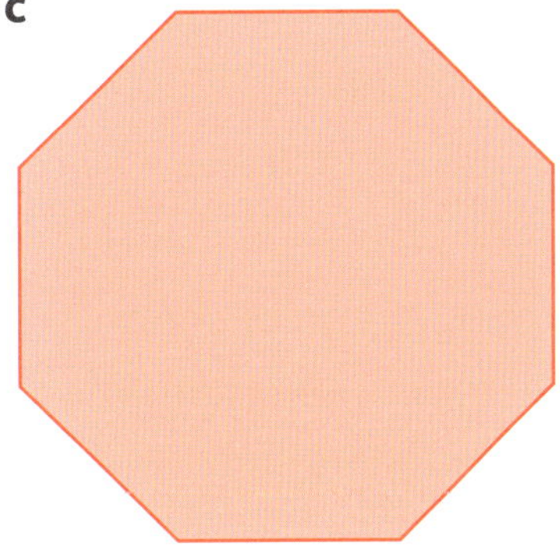

2 Find the perimeter of each rectangle. Measure to the nearest millimetre. How can you make calculating easier?

a

b

c

3 Here are some 2-D polygons. These are all irregular.

A

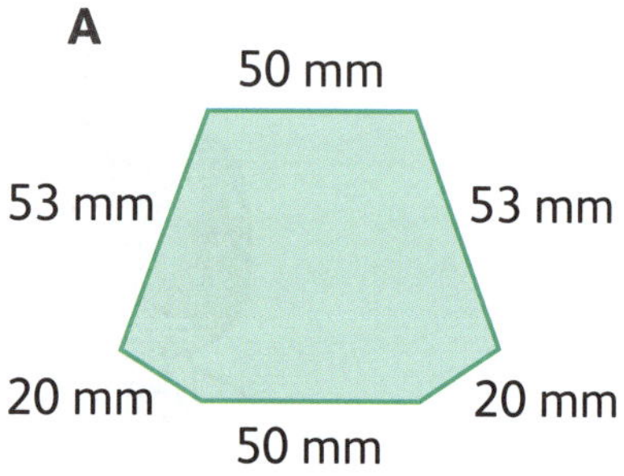

B

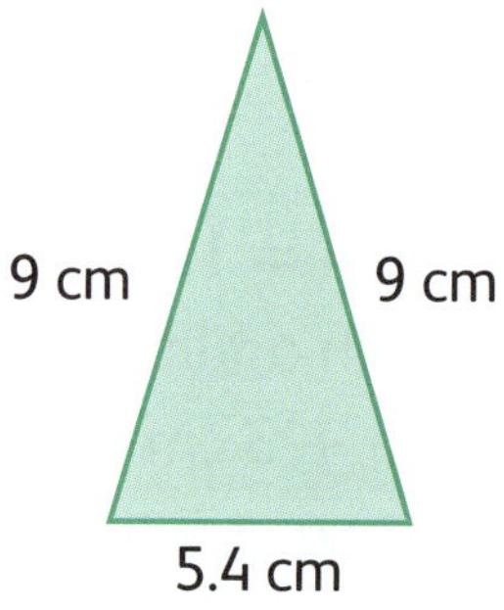

C

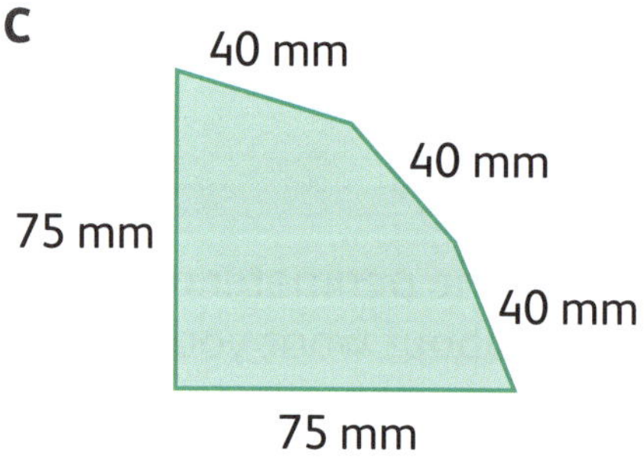

a Name each shape.
b Calculate the perimeters.
c Explain how you made calculating easier each time.

4 The perimeter of a regular shape is 30 metres.
What is the length of one side when the shape is:

a an equilateral triangle
b a square
c a pentagon
d a hexagon?

Working with distance

Key maths idea

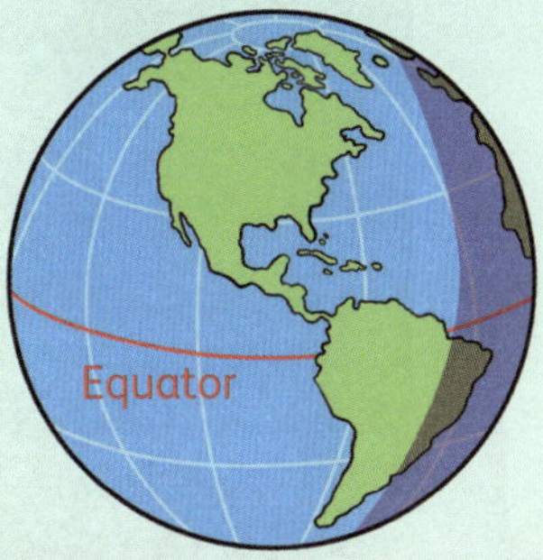

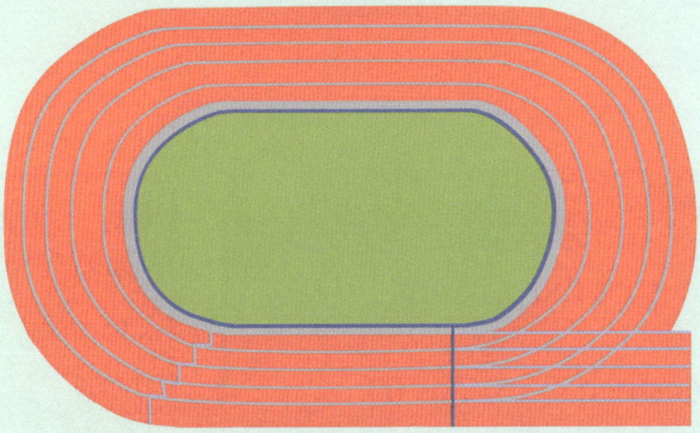

We use **metres** to measure shorter **distances**.

We use **kilometres** to measure longer distances.

There are 1 000 metres in a kilometre.

The imperial measure of **miles** is often used for distance.
Where have you seen distances shown as miles?

1 mile is approximately 1.6 kilometres.

Key words

metre
distance
kilometre
miles

The total distance around the equator is 40 075 kilometres or 24 901 miles.

The total distance around the running track is 400 metres or just under a quarter of a mile.

1 Say which is further:

- **a** 1 200 metres or 1 kilometre
- **b** 7 500 metres or 7.6 kilometres
- **c** 200 000 metres or 201 kilometres
- **d** 10.5 kilometres or 10 500 metres?

2 Elizabeth filled in this table to show how far she walked in five days.

- **a** Write the distances so that they are all in kilometres.
- **b** What distance did she walk altogether in kilometres?
- **c** On which two days did she walk exactly 5 km altogether?
- **d** On Thursday Elizabeth recorded $\frac{1}{2}$ km too much by mistake. What is the correct distance for Thursday?
- **e** What is her new total distance in kilometres?
- **f** How many metres did Elizabeth walk in the five days?

Day	Distance walked
Monday	3.5 km
Tuesday	2 400 m
Wednesday	2.6 km
Thursday	3 200 m
Friday	850 m

3 This road sign shows the distance in miles to different locations in the world.

a How much further is it to New York than to Miami?

b How much further is it to Paris than to London?

c What other comparisons can you make?

d Amy says that the distance to Paris is about 5 000 km. Do you agree? Explain why you think this is right or wrong.

Try this conversion challenge!

Working with mass

Key maths idea

We use **grams** and **kilograms** to measure the **mass** of everyday objects.

Objects that are really heavy are measured in **tonnes** (t).

1 tonne = 1 000 kilograms

Trucks, ships and aeroplanes are measured in tonnes. So are some animals, for example, elephants.

This is an equal arm balance. It shows that 1 kg is equal to 1 000 g.

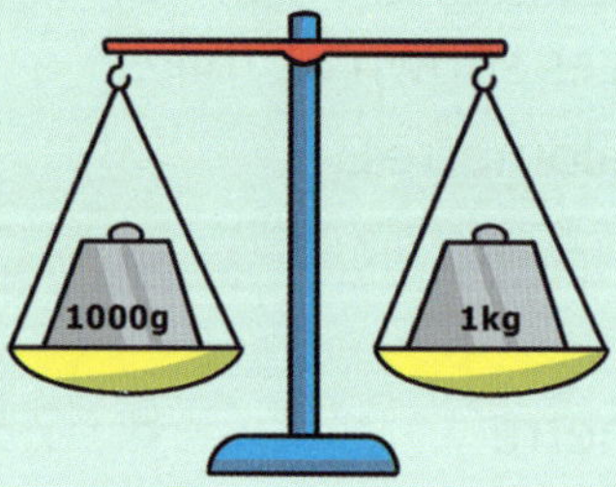

Items with a mass of 1 kg		Items with a mass of < 1 kg	
			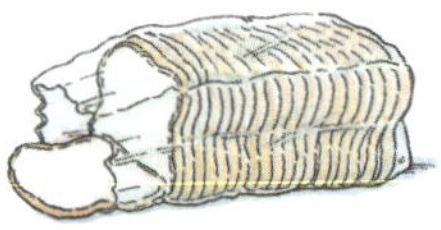
1 litre	1 kg	250 g	$\frac{3}{4}$ kg

Remember that 'kilo-' means 'thousand'. So a kilogram is a thousand grams.

We multiply by 1 000 to convert kilograms to grams.

We divide by 1 000 to convert grams to kilograms.

Key words

gram
kilogram
mass
tonne

1 Use the items from the previous page to decide which balance scales are correct.

a
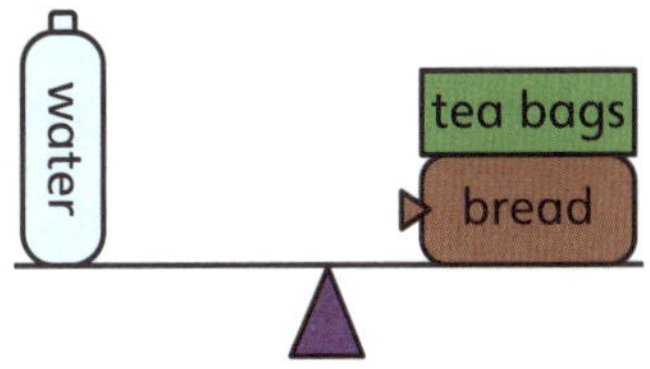

b

c
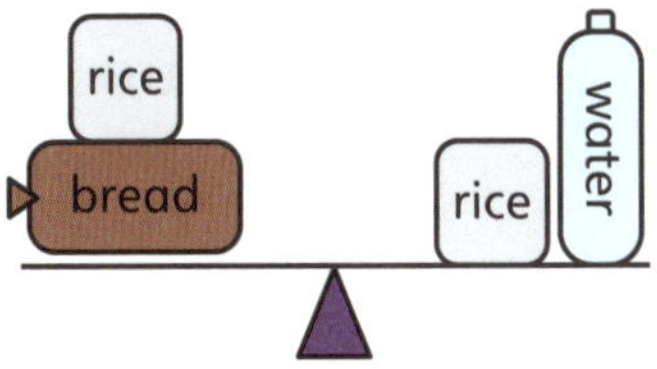

d

2 What mass is shown on each scale?

a
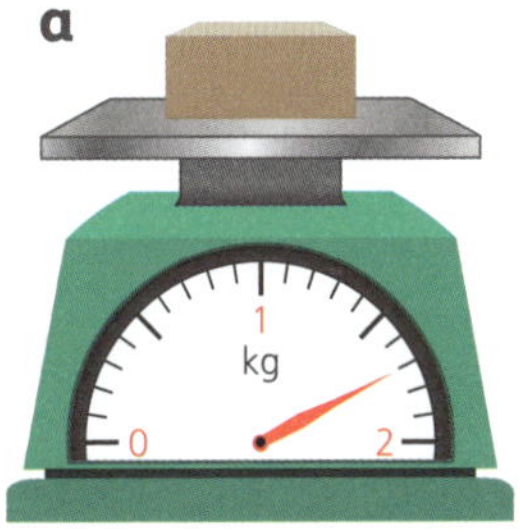

b
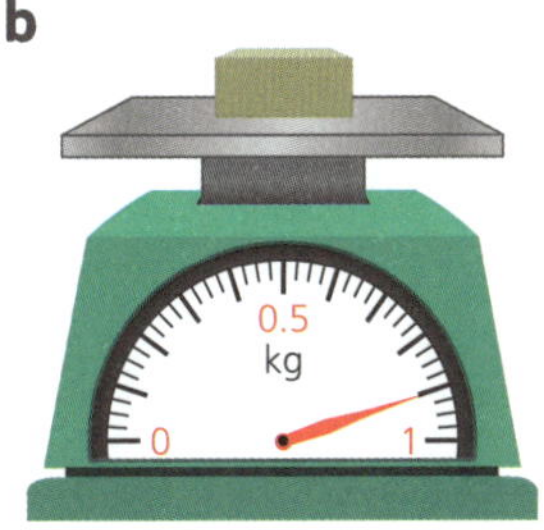

c
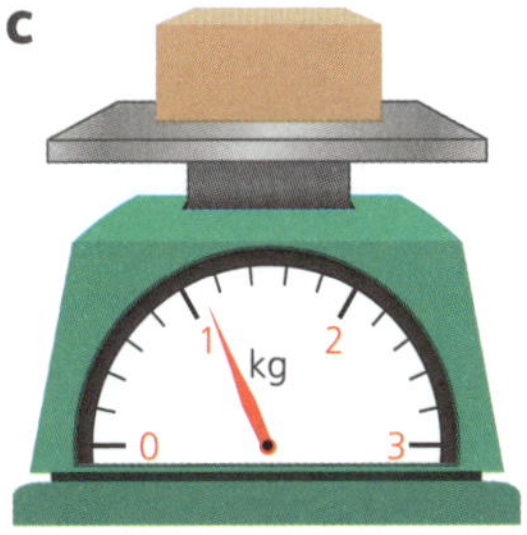

3 Write these sets of masses in order from lightest to heaviest.

a 1 000 g $\frac{1}{2}$ kg 2 kg 1.75 kg $\frac{3}{4}$ kg

b $1\frac{3}{4}$ kg 1 t 3 000 g 2.5 kg $2\frac{1}{4}$ kg 750 g

4 Place a litre bottle filled with water on one side of a balance scale. Compare its mass to six objects in the classroom.

a Copy the table below and complete the column for your estimates.

Object	Estimate (write < 1 kg, $= 1$ kg or > 1 kg)	Actual mass in grams	Actual mass in kilograms

b Now measure accurately to find the mass of each object. Complete the table.

5 Solve these problems.

a A truck has a mass of 11 tonnes. What is this in kilograms?

b Three elephants each have a mass of 6 000 kg. What is the total mass of the elephants in tonnes?

c A trailer has a mass of 50 tonnes, including the load. The load on the trailer is 38 000 kg. What is the mass of the trailer in tonnes when the load is removed?

Working with capacity

Key maths idea

Amanda has a collection of jars. When they are empty, she wants to clean them and refill them with homemade jam.

Which jar will take up most space on a shelf?
Which jar will hold the most jam?
Which will hold the least jam?
Which will hold about the same amount?

Containers come in different shapes and sizes, and they can hold different amounts.
When you measure how much space a **container** takes up, you are measuring **volume**.
When you measure how much the container holds, you are measuring **capacity**.

We use the metric unit of **litres** to measure capacity. There are 1 000 **millilitres** in a litre. We can also use imperial units to measure capacity, for example, **pints** and **gallons**.

Key words

container
volume
capacity
litre
millilitre
pint
gallon

1 **a** Find the values for A, B, C and D in litres and then in millilitres.

b Round each value in litres to the nearest litre.

c Another value, E, rounds down to 3 litres when we round it to the nearest litre. Give a possible value for E.

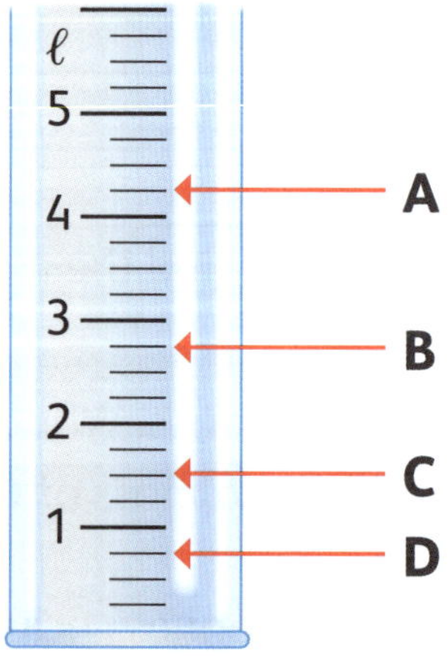

2 **a** Bring some bottles, cups and jars from home.

b Order the bottles from the one that holds the most to the one that hold the least. If the bottles don't have labels, you will have to estimate.

c After you have estimated, use measuring cups and water to help you find the capacity of bottles that do not have a label.

d Write each a letter (A, B, C, D and so on) on each container, in the correct order.

e Write the capacity (in ml or ℓ) on each container.

f Write all the capacities first in millilitres (ml), then in litres (ℓ).

g Copy and complete the table to show your estimates and measurements.

Container	Estimated capacity	Actual capacity (ml)	Actual capacity (ℓ)
A			
B			
C			
D			
E			
F			

3 A pint is approximately 568 millilitres. There are 8 pints in a gallon. Copy and complete this conversion table.

	Imperial measure	Approximate metric measure (ml)
a	2 pints	
b	1 gallon	
c	$\frac{1}{2}$ gallon	
d	10 pints	
e		56 800 ml

Try this conversion challenge!

4 Solve these capacity problems.

a A container holds 750 ml of water. A bucket holds 10 times as much. How many buckets can be filled from a 30 litre tank of water?

b A recipe for 2 people uses 0.25 litres of water. How much water is needed to make the recipe for 8 people?

Problem solving

Key maths idea

When you do problem solving, you will have to work very systematically. Try to solve this puzzle. Draw, write or use objects to work out the solution.

This puzzle dates back to the 9th century!

Once upon a time a farmer went to town to sell his vegetables. When he'd sold all his vegetables, he bought a chicken, a fox and a bag of corn. On his way home, he had to cross a river by boat, but he suddenly found he had a problem! The boat could take only the farmer and one of the things he had purchased. If he left the fox and chicken alone, the fox would eat the chicken. If he left the chicken and the corn, the chicken would eat the corn.

How can the farmer get himself and all the things he bought to the other side of the river without any damage?

farmer = 84 kg, fox = 14 kg, chicken = $4\frac{1}{2}$ kg, bag = $2\frac{1}{2}$ kg

1 **a** What is the total mass in the boat on the first crossing?

b What is the total mass in the boat on the other six crossings?

2 The distance across the river in one direction is 25 m. How far does each of the following travel in the boat in total?

a the fox

b the chicken

c the bag of corn

d the farmer

3 Each crossing takes 12 minutes.

a How long do the seven crossings take in total? Give your answer in hours and minutes.

b The farmer rests for five minutes between crossings. He started the first crossing at 14:30. What time did he finish the last crossing?

4 The farmer discovers that the boat has a small leak! With each crossing, 400 ml of water enters the boat. The farmer empties out 250 ml of water each time he reaches the river bank.

How much water is in the boat at the end of the last crossing? Explain how you worked this out and show the methods you used to calculate this.

Talk about

Think about how you can represent the problem to help you solve it. Discuss in groups how drawing a diagram can help you keep track of the crossings and the water in the boat each time.

Maths detective: At the airport

Go on a maths detective journey to solve these cases.

Case 1

a

b

c

d

e

You are going on a work trip. The airline charges $8 000 for the first 22 kg of luggage and then $2 000 for every additional 1 000 g of luggage. But they made a mistake with the cost of one of these pieces of luggage. Which piece is it?

Case 2

You are watching the bag handlers fill trolleys with luggage to take to the aeroplane. They tell you that the total mass of the luggage is 2.4 tonnes. If the average mass of one piece of luggage is 20 kg, approximately how many pieces of luggage are on the trolleys?

Case 3

Finally, you are on the plane. You have calculated that the aeroplane has a take-off weight of 80 tonnes. This includes the mass of the plane, which is 41 tonnes, and fuel, which totals 18 000 kg. The rest of the mass is made up of passengers, cargo and crew.

1 Calculate the total mass of the passengers, cargo and crew. Give your answer in kilograms and in tonnes.

2 You know that the aeroplane burns approximately 9.5 kg of fuel per kilometre. Approximately how many tonnes of fuel will be left after a 1 000 km flight?

Real-world maths: Be water wise!

We need water to live, but approximately how much water do we use each day?

1 Collect information about your own use of water for a period of 5 days and record it in a table.

2 Then draw a graph that shows your water usage. Remember to think about the water you use to wash, and any other water you use in your home.

3 How can you find the average amount of water you use each day? Think about your work in statistics to help you. Will you need to find the mean, mode or median?

4 Read the text and answer the questions.

In many countries, people use about 130 litres of water each day. Because water is heavy, spacecrafts carry as little water as possible. Water is a by-product from fuel cells on a space shuttle. Fuel cells make electricity when hydrogen and oxygen combine. When the two substances combine, they make electricity, and they make water. The astronauts can recycle this water.

Electrical power for the International Space Station (ISS) is generated from solar panels rather than from fuel cells, so no water is produced. The ISS recycles small amounts of water from cabin air. This is why the ISS astronauts limit their water use to about 11 litres a day.

Adapted from: NASA - Human Needs: Sustaining Life During Exploration

a If you can, find out more details about the amount of water each astronaut (or cosmonaut) is allotted each day for a space mission. How does that compare with the amounts you use for different purposes in daily life?

b Make a table to show the amount of water needed per astronaut for different length missions, for example, 200 days. What calculations can you use?

c Use what you know about the mass of 1 litre of water to think about the mass of a daily amount required on the space shuttle. Explain why the amount of water astronauts take on a mission is limited.

d How much space will be needed for water for each astronaut per week? Measure the space various containers of water take up to help you make decisions.

5 Think about ways to conserve water. What could the astronauts do to conserve water?

Measurement

What I have learned

Th	H	T	O	•t	h	th
				•		
				•		

Explain how you can use a place value chart and what you know about multiplying and dividing by 10, 100 and 1 000 to convert between these units of measurement:

- millimetres ↔ centimetres
- centimetres ↔ decimetres
- centimetres ↔ metres
- metres ↔ kilometres
- milligrams ↔ grams
- grams ↔ kilograms
- kilograms ↔ tonnes
- millilitres ↔ litres

Talk to a partner about the way the prefixes 'milli-', 'deci-', 'centi-' and 'kilo-' help you to decide how to calculate between units of measurement.

Practice questions

1 Copy and complete. Use <, > or = .

a 18 cm ☐ 180 mm **b** 4.5 m ☐ 425 cm **c** 33 km ☐ 30 000 m

d 5 000 g ☐ 4.9 kg **e** 2 000 kg ☐ 2 tonnes **f** 750 ml ☐ 0.75 ℓ

2 A store sells planks that are 32 mm thick. Shelly-Ann buys 7 planks.

a She stacks the planks on a shelf. How high is her stack?

b She places the stack on a shelf that is 30 cm high. How many more planks can she fit on the stack?

3 Look at this line graph. It shows the amount of rainfall in millimetres during a week.

a How many millimetres of rain fell on the wettest day of the week?

b Which day saw the equivalent of 3 cm of rainfall?

c A decimetre of rain fell on one of the days. Which day was it?

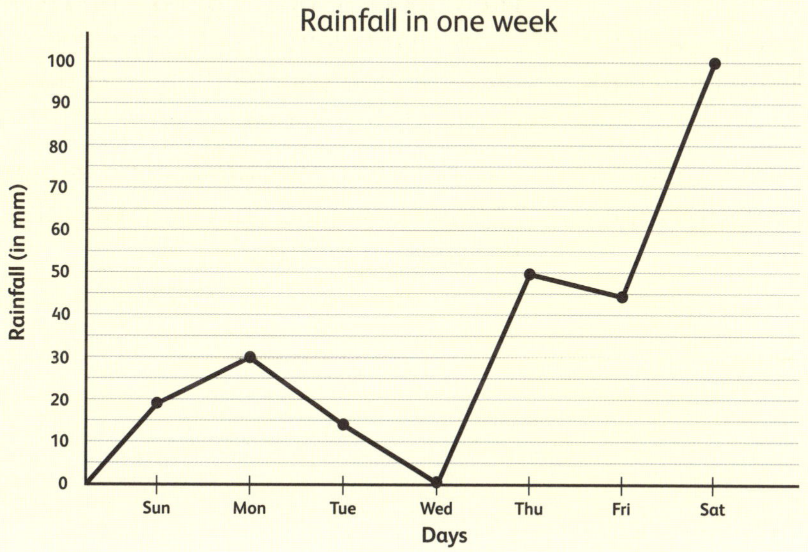

4 Audrina has collected some books for the school library. The total mass of the books is 55 kg. She has a box that can hold no more than 9 kg at a time, and she needs to carry the books from her classroom to the library. What is the smallest number of trips she will need to make?

5 A packet of sugar weighs 2.5 kg. You need 350 g for a batch of muffins. How many batches can you make from one packet?

6 The total mass of two identical cartons of juice and a cake is 950 g. The cake is 0.2 kg heavier than one carton of juice. What is the mass of the cake?

5 Self check

Check you can do this:	If you have difficulty, try this:
1 Explain the relationship between different units of measurement.	Revisit pages 73–74. Remind yourself of the meanings of the prefixes 'centi-', 'milli-' and 'kilo-'and how to convert between them.
2 Estimate and measure length, perimeter and distances.	Revisit pages 75–76 and remind. Remind yourself how to use a tape measure or ruler. Look again at pages 77–78, and remind yourself how to measure the distance around a 2-D shape.
3 Estimate and measure mass, using the relationships between grams, kilograms and tonnes.	Refer back to pages 80–81, and remind yourself how to measure mass using scales, and how to convert between different units of mass.
4 Estimate and measure capacity, using the relationship between millilitres and litres.	Refer back to pages 82–83, and remind yourself how to measure how much a container holds. Also remind yourself how to convert between millilitres and litres.
5 Solve measurement problems involving the four operations.	Go back to pages 84–85 and remind yourself how you worked out which operation to use to solve each question.

Unit 3: Geometry

Chapter 6 Lines and line segments

In this chapter you will explore this focus question:

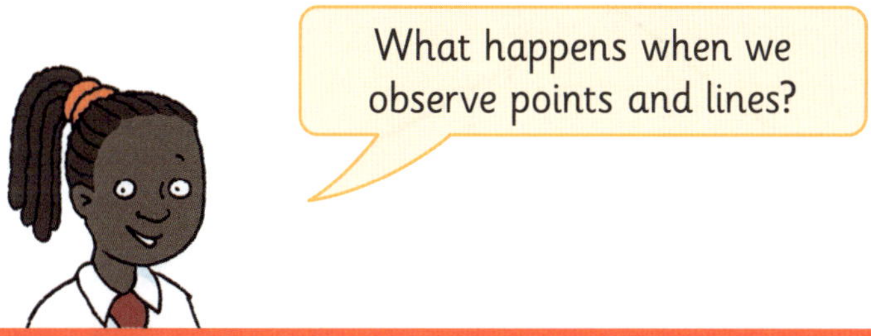

To answer this question, you will:

- recognise horizontal, vertical and intersecting line segments
- describe and draw parallel, perpendicular and intersecting line segments within an environment
- differentiate between models of parallel and perpendicular line segments when drawn or seen in the environment.

Starting point

1 a Work with a partner. Describe the **geometric shapes**, lines and angles you can see in this picture. What words do you use to describe the different lines?

b Pick two lines that are the same in some way. Then, pick two lines that are different in some way. What makes them the same or different?

Key word
geometric shape

Lines and line segments

Key maths idea

A **line segment** joins two points. We name a line segment by labelling its end points.

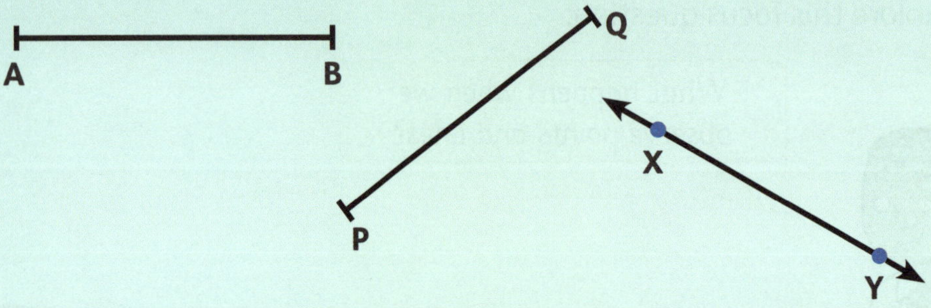

A **line** passes through two **points** and continues. We name a line by any two points along it. A line is shown with two arrows. The points on a line are often shown as **dots**. XY is a line.

Key words

line segment
line
point (dot)
slanted
oblique

Line segement PQ and line XY are both **slanted**. We can also say they are **oblique**.

1 Name each line or line segment.

a

b

c
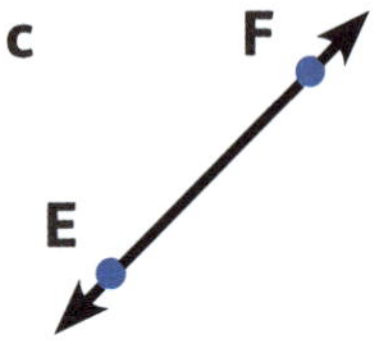

d

e

f
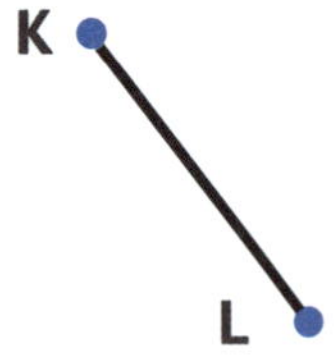

2 Here are two examples of horizontal, vertical and oblique line segments in real life. Sketch and label the types of lines you can see.

a
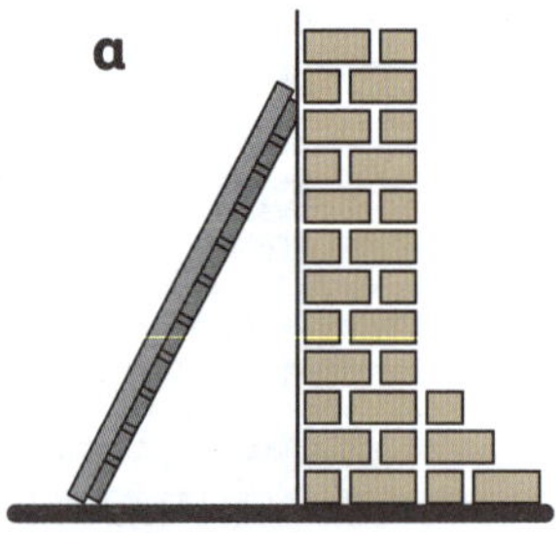

b

3 Find two examples of horizontal, vertical and oblique line segments in your class or school. Now sketch and label the types of lines you can see.

Horizontal and vertical lines

Key maths idea

Horizontal lines are named after the horizon. They go from left to right, or right to left. **Vertical lines** are directed straight up and down. Oblique or slanted lines are neither horizontal nor vertical.

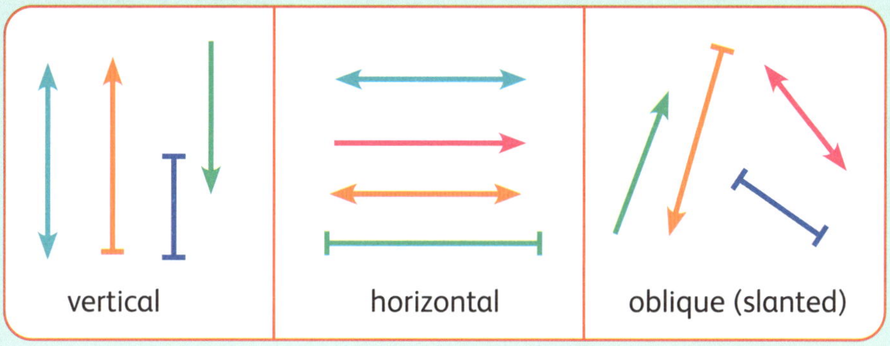

Lines may enclose a geometric shape, such as this triangle.

Can you identify horizontal, vertical and oblique lines in this photograph?

Key words
horizontal
vertical

1. Use a geoboard or pinboard to create different horizontal and vertical line segments and polygons.
2. Is it possible to create a closed shape with only horizontal lines, or only vertical lines? Explain your answer.
3. Can you create a rectangle with zero horizontal lines?
4. Can you create a triangle with one vertical and one horizontal line?
5. Can you create a shape with two horizontal lines but no vertical lines?
6. Create a pentagon or hexagon using only horizontal and vertical lines.

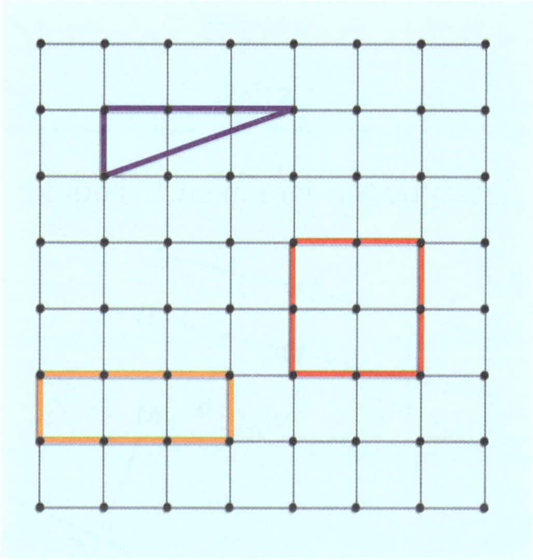

Perpendicular and parallel lines

Key maths idea

Intersecting lines are two lines or line segments that cross or meet (touch) each other.

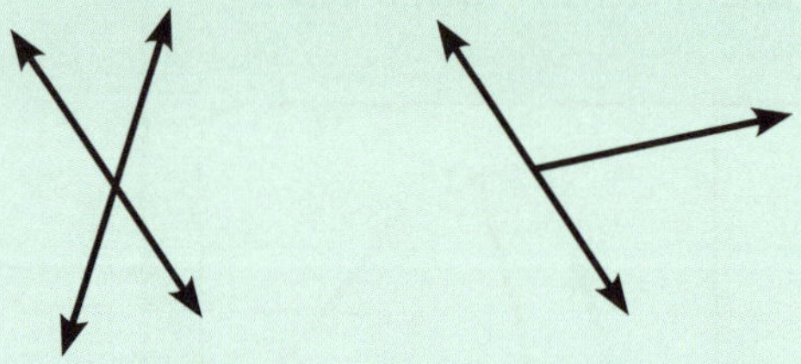

Perpendicular lines are two lines or line segments that **intersect** or meet at **right angles**. The small square at the intersection of the lines shows that they are perpendicular.

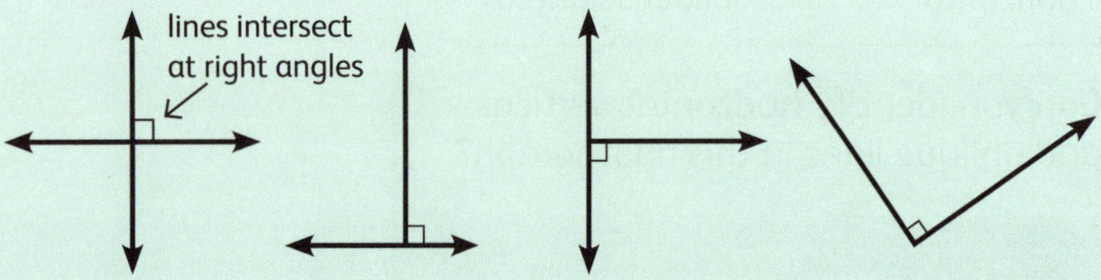

Parallel lines are like railway tracks: they are the same distance apart and they point in exactly the same direction.
A pair of parallel lines never intersect.

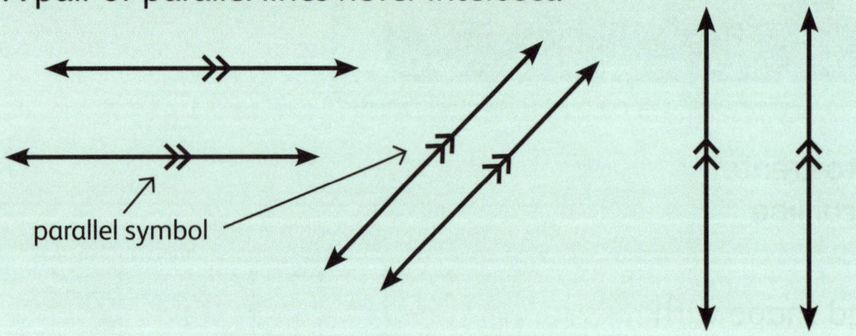

Key words

intersecting
perpendicular
intersect
right angle
parallel

1 Describe the lines in each diagram.

a
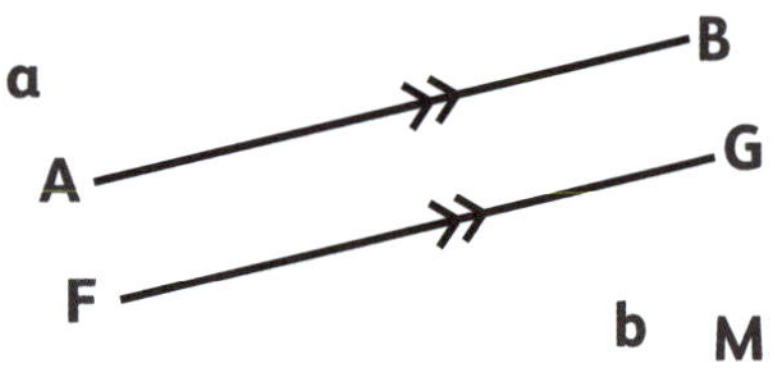

b
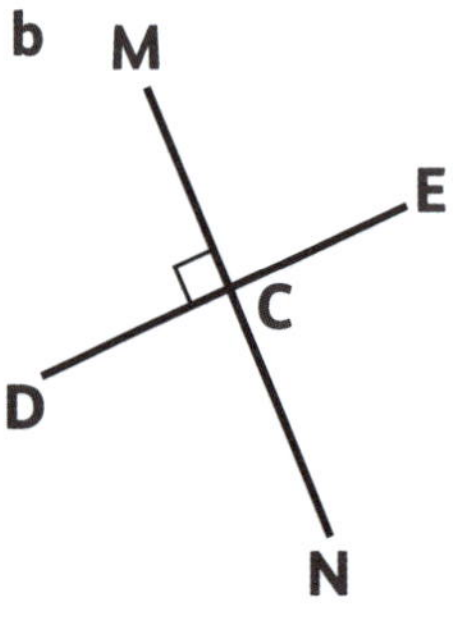

c
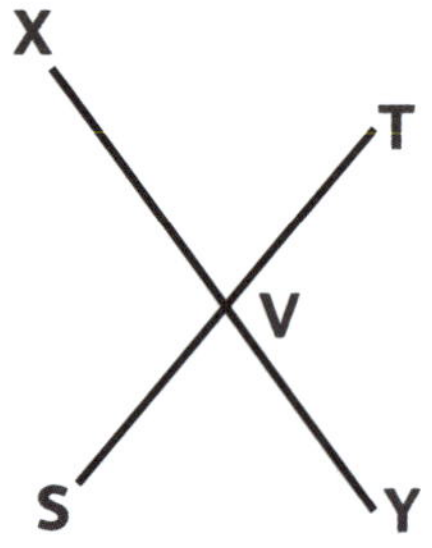

d
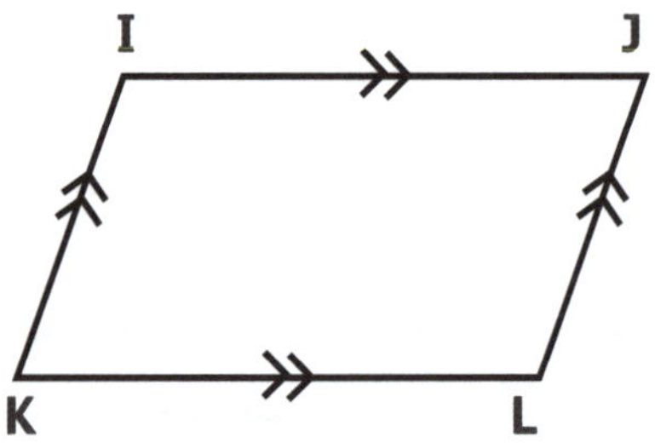

2 Work in groups to make some parallel and perpendicular lines. Draw diagrams of the shapes you made. Here are some examples.

3 Look at these shapes. Copy them and show all the parallel and perpendicular lines. Remember to use the symbols for parallel lines.

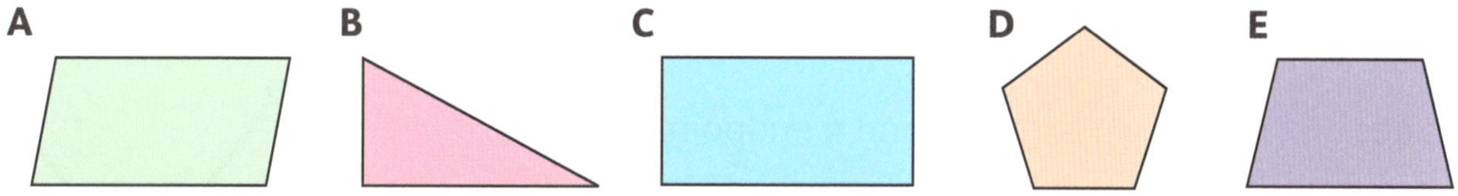

a Which shapes have parallel lines?
b Which have perpendicular lines?
c Which have parallel but no perpendicular lines?

Real-world maths: Architectural design

1 Architects and artists use horizontal, vertical and oblique lines in their drawings of buildings and in **art**.

a Draw your own architectural drawing using different types of lines. You can experiment with isometric graph paper if you have any available.

b Investigate art in your community or **environment** that makes use of different types of lines. Draw your own artwork in which you use different lines.

2 a Describe the different sets of parallel lines can you see in this picture.

b Describe the perpendicular lines you can see.

Key words
art
environment
pattern

Geometry

Maths detective: Patterns in line segments

Use the clues to discover some ways of joining line segments. You can use sticks and plasticine, or draw lines and dots.

Case 1

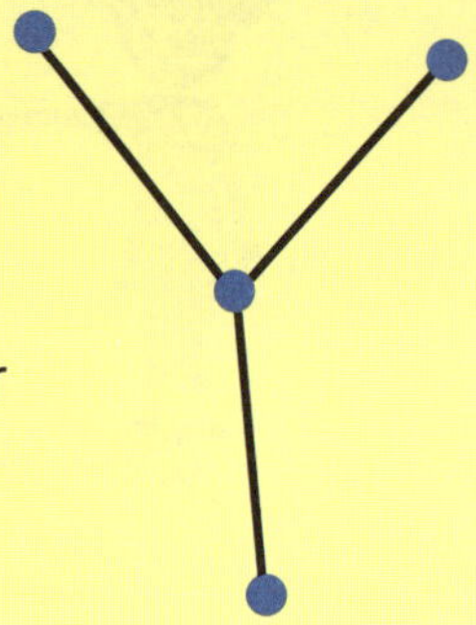

This design uses 3 sticks as line segments. They are joined with a blob of plasticine at each end. There are 3 line segments and 4 end points.

1 Use 3 line segments to make different designs. Record the number of end points for each design.

2 What is the maximum number of end points?

3 What is the minimum number of end points?

Case 2

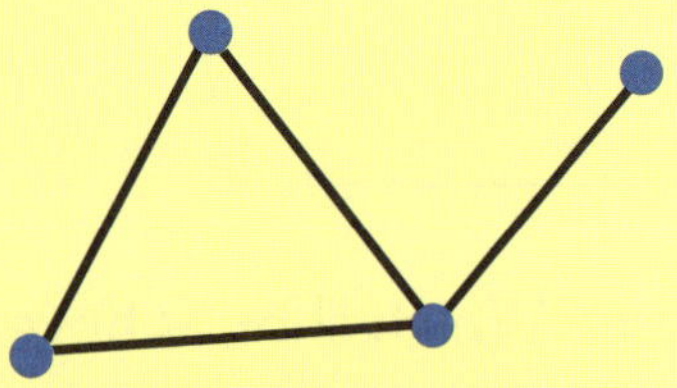

This design uses 4 line segments and 4 endpoints.

1 Investigate some different arrangements you can find.

2 What are the maximum and minimum number of endpoints for designs that use 4 line segments?

Case 3

Complete the table to show the maximum and minimum endpoints possible for different numbers of segments.

Number of line segments	Maximum endpoints	Minimum endpoints
2		
3		
4		
5		
...		

Do you see a **pattern**? Describe it.

What I have learned

Draw and label your own diagrams to show the difference between these terms.

1 lines and line segments

2 horizontal, vertical and oblique lines

3 parallel and perpendicular lines.

Practice questions

1 Copy the Venn diagram. Then draw or write an example of a polygon for each section of the sorting diagram.

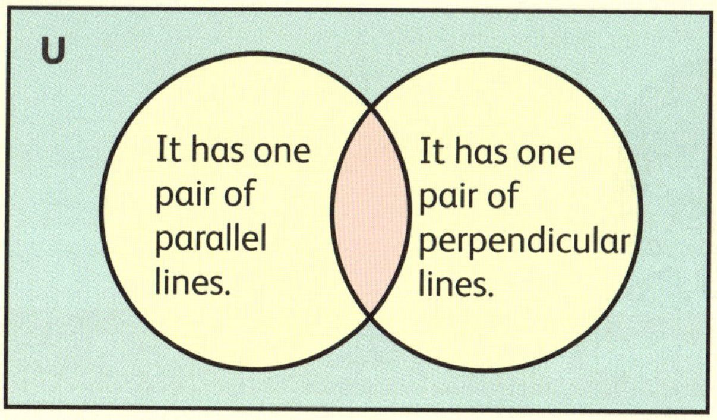

2 Sketch diagrams of objects or parts of the building in your school environment to show an example of each of these lines and angles.

horizontal line	vertical line
right angle	parallel lines
perpendicular lines	oblique line

6 Self check

Check you can do this:	If you have difficulty, try this:
1 Recognise and draw horizontal and vertical line segments.	Go back to page 90 and remind yourself how to identify each type of line. Practise drawing each type of line.
2 Recognise and draw intersecting lines, and identify and draw parallel and perpendicular lines.	Go back to pages 92–93. Remind yourself what we mean by intersecting lines, and draw some pairs of lines that intersect at right angles, and some that do not. Remind yourself how we identify parallel and perpendicular lines, and draw some examples.
3 Identify parallel, perpendicular and intersecting line segments in diagrams and in your environment.	Look around and find examples of each type of line. You can look at buildings, furniture and the lines created by objects such as shelves and cupboards.

Chapter 7 Types of angles

In this chapter you will explore this focus question:

To answer this question, you will:

- recognise acute, right, obtuse and reflex angles
- use 45°, 90° and 180° as benchmarks to estimate the size of other angles
- use a protractor to measure and draw angles accurately to the nearest degree.

Starting point

1 What kinds of lines can you see in this picture? Identify examples of:

  a vertical lines
  b horizontal lines
  c intersecting lines
  d oblique lines
  e parallel lines
  f perpendicular lines.

2 Use a ruler. Draw your own picture of a city using straight lines.

Angles

Key maths idea

An **angle** is a measure of the degree of **turn** between two lines that meet or intersect at a point. The point of intersection forms the **vertex** of the angle, and the two lines form **arms**.

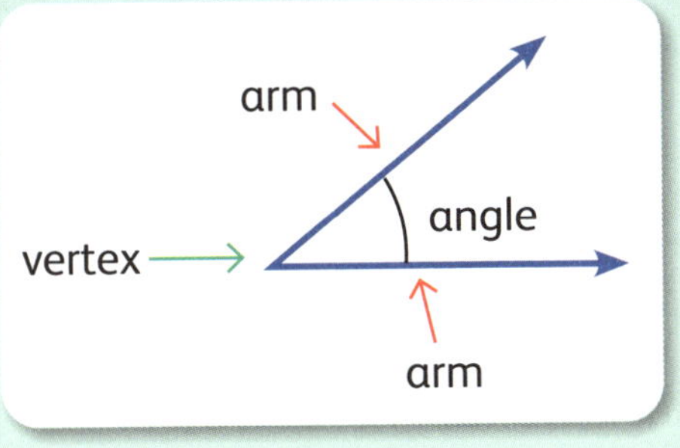

We measure angles in **degrees**.
A whole turn is 360 degrees. We write this as 360°.
A right angle is a quarter turn. It is 90°.
A half turn is 180°. It looks like a straight line.

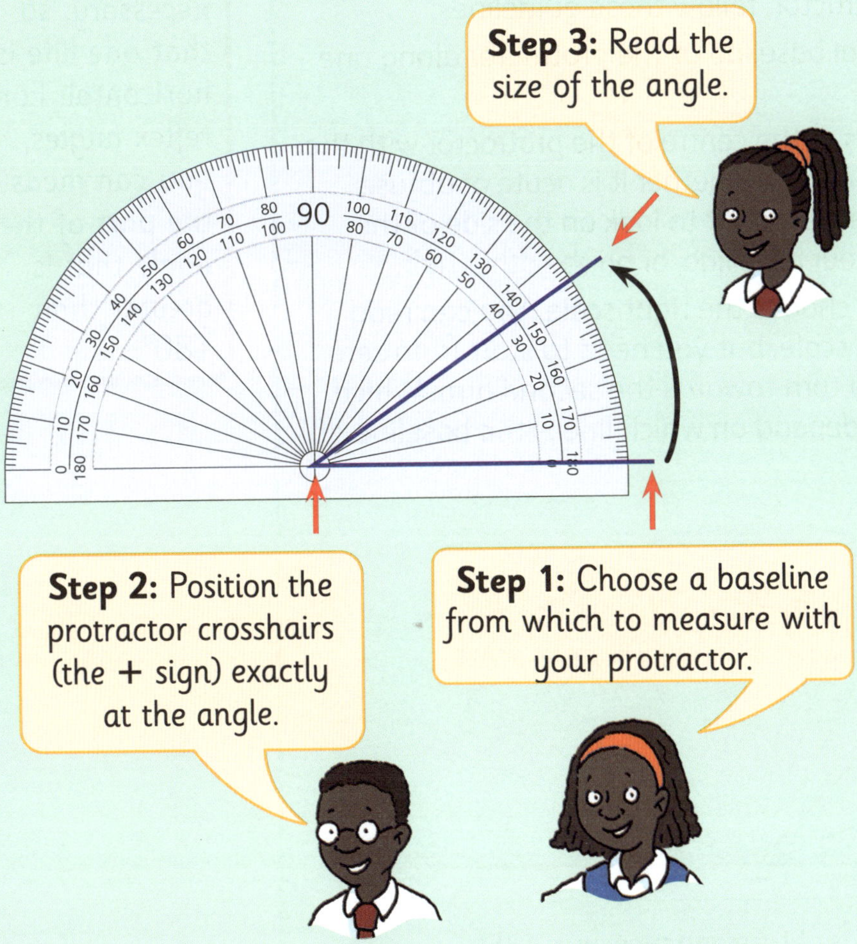

A **protractor** is a tool for measuring and drawing angles.

Key words

angle
turn
vertex
arm
degree
protractor

Talk about

What angles can you see on your classroom objects?

1. Practise drawing and measuring your own angles of different sizes.
2. First estimate the angles you can find on objects in your classroom. Then measure them with your protractor to check your estimates.

Identifying and measuring angles

Key maths idea

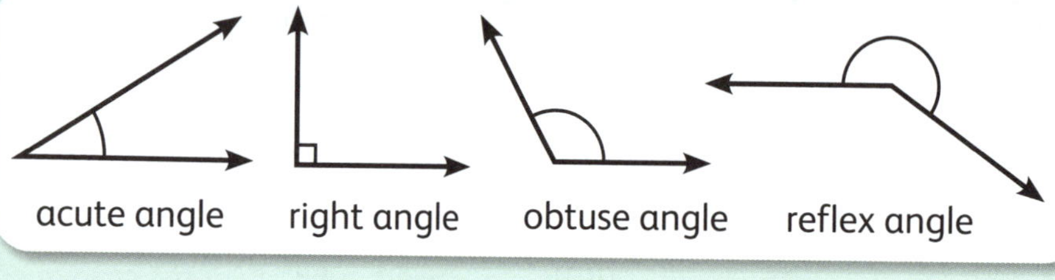

An **acute angle** is an angle between 0° and 90°.
A **right angle** is exactly 90°.
An **obtuse angle** is greater than 90° and less than 180°.
A **reflex angle** is greater than 180° and less than 360°.

When you use your protractor, follow these guidelines:

- Line up the horizontal baseline of the protractor along one of the arms.
- Line up the crosshairs at the centre of the protractor with the vertex of the angle. Identify whether it is acute or obtuse. That will tell you whether or not to look on the side of the protractor closer to your base line, or on the other half.
- Then make sure you choose the right scale. You can read the inside or outside scale, but you need to start from zero at the base arm, and turn towards the second arm. Which one you choose will depend on which line is the baseline of the angle.

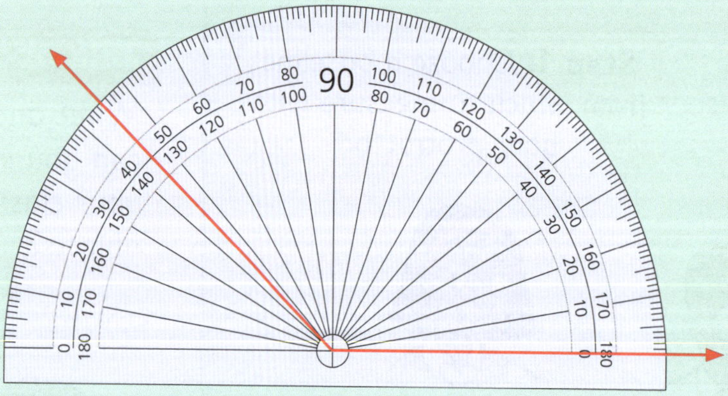

The angle above is obtuse and measures 135°.

Key words

acute angle
right angle
obtuse angle
reflex angle

When you estimate an angle, you can tilt your head or the page if necessary, so that one line is horizontal. For reflex angles, you can measure the part of the angle that is greater than 180°.

1 First look at each angle to decide whether it is closer to 90°, 180° or 45°. Then use a protractor to measure each angle. Write your answers like this:

a ∠ABC = ______ **b** ∠DEF = ______ **c** ∠GHI = ______ **d** ∠JKL = ______

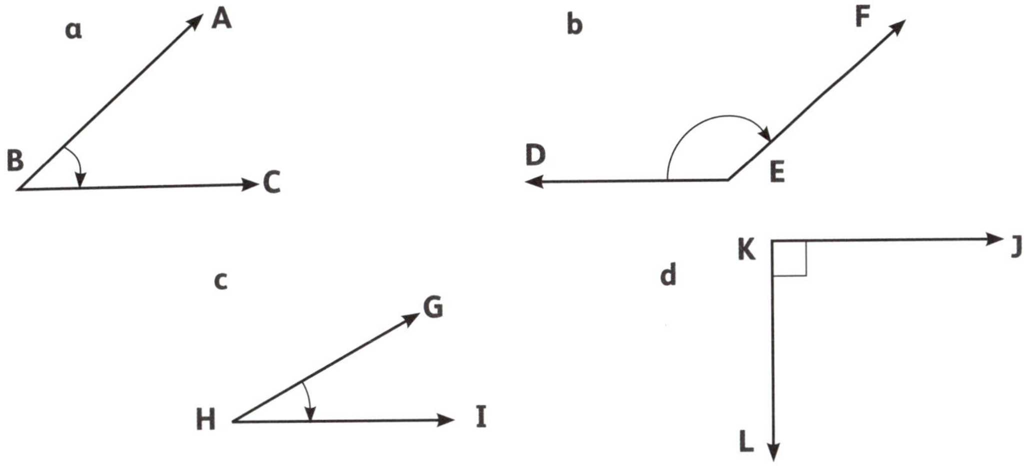

2 Use these words to help you write accurate definitions for these terms.

turn	between	line	point	size	amount	degrees

a The definition of an angle is ________

b An acute angle is ________

c A reflex angle is ________

3 Follow these steps to draw a 50° angle.

Step 1 Draw the baseline of the angle.
Make sure the protractor is accurately placed on the baseline and that the centre is at one end of the line.

Step 2 Mark the angle on the correct scale. Choose the angle scale that shows 0° on the baseline.

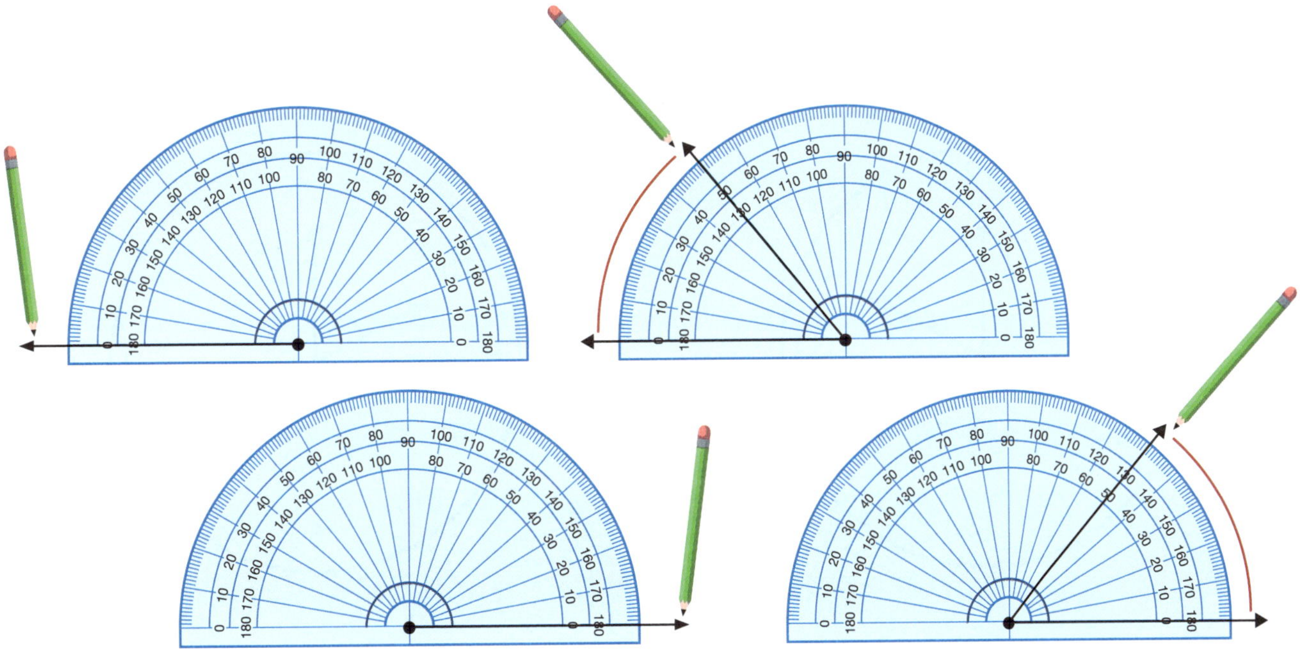

Step 3 Remove the protractor. Use a ruler to join the end point of the line to the mark you made. The point at which the two lines of the angle meet is called the vertex of the angle.

Step 4 Check your angle. Is the angle acute, right, obtuse or reflex?

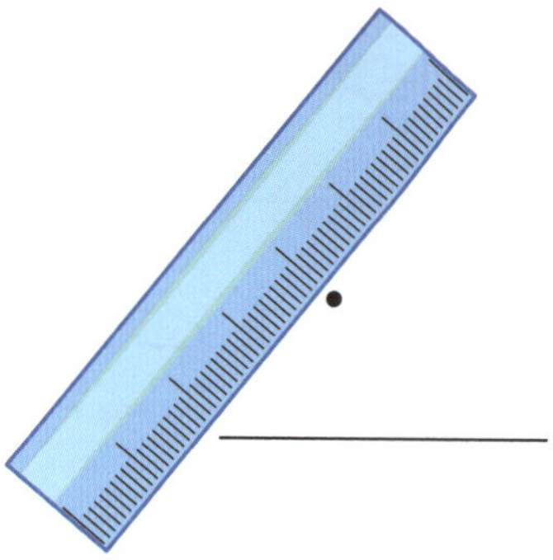

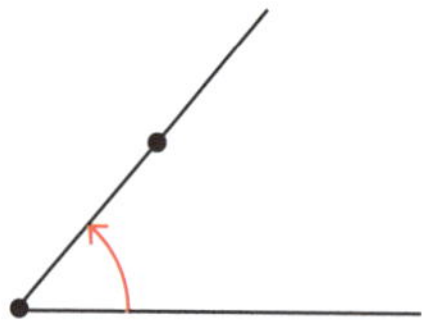

4 Now use the steps in 3 to draw the following angles.

a 120° **b** 55° **c** 89°

d 100° **e** 175° **f** 10°

Then swap your work with a partner and check each other's angles by measuring them.

5 Draw three different versions of a 45° angle by starting with different baselines.

Maths detective: Investigate angles on a clock

Use your maths detective skills to investigate the angles created when the hands turn on a clock.

> Remember that angles measure turn. So the angles here show how far the hour has turned from 12 o' clock.

Case 1

Measure the angles between the hands on these clocks.

What do you notice?

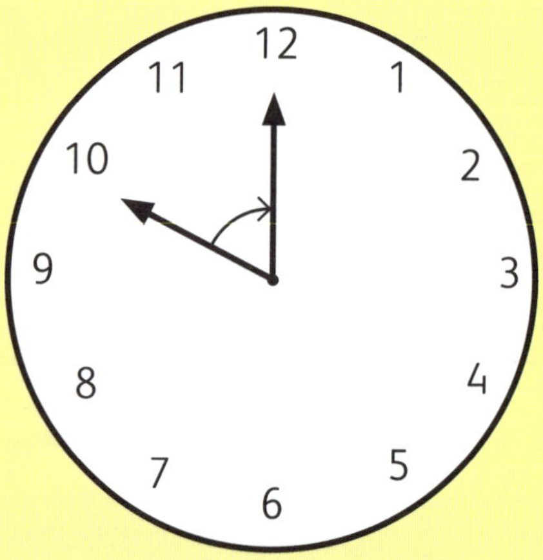

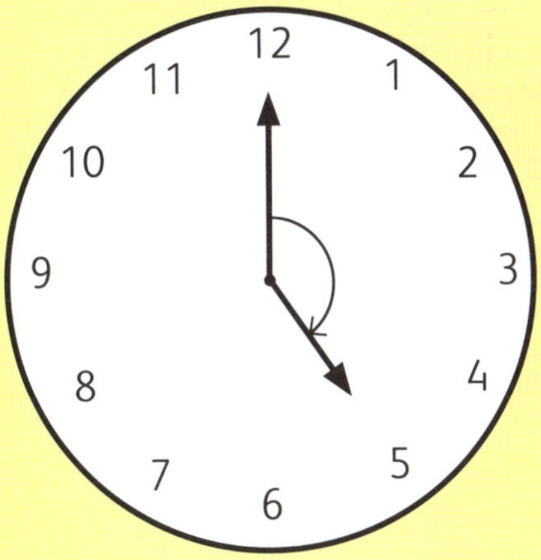

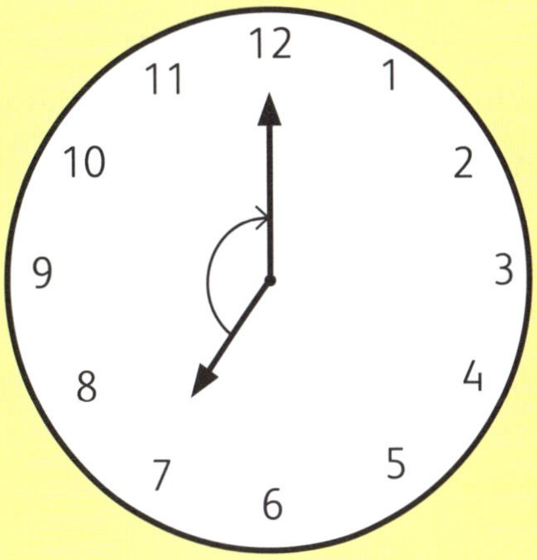

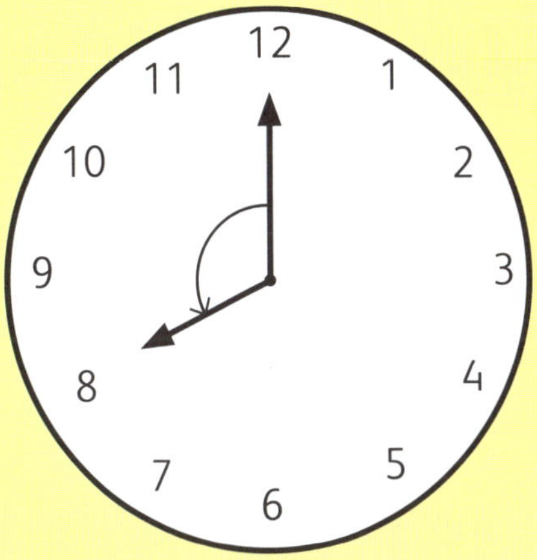

Case 2

The arms on this clockface create a right angle.
Draw four other clockfaces to show different times that make right angles.

Case 3

The arms on a clockface always create two angles.

Investigate these statements and decide which are always true, which are sometimes true and which are never true.
Draw the angles to show your thinking.

When there are two angles on a clock:

1 one angle is reflex and one is obtuse
2 both angles are obtuse
3 one angle is acute and one is obtuse
4 two angles are reflex.

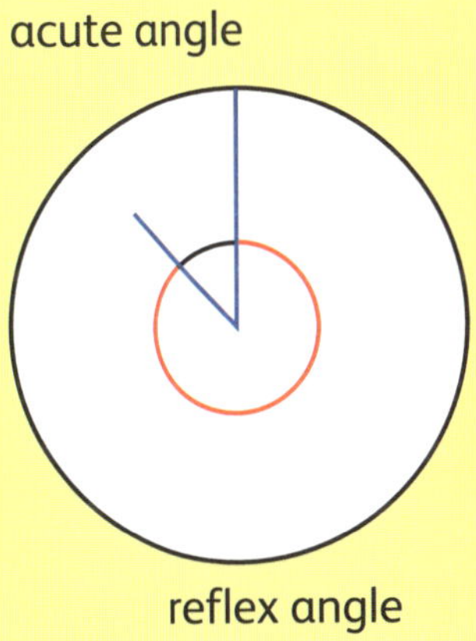

Case 4

Investigate the following possibilities and decide whether they are true or false.

1. When you add any two acute angles, the result is obtuse.
2. When you add any two obtuse angles, the result is reflex.
3. The difference between two acute angles is an acute angle.

Write down your findings about each statement. Draw diagrams to show your thinking.

Real-world maths: Angles in construction

1. In construction, measuring angles precisely is very important. A slight error of even one degree can lead to weakness or dangerous safety problems. Look at the picture and answer the questions.
   a. Why is the angle of the drill important? Explain what would happen if the woman held it at a different angle?
   b. What examples of right angles can you see in the picture? Why are these angles important for the functions of the frames?
   c. Why do you think most doors, walls and windows have to be at right angles to the ground and the ceiling?

2. If your teacher can arrange for a builder, carpenter or cabinetmaker to visit your class, find out which tools they use to make sure the lines in their work are straight, and that the angles are accurate right angles.
3. Play a game with a partner. One student shapes their body into a pose with any two of the following angles:
   - right angle
   - obtuse angle
   - reflex angle
   - acute angle
   - straight line.

   The other student must say which angles are being formed, and draw a stick person to show how they have been made. You can put your drawings together in a booklet to display in class.

What I have learned

Dancers and athletes also use angles in their work.

1 Work in pairs to identify the different angles you can see in the picture.

2 Work in pairs or groups to make up a dance and a song to demonstrate what you have learned about angles in this chapter.

Practice questions

1 Identify whether each angle is acute, obtuse or reflex.

a

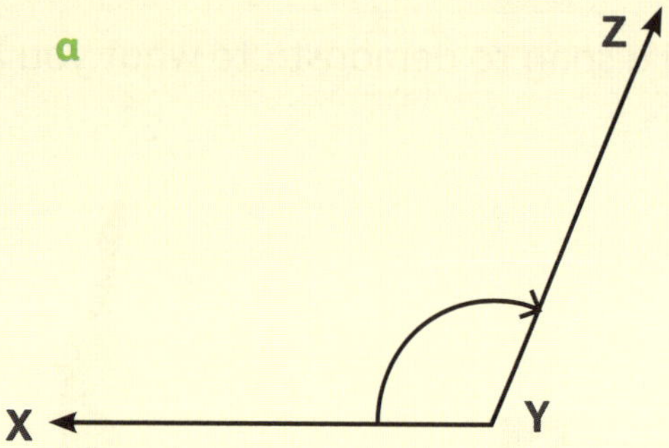

b

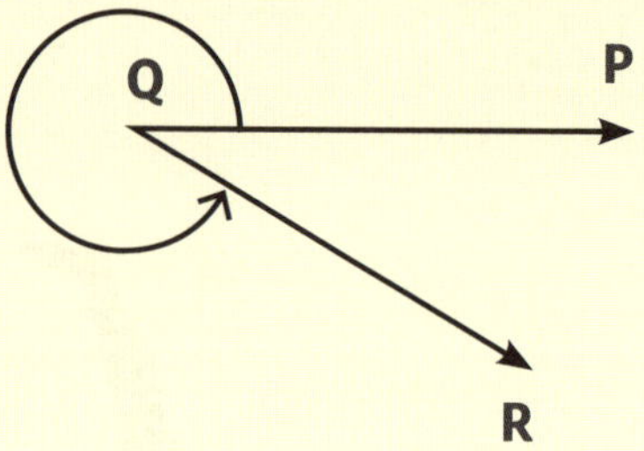

c

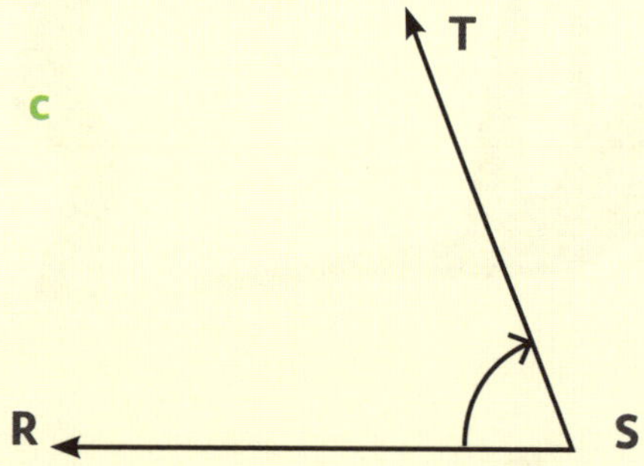

d

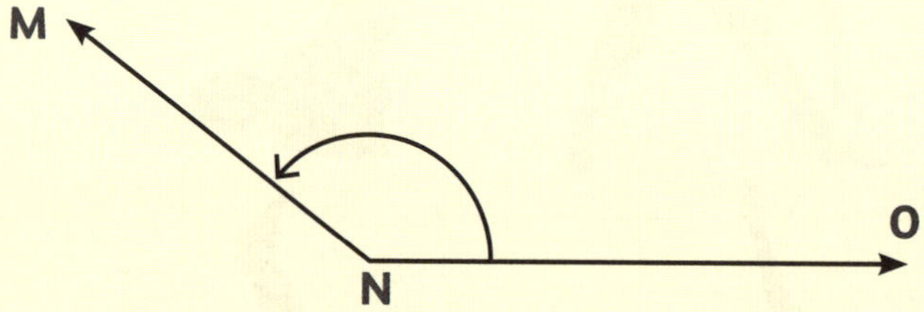

2 Measure these angles accurately.

a

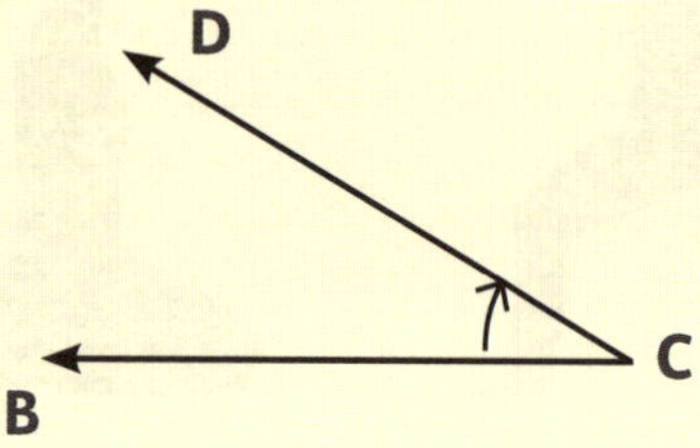

b

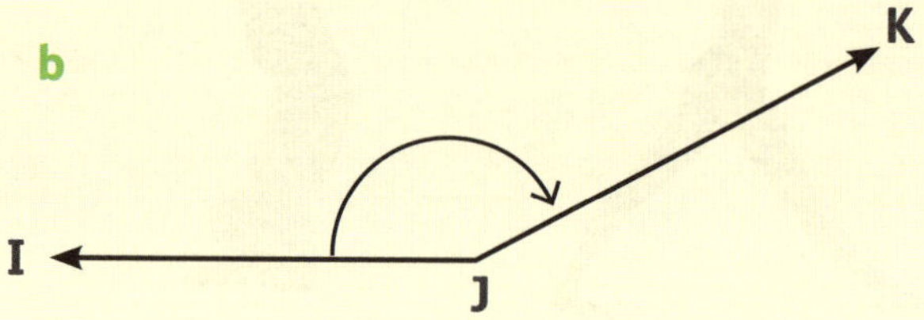

c

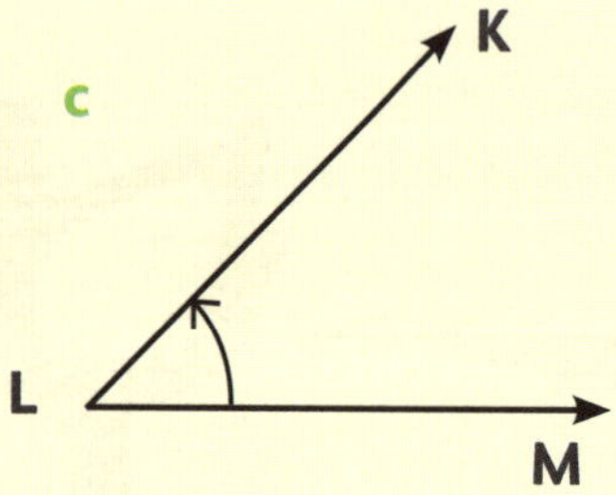

d

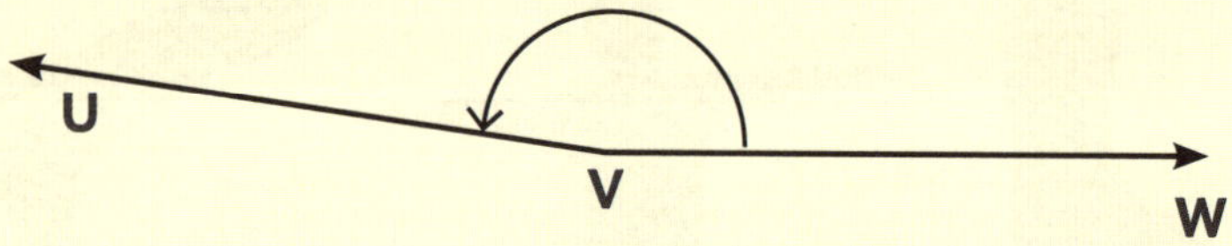

3 Form parallel and perpendicular lines by joining two vertices of a regular hexagon.

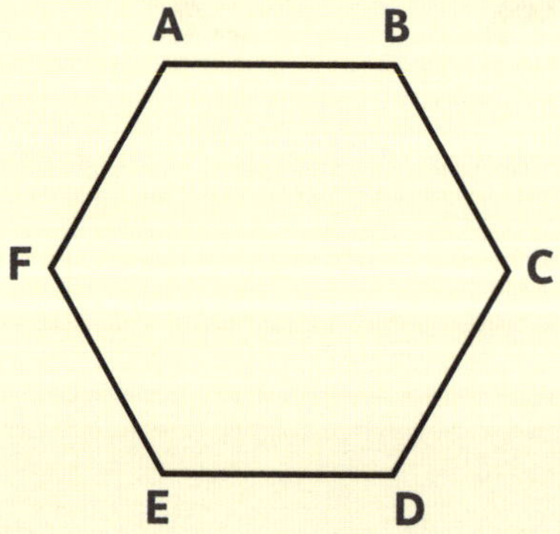

7 Self check

Check you can do this: 🙂	If you have difficulty, try this: 😐
1 Identify an angle as acute, right, obtuse or reflex.	Go back to page 98. Remind yourself of all the terms. Start with right angles, and make sure you understand the other terms and how they relate to right angles.
2 Use 45°, 90° and 180° as benchmarks to estimate an angle.	Fold a piece of paper in half, then in half again, to make a right angle. Fold your angle in half to make a 45° angle, then open it out. Use your folded paper to help you estimate some angles around you.
3 Use a protractor to measure an angle to the nearest degree.	Go back to pages 98–99. Remind yourself of the steps to follow when you use a protractor. Practise measuring some different drawn angles.

Unit 4: Statistics and probability

Chapter 8 Collecting information

In this chapter you will explore this focus question:

To answer this question, you will:

- understand what we mean by a population and a sample
- explain what a fair sample is and understand the concept of bias
- use a sample to make claims about a larger population
- develop questionnaires and use them to collect data
- use technology to organise and store data.

Starting point

1 Discuss how this picture illustrates the idea of a **population** and a **sample**.

 a Identify the total population.

 b Identify the sample group.

 c How many more people would you include in the sample group if you were studying this population?

2 Imagine that you run a supermarket. Think of some questions you might like to ask the people who shop at your supermarket. In groups, discuss:

- What information would be useful to someone running a supermarket?
- What would be a good way for them to find out more about their shoppers?
- What size of sample would be suitable for this kind of study?

Key words

population

sample

Populations and samples

Key maths idea

When you conduct **research**, you collect data about a population group. If the population group is small, you can survey the whole group. If it is large, you can survey part of the whole group. This is called **sampling**.

Researchers usually use a sample size of around $\frac{1}{10}$ of the whole population. In a very large population, the sample size may be a smaller fraction.

Key words

research
sampling

At a school of 1 200 students, the parent committee can offer three products as snacks at the tuckshop. They ask a sample of students to choose their favourite snacks from a list of options.

Snack option	Number of votes
banana chips	☺☺☺☺☺☺☺☺
spiced bun	
patties	☺☺☺☺☺☺☺☺☺☺
crispy cheese snack	☺☺☺☺☺☺
chopped fruit	☺☺☺☺☺☺☺☺☺
coconut biscuits	☺☺☺
ginger cake	☺☺☺☺
tamarind balls	☺
grater cakes	☺☺☺☺

1 a What was the sample size?
  b Do you think this was a fair sample size or not?
  c What sample size would you suggest for this size of population?

2 Which three options do you think the parent committee should offer? Give reasons for your answer.

3 What sample size would be suitable for schools with the following populations:
  a 50 students b 190 students c 830 students?

4 Study this chart.

a What question do you think the researchers wanted to answer?

b What was the sample size?

c Write three statements about the data shown on the graph.

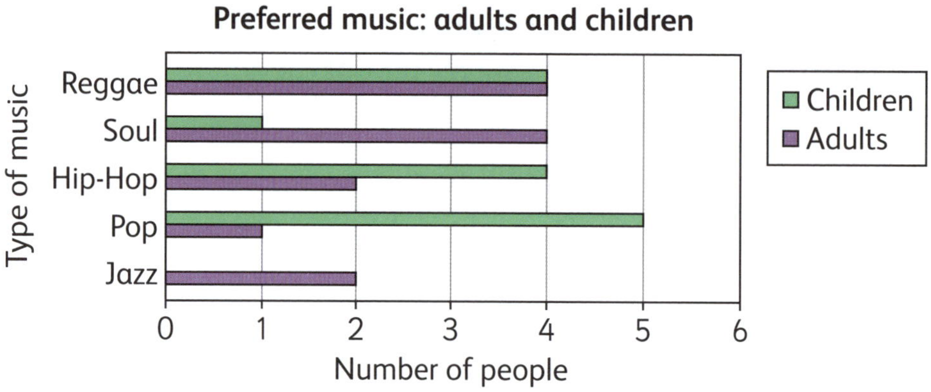

Challenge: What do you think this graph would look like if you asked this question to your family? And if you asked your class? What about if you asked everyone in the school?

Fairness and bias

Key maths idea

A **fair** sample is big enough for the group it represents. We can say it is **representative** of the whole group, because it represents the whole population without favouring one group over another group. If the sample is too small, the results may not capture the information about the whole group. If the sample is too large, it will be very difficult to collect the data (information) accurately.

One of the most important principles in data collection is that the data should not be **biased**. It is not only sample size that causes bias. Here are examples of other factors that researchers need to consider when they select their samples:

- Sample selection method: The sample should be chosen at **random** from the whole population group.
- Gender: It is important to reflect the genders of the whole population, and not to favour one group over another, unless you are only studying one particular gender group.
- Race: The sample should reflect the race groups in the whole population, without excluding any specific group.
- Language: If a population group speaks many languages, researchers may need to think about how to make sure that each group is represented.

Can you think of other factors that researchers need to consider to avoid bias in their research?

Key words

fair

representative

biased

random

1 Explain how each of the following situations might cause bias or might fail to provide a representative sample group.

- **a** A group of researchers wants to find out how much people generally understand about Covid-19. They decide to go to a hospital and interview the staff.
- **b** A student wants to find out about favourite sports at the school. He asks five boys in his football club their opinions.
- **c** A teacher wanted to find out about whether there was a need for an anti-bullying programme at school. She asked students to stand up at assembly to share their views.

2 A class of Grade 5 students were trying to find out the most popular type of music among students at their school. Explain the problem with each of their samples.

3 Design a sample that you think would provide accurate results.

- **a** What would your sample size be?
- **b** How would you make sure your sample is not biased?

4 You will now design a sample to answer a question about your school population.

- **a** Decide how to use a sample to find the information you need. Think about:
  - What size is the total population of your school?
  - How will you make sure your sample is not biased?
  - What options will you give participants to choose from?
- **b** Collect the sample information, and describe your findings and reasoning.
- **c** There are many different types of music. Which will you offer as options?

Collecting information

Key maths idea

A **survey** or **questionnaire** is one way to collect information.
Here is an example of how to take an **opinion poll**.

Dear parents,

We are reviewing our homework policy.

This is a survey to find out your opinion about homework in this school.

Please tick one option only.

I think my child should get:

- ☐ much more homework
- ☐ more homework
- ☐ the same amount of homework
- ☐ less homework
- ☐ much less homework

Please return this questionnaire to your child's teacher.

This survey uses tick boxes so that the results can be categorised.

Key words

survey
questionnaire
opinion poll

1 Design a questionnaire to be sent home to your parents to find out more information about homework. Collect information about:

- the amount of time spent each week
- which subjects have the most homework
- opinions about the amount of homework
- the types of tasks that should be set
- what problems get in the way of completing the homework.

2 Try out your questionnaire on students from a different class.

  a Collect and categorise your results. Think carefully about how you will do this.

  b Summarise your main findings in a short paragraph.

3 Compare and evaluate your questionnaires with other students.

  a What types of questions did you use?

  b Which questions gave you clear results?

  c How can you improve your questions to get clearer answers?

Real-world maths: Conduct an interview

1 Researchers can interview small samples of a population to collect in-depth data for opinion polls.

  a Write some interview questions to find out more about the opinions of students in your school about homework. Remember that your questions might need more than a one-word answer or a tick box. Think about questions like these:

  - What are the advantages of homework?
  - How can homework help students learn better?

  b Decide where and how to do your interviews. Then interview a small sample of students. If possible, use ICT equipment to record any detailed answers from the students you interview. (Make sure you obtain permission from each interviewee first!)

  c Present your findings to the class. Think about how to represent the data clearly and accurately for everyone to understand.

2 How many people in Jamaica can drive a car?

  a Discuss what is wrong with this reasoning.

  b What would be a suitable sample to find the answer to this question?

Maths detective: Samples and populations

Use your reasoning skills to draw some conclusions about data.

Case 1

Your school wants to find out how many people play different sports. There are 1 000 students in the school. They asked a sample of 50 students. Here are the results:

Use the information in this sample to predict the total number of students who play each sport in the whole school.

Students playing different sports

Basketball	12
Cricket	15
Soccer	18
Tennis	5

Case 2

Marva is a town official in a town that has a population of 20 000. Her team needs to find out the total number of people who believe the town's roads need repair. Here are the results of a sample of 500 people:

Use this sample to predict the total number of people in town who would give each answer.

Condition of roads

Need repair	180
Do not need repair	200
Not sure	120

Case 3

A pet food factory wants to research the total population of Jamaica. Use your class as a sample to predict the total number of people who own at least one pet in Jamaica.

What I have learned

1 Explain how you could use what you learned in this chapter to draw conclusions about each of these questions.

2 Copy and complete each sentence:

a A biased sample may ____ .

b A fair sample includes ____ .

Practice questions

1 Suggest a sample to find conclusions to these questions.

a What is the most common method of transport students in your school use to get to school?

b How many students in your school are left-handed?

c What is the most popular ice-cream flavour in Jamaica?

2 A town has a population of 8 000 people. Out of a sample of 500 people, 400 support a pay rise for nurses. Predict the total number of people in the town who support a pay rise for nurses.

3 Design a survey questionnaire to find out how much exercise people in your street or your class do each week.

8 Self check

Check you can do this:	If you have difficulty, try this:
1 Decide on a suitable sample size.	Go back to page 107. Remind yourself of the meanings of the terms 'population' and 'sample' and how researchers choose a sample size.
2 Explain why some samples may be biased.	Go back to page 108, and remind yourself what we mean by a representative or biased sample.
3 Make predictions based on the results of a sample.	Look at the information you collected in your questionnaire. Make statements about what your sample shows.
4 Design and use surveys and questionnaires to collect information.	Think of a question you might like to research in your class or at your school. Design a list of questions that you could use to find out the answer to this question.

Chapter 9 Presenting and interpreting data using charts and graphs

In this chapter you will explore this focus question:

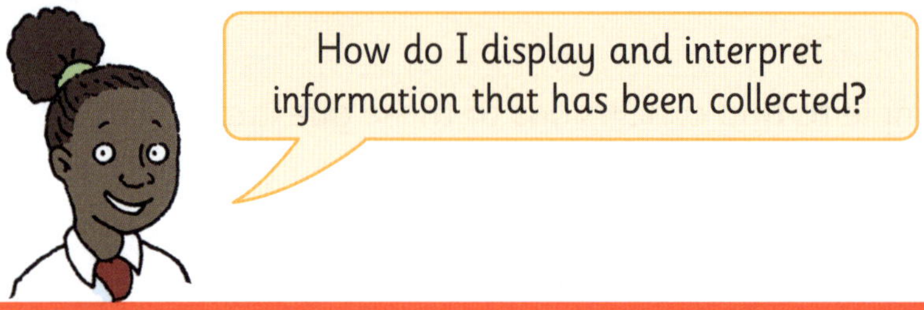

To answer this question, you will:

- discuss how we use tables and graphs
- draw simple graphs and interpret data from graphs
- draw and interpret data in pictographs, line graphs and bar graphs
- report on the mathematical content and interpretation of data
- represent data using bar graphs, double bar graphs, pictographs, pie charts and line graphs.

Starting point

1 What does this **pictograph** show?

Litter on a beach before, during and after environmental awareness week

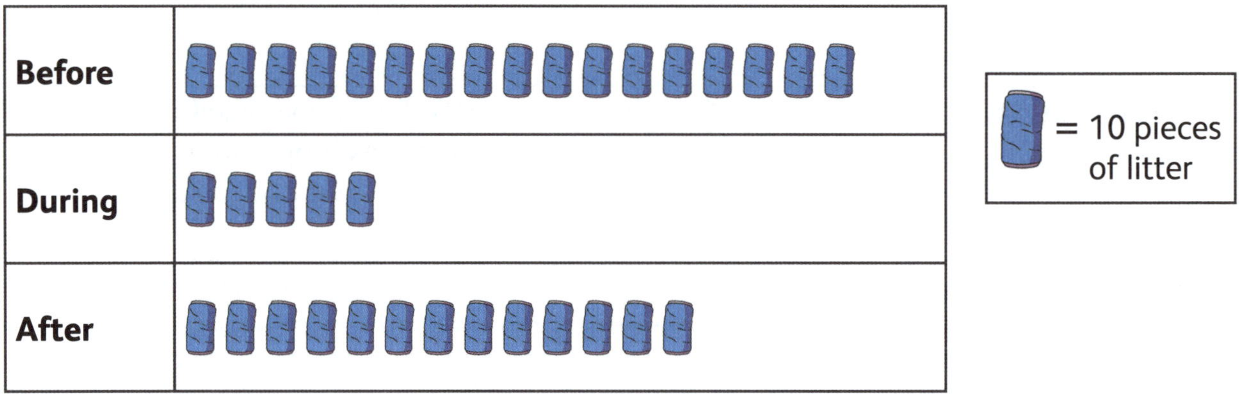

- Work out how many items the table shows for each category.
- What does the table show about the effect of the awareness week?

Key word
pictograph

Using and interpreting pie charts

Key maths idea

A **pie chart** is a graph in the shape of a circle. It is also called a **circle graph**.

The whole circle represents the whole sample.

The sections show the different categories.

The **table** alongside shows how often each piece of **data** was recorded (also called the **frequency**).

Favourite subjects

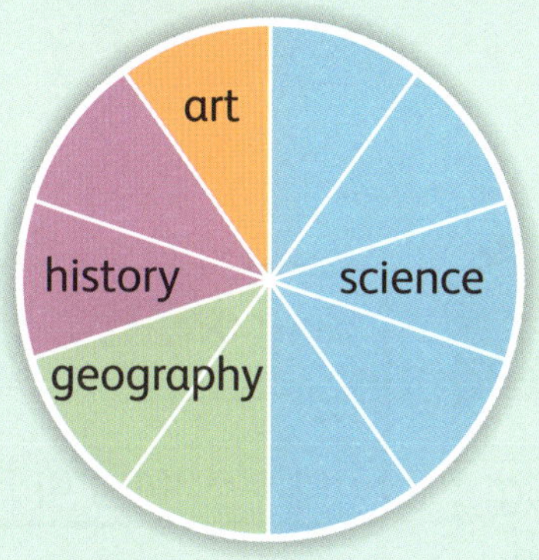

Favourite subjects	Number of students (frequency)
Science	5
History	2
Geography	2
Art	1

In this pie chart, the whole sample is made up of 10 students.

Each segment of the circle represents the choice of one student.

You can make simple pie charts by folding a circle.

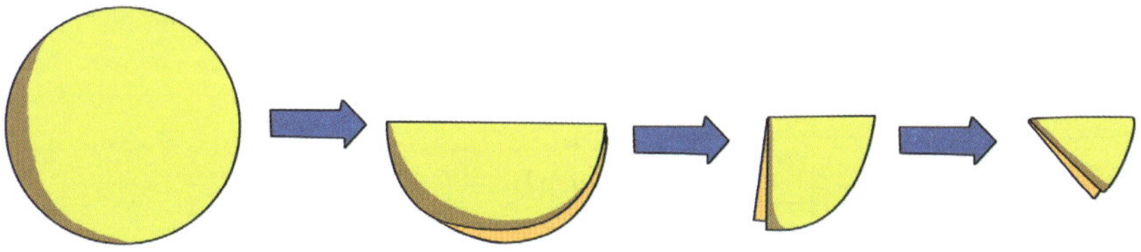

When you open up the circle, your pie chart will have eight sections. You can colour the sections to make a simple pie chart with eight items of information.

Key words

pie chart
circle graph
table
data
frequency

1 Look at the pie chart and the table again. What fraction of the students chose each subject?

2 Collect data for 16 students about their favourite subjects. Create a pie chart by folding and unfolding a circle to show the results.

Statistics and probability

3 Collect data for 32 students. Use this questionnaire:

> **SURVEY**
>
> **Importance of environmental issues**
>
> Tick one box to choose the environmental issue you think is most important:
>
> ☐ Recycling ☐ Extinctions
>
> ☐ Pollution ☐ Climate change
>
> ☐ Deforestation

a Present your results as a pie chart.

b Choose a suitable heading for your pie chart.

Bar graphs and double bar graphs

Key maths idea

Bar graphs show information by arranging the data in bars along a set of axes. A horizontal bar graph has the bars running from left to right, parallel to the x-axis, and a vertical bar graph has the bars perpendicular to the x-axis.

A bar graph must always have a heading, and the axes and bars must be labelled. There are equal spaces between the bars.

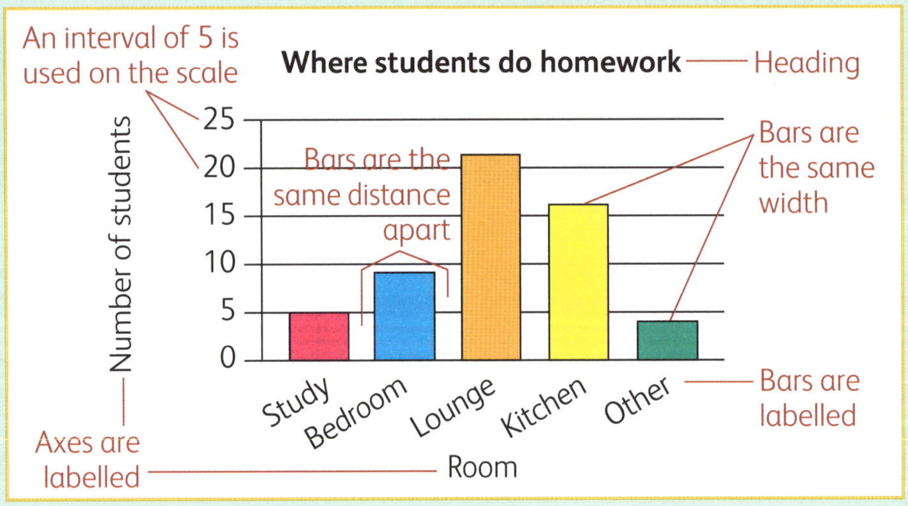

Key words

bar graph

double bar graph

Sometimes we can split the data to show more detail about it. The bar graph on the left shows the sports that a number of students chose as their favourites. The **double bar graph** on the right splits the data to show how many boys and how many girls chose each category.

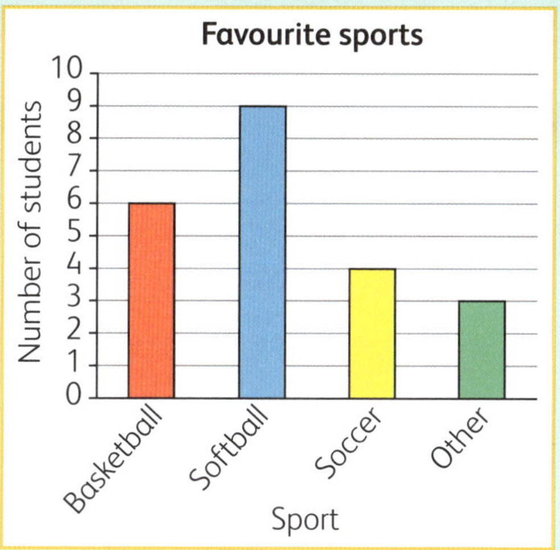

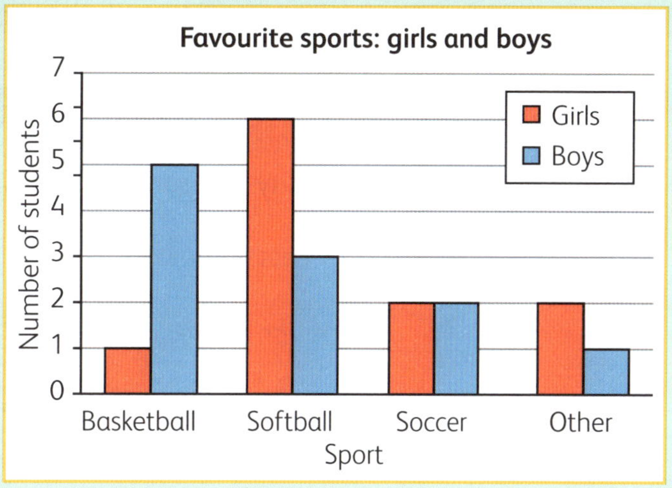

1 Look at this bar graph for favourite subjects: girls and boys.

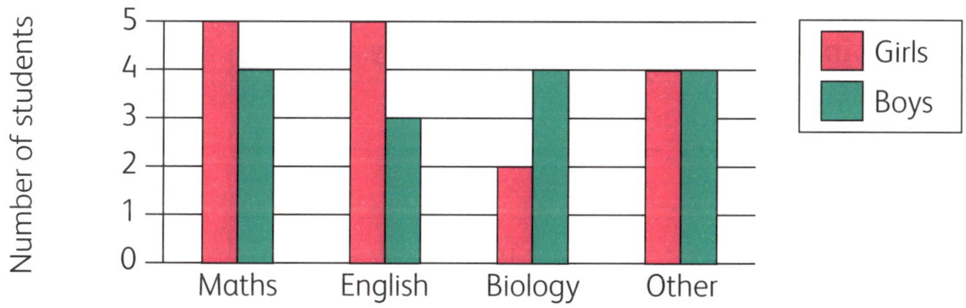

a How many students were surveyed in total?

b How many of the students were girls?

c How many were boys?

2 Now collect information about favourite sports for your class, and create your own double bar graph to show your results.

3 These graphs show how much time two students spent exercising over four weeks.

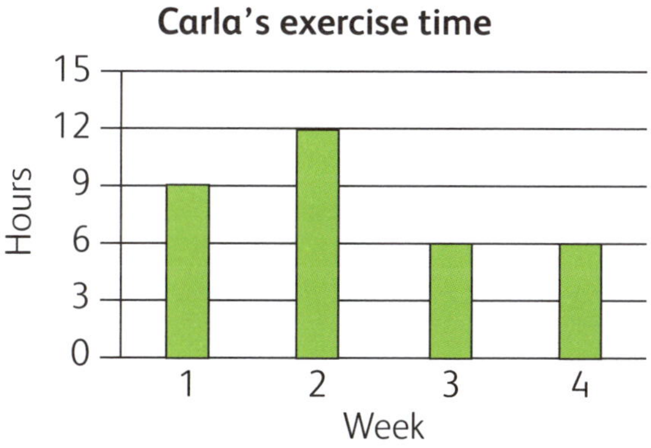

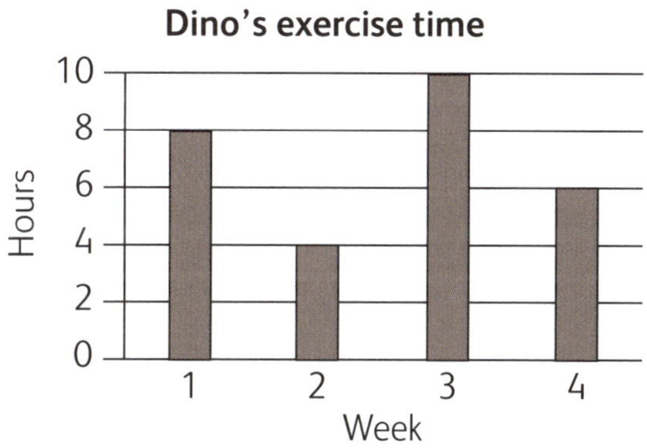

Combine the results of these two sets of data into a double bar graph. Look at the vertical scales carefully!

Statistics and probability

4 This graph shows the results of the parents' answers to the homework survey on page 111 in Chapter 8.

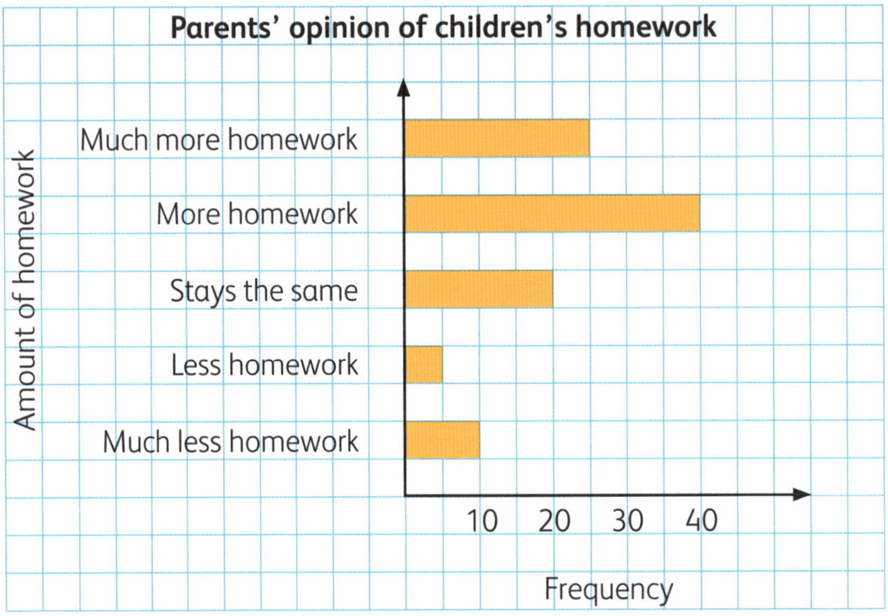

a This table shows the results from asking students how much homework they should get:

Opinion	Frequency
Much more homework	1
More	4
Stay the same	10
Less	35
Much less	50

Create a double bar graph to show the information for students and parents on the same graph.

b Describe and explain the trends you see when you compare the two sample groups.

Line graphs

Key maths idea

Line graphs are used to show **continuous data**, for example, data that changes over time.

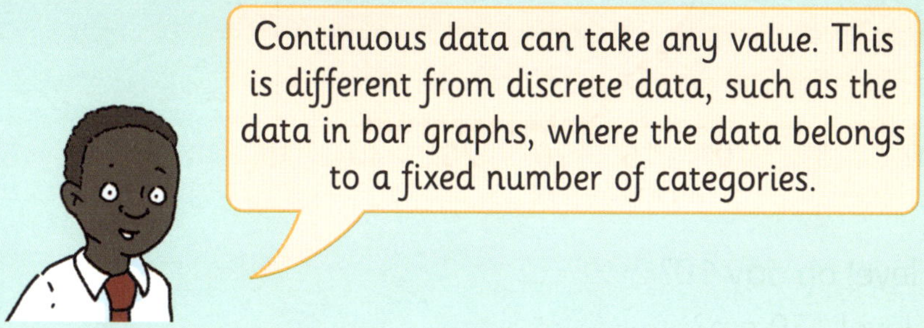

This graph shows the total number of runs two different cricket teams made.

Runs scored in a cricket match

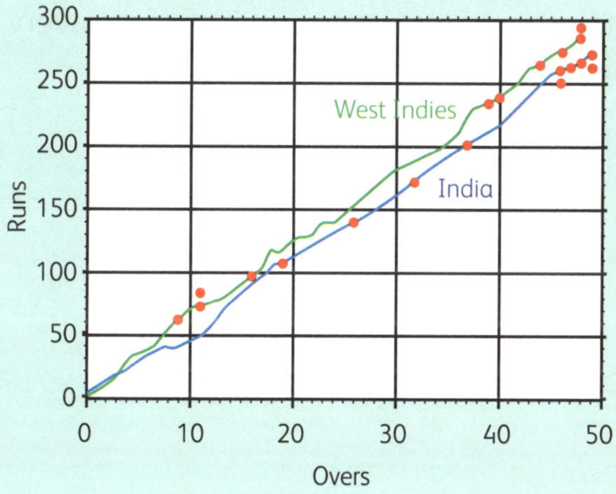

You can read and estimate values from the graph, like this:

Runs scored in a cricket match

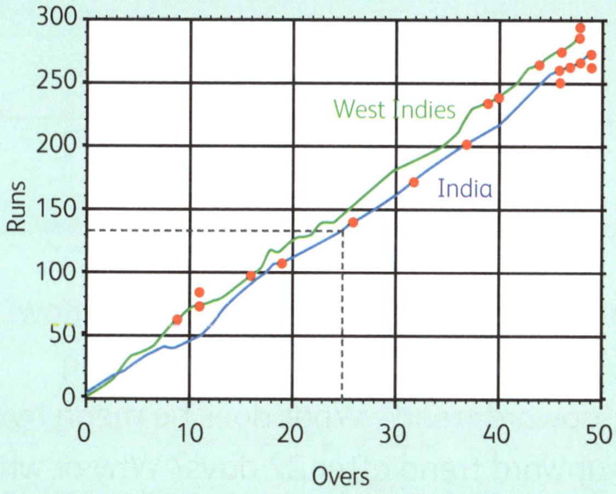

The dashed lines show that the Indian team's score was approximately 130 runs after 25 overs.

Key words

line graph
continuous data

Statistics and probability

1 This graph shows the water level in a reservoir.

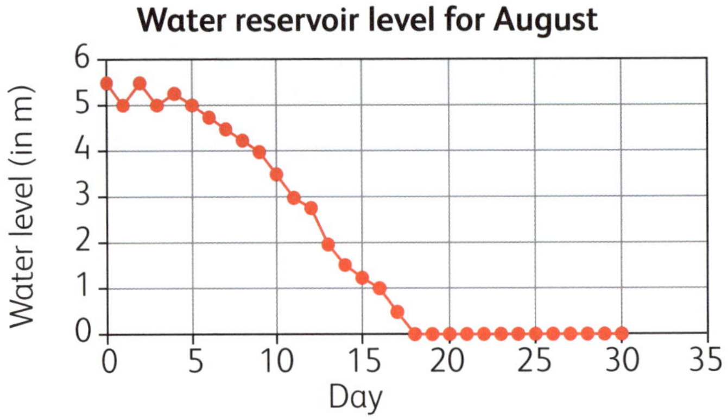

a What was the water level on day 10?

b When was the water level 150 cm?

c What happened around day 18?

d Alex says: 'This graph displays a downward trend.' Explain what you think he means. Also describe what an upward trend would look like.

2 Devon planted a bean and tracked its growth. He measured its height every day for 3 weeks. This line graph shows his results.

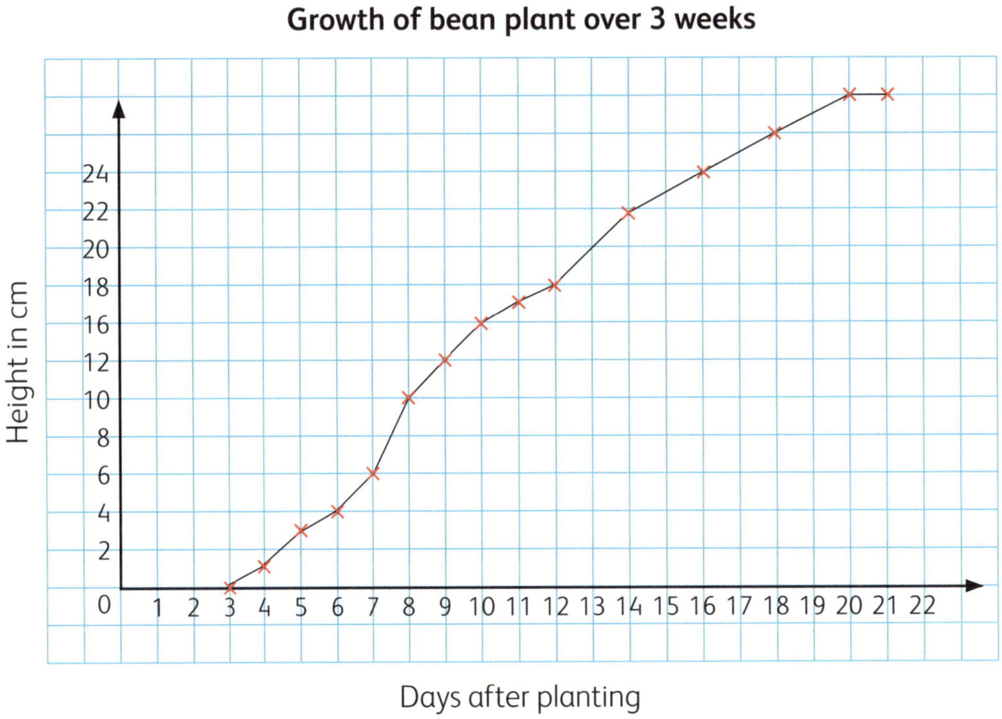

a On which day did the bean shoot start growing? Explain how you know.

b On which day did the bean grow the fastest? Explain your reasoning.

c Devon says the graph shows an 'upward trend'. What does he mean by this?

d Do you think it will continue the upward trend after 22 days? Why or why not?

3 Use the line graph in Exercise 2 to read or estimate the height of the plant on the following days:

a day 4 **b** day 7 **c** day 12 **d** day 16 **e** day 20

Choosing graphs

Key maths idea

It is important to choose the right type of graph to show the data. Always think about the purpose of your graph.

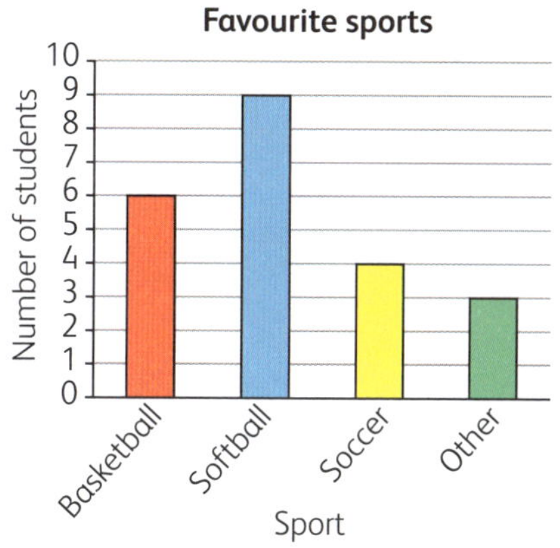

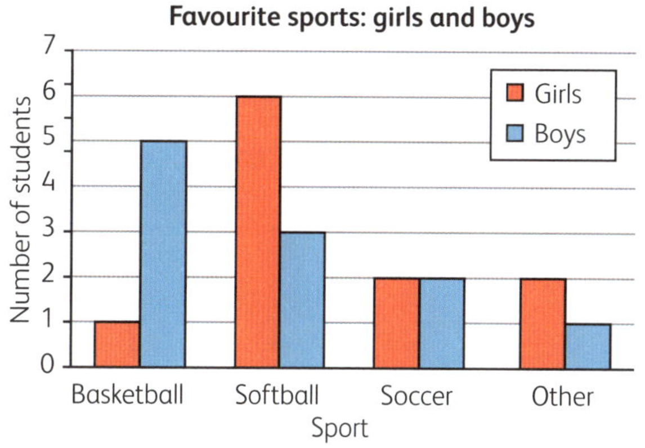

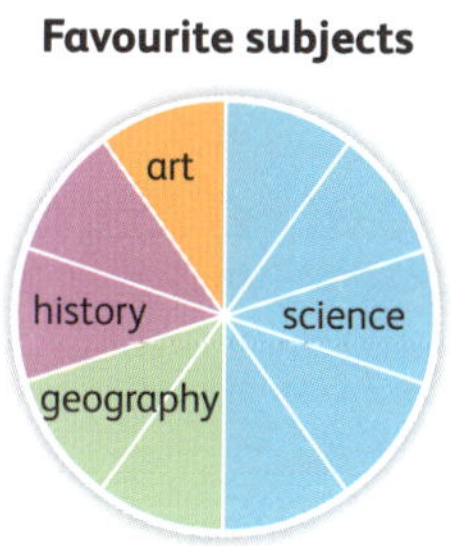

Favourite subjects	Number of students (frequency)
Science	5
History	2
Geography	2
Art	1

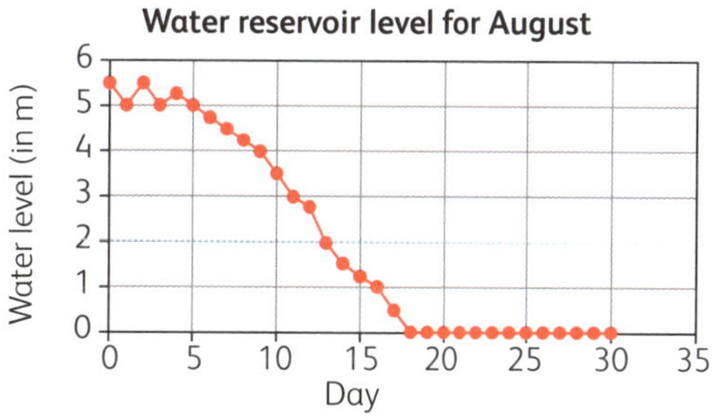

Bar graphs are useful for comparing data from different categories.
Pie charts are useful for showing categories of data as fractions of the whole.
Line graphs show data that changes over time.
Tables show exact numbers, but may be hard to read.

1 **a** Look at your survey data about the importance of environmental issues on page 116. Decide if you wish to collect further data, such as opinions on different aspects of the environment.

b Discuss whether a pie chart is the best way to present the data you collected.

c Work with a partner or in a small group to present your data. Choose an appropriate graph, and include accurate labels and scales, and write a suitable title for your graph.

Real-world maths: Spreadsheets

Many researchers use **spreadsheet** software to collect and process data. A spreadsheet looks like a grid with rows and columns. Each block in the grid is called a **cell**. You can enter data into the cells. This is called **data capture**. Once all your data is in the spreadsheet, you can use some of the tools in the software to find out more about it. Spreadsheet software makes it easy to add up values in rows or columns, to apply a formula, or to rearrange your data from tables to graphs.

This spreadsheet is being used to collect test scores for a group of students.

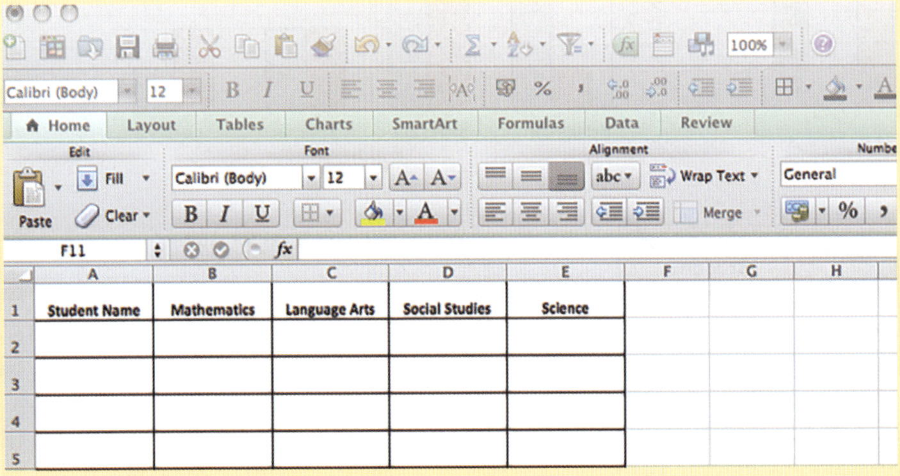

1 Opinion polls can have an impact on decisions and policy. Write a few more questionnaire questions about environmental issues. Here's a suggestion for the first question:

> **Recycling could be improved at our school**
>
> ☐ Strongly agree ☐ Disagree
>
> ☐ Agree ☐ Strongly disagree
>
> ☐ Don't know

Collect 16 or 32 pieces of data for each question. Present your results as pie charts. Present your findings in a school assembly to start a discussion about how the students at your school can improve their environment.

Key words
- spreadsheet
- cell
- data capture

2 This bar graph presents data about the length of time students spent washing their hands.

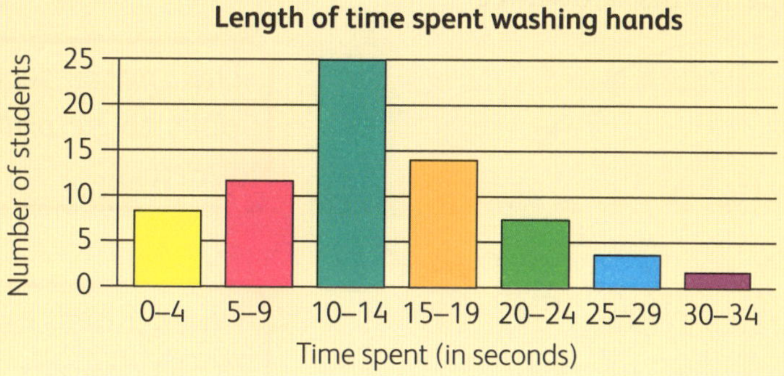

- **a** This bar graph shows grouped data. How many students were surveyed?
- **b** Conduct a similar survey in your school. Collect information for two different classes, of different ages.
- **c** Present your findings as a double bar graph.
- **d** Interpret and explain the findings of the graph in a paragraph.
- **e** **Challenge:** Use spreadsheet software to store and enter the results of your survey about homework. Experiment with using the software to present the data as different graphs.

Maths detective: Pie charts

Let's investigate some pie charts that are not split into equal sections. You can estimate the fractions in the pie charts, or you can measure the angles.

Case 1

What fraction of each pie chart is shaded?

Measure the angle that each shaded section makes in the centre.

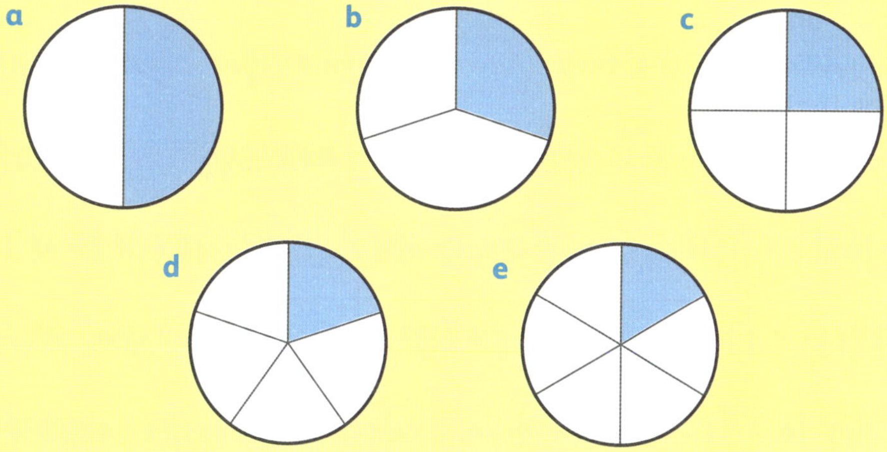

Case 2

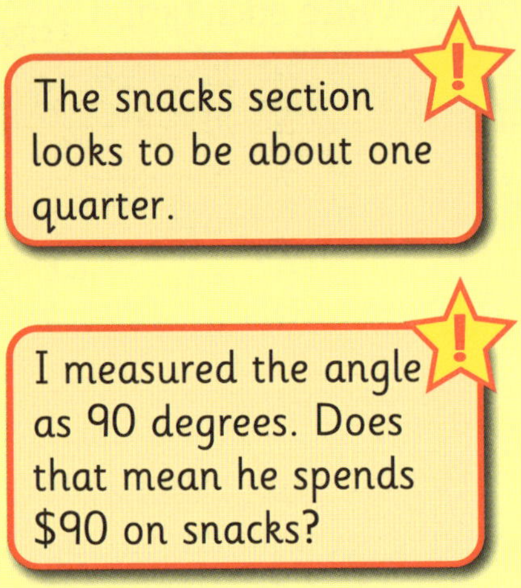

1 Measure the angle of each section and give the fraction for each category.

2 Work out how much Hakeem saves each month if his total allowance is $1 800.

Case 3

Look at the pie chart.

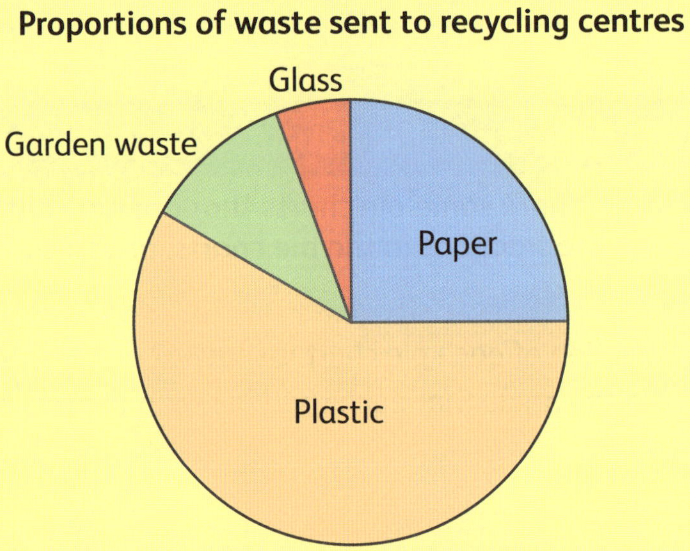

Estimate the fraction of each section of the pie chart.

Then measure the angles to check your estimates.

Copy and complete the table.

Pie chart section	My estimate	My partner's estimate	Measured angle	Who was closest?
Plastic				
Paper				
Garden waste				
Glass				

What I have learned

Read the sentences. Then choose one description from A to D from the box to match each type of graph. Write each graph and its description in your notebook to summarise what you learned in this topic.

1 When would you use each graph or chart?

a pie chart **b** bar graph

c double bar graph **d** line graph

A to compare data from two groups with the whole population
B to show the data as fractions of the whole group
C to present data that changes over time
D to compare different categories of data

2 Name the features missing from this bar chart. Why are they important?

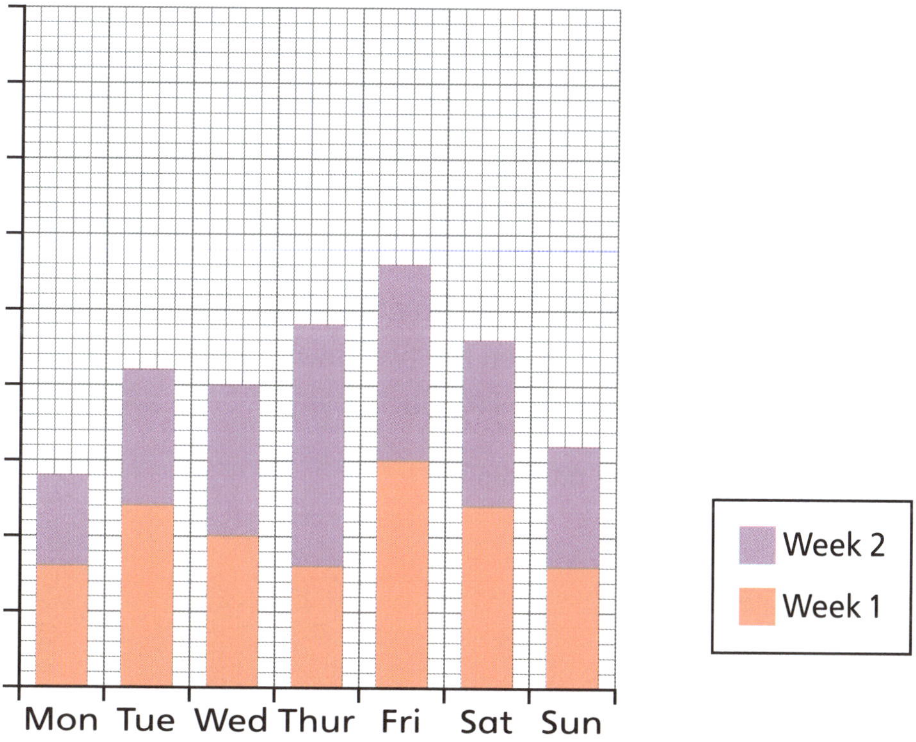

Practice questions

1 Which pie chart has a blue section that is greater than one half?

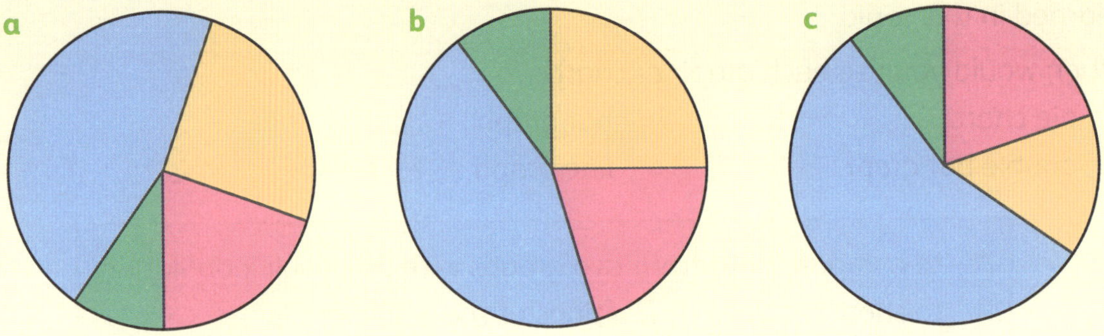

2 Describe in your own words what this double bar graph shows about the effect of preparing for a test.

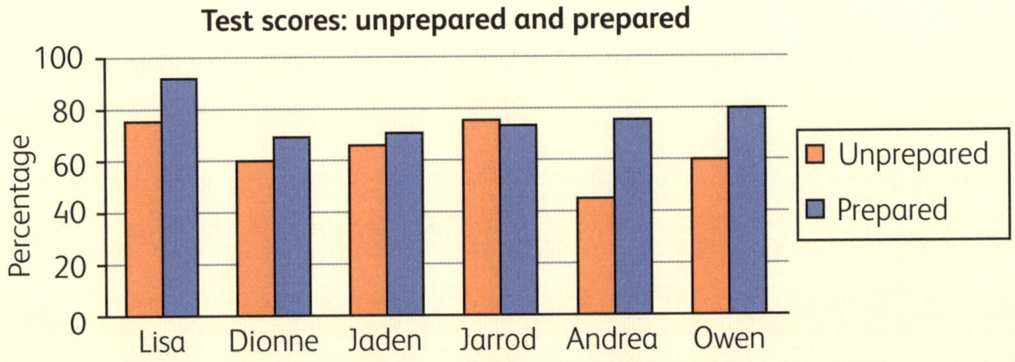

3 These graphs show the recorded temperature and rainfall in two different towns.

a What was the difference in rainfall between the two towns in July?

b Estimate the temperature recorded in April for each town.

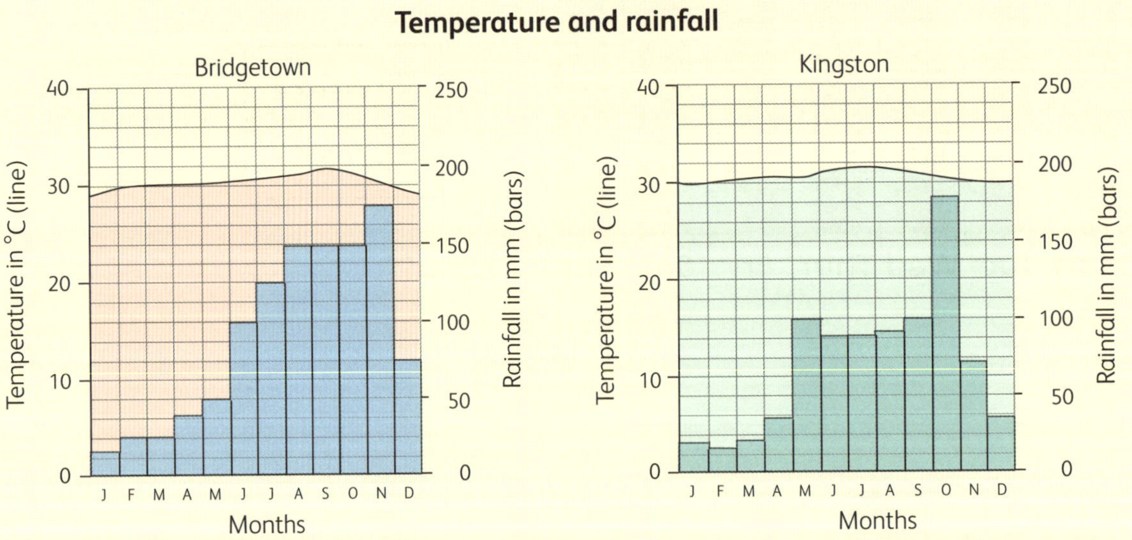

4 This line graph shows the growth of a tree over time.

Graph showing growth of a tree

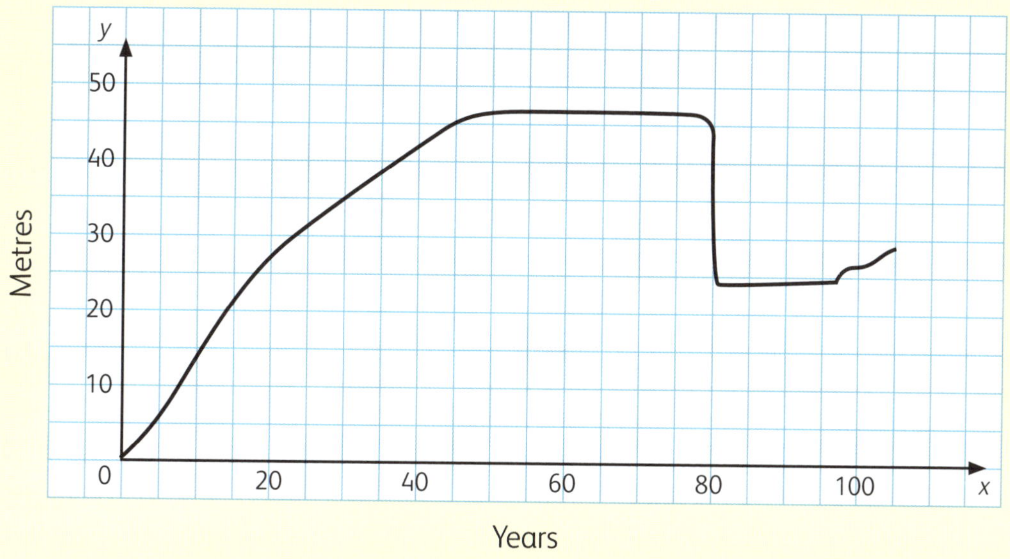

a What was the height of the tree at 35 years?

b When was the tree 20 metres tall?

c Explain what happened when the tree was 80 years old.

9 Self check

Check you can do this:	If you have difficulty, try this:
1 Draw and interpret pie charts.	Go back to page 115. Remind yourself how to work out a fraction of a full turn, and how to use angles to show the correct portion of the circle.
2 Draw and interpret bar graphs, including double bar graphs.	Go back to pages 116–118 and revise the work you did on these types of graphs.
3 Interpret line graphs.	Go back to page 119 and revise how to read data off this type of graph.
4 Choose which graphs to use to show data.	Go back to page 121 and remind yourself how each type of chart or graph is suited to particular types of data.

PEP task: Fitness fanatic

This performance task has 3 parts. Your teacher will tell you when to do each part.

Part 1

The owner of a fitness centre wants to lease new treadmills for the centre. The green shaded area on the plan below shows how much space is available for the treadmills.

You will need:
- a ruler
- a pencil
- squared paper
- a pair of compasses
- a protractor
- a calculator.

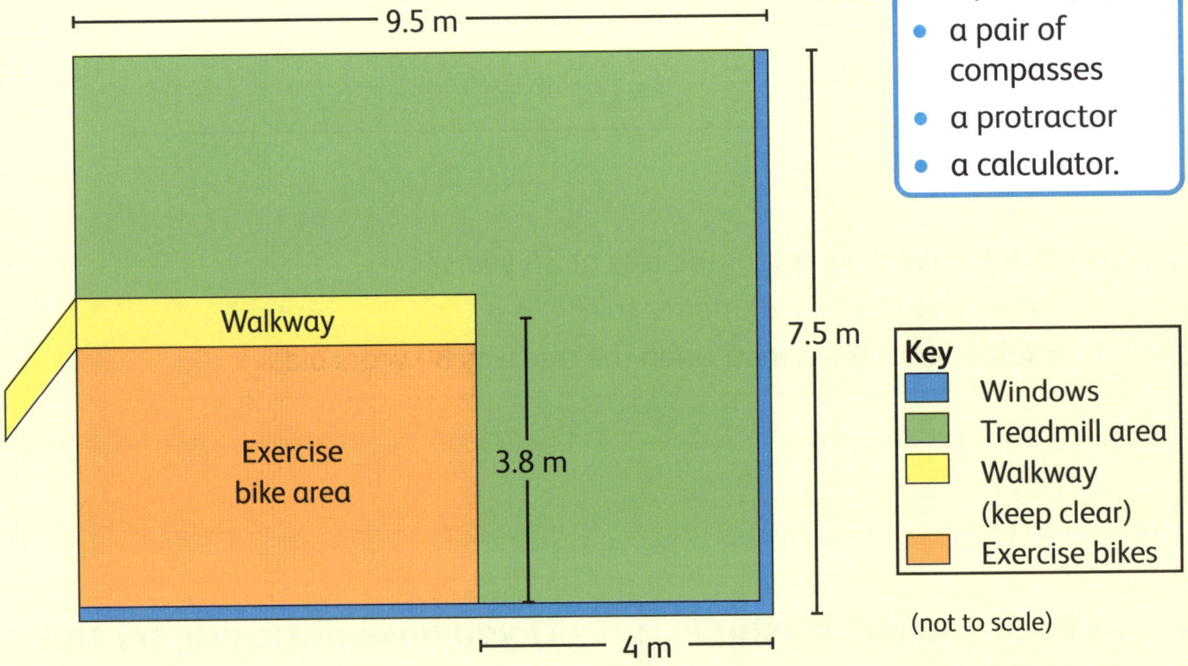

1 Draw a scaled plan on squared paper. Use a scale of 2 cm to 1 metre.

2 The owner wants to fit in as many treadmills as possible. This is the information that she is working with:

- The treadmills have a footprint that is 80 cm wide and 175 cm long. This is the amount of space they take up on the floor.
- There must be at least $\frac{3}{4}$ of a metre space between treadmills.
- Treadmills need to be at least 0.75 m from any windows for safety reasons.

a Calculate the footprint that a treadmill will have if it is drawn to scale on your plan. You can work this out by multiplying lengths in metres by 2 to get a centimetre length for the plan.

b Work out how many treadmills the owner can fit into the space she has. Show the position of the treadmills on your plan.

c Write two sentences explaining why you arranged them as you did.

Part 2

Elaine has been running on a treadmill. This is what the display looks like.

Jeffrey is also using a treadmill. This is what the display on his treadmill looks like.

Use the information on the treadmill displays to answer these questions.

1 BPM stands for beats per minute and tells the user what their heart rate is while they are training.

- **a** Whose heart rate is higher, Jeffrey's or Elaine's?
- **b** What is the difference between the two heart rates?

2 The time display shows how long each person has been training on the treadmill.

- **a** Who has been training for half an hour?
- **b** If it is 6:15 a.m. now, when did Jeffrey start using the treadmill?
- **c** If Jeffrey plans to train for $\frac{3}{4}$ of an hour, how much longer will he spend on the treadmill?

3 The speed on the treadmill tells people how fast they are walking or running in kilometres per hour.
You can use a special formula to work out speed, distance and time.

- Speed = distance (km) ÷ time (hours)
- Distance = speed (km/hr) × time (hours)
- Time = distance (km) ÷ speed (hours)

- **a** Express each person's time as a fraction of an hour.
- **b** Calculate Elaine's speed in kilometres per hour.
- **c** Work out how far Jeffrey has run.
- **d** If Jeffrey continues to run at the same speed for 45 minutes, how far will he have run?
- **e** Elaine runs for another half an hour. At the end of her training session, the time shows 60:00 and the distance covered is 13.2 km. Did she run faster or slower in the second half hour of training? Explain how you worked this out.

4 Treadmills can be set so that it feels like you are running up a slope. The gradient or incline tells you how many metres you go up, for every 100 metres you run.

Gradient can be defined as $\frac{\text{rise}}{\text{run}}$.

A gradient of 12% means that you go up 12 metres for every 100 metres that you are running. This is a steep gradient that would feel like you are going up a very steep hill.

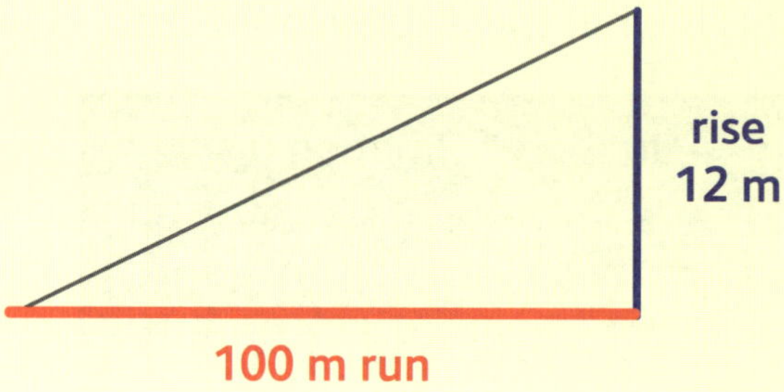

Draw a triangle like this for each athlete and show how many metres they would have risen if they were running on a real hill for half an hour.

Part 3

When athletes compete in the Olympic Games, they stay in an Olympic Village and the organisers prepare and serve food.

During the 2021 Olympics, the organisers served:

- 25 000 loaves of bread
- 330 000 kilograms of fruit and vegetables
- 75 000 litres of milk
- 1.2 million meals.

1 Look at this pictograph showing the data above.

Loaves of bread	
Litres of milk	
Kilograms of fruit and vegetables	
Meals served	

Key

 =	 =
 =	 =

a Use the information above to work out what each symbol represents.

b Copy and complete the key.

c If there were 9 375 athletes, how many loaves of bread is that per person?

d Is it reasonable to say that the organisers served approximately 130 meals per person? Explain your answer.

2 You already know the importance of eating a healthy balanced diet to provide the energy your body needs to function and perform well. A balanced diet includes all of the foods your body needs in the correct proportions. When you exercise, your muscles burn more energy, so the proportion of different foods in your diet may change.

Read this information for competitive athletes carefully.

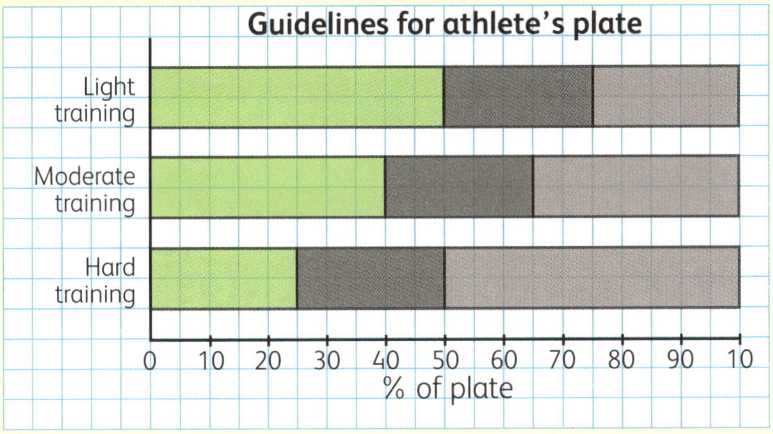

Key	
Fruit/Vegetables	(light green)
Lean protein	(dark grey)
Grains/Carbohydrate	(light grey)

Light training	Moderate training	Hard training
1 teaspoon healthy fats	1 tablespoon healthy fats	2 tablespoons healthy fats
	Extra 1 side of fruit	Extra 1 side of fruit

Complete these statements. Write down only the word that completes the statement.

a When you are doing light training, $\frac{1}{2}$ your plate should be ____ .

b Lean protein should always be ____ of your plate.

c When you train harder, you can eat more ____ .

d You need the least amount of grains and carbohydrates when you are doing ____ training.

e You can add ____ fats when you are training hard.

3 This plate diagram shows the proportion of different types of foods when you are doing light training.

Draw a similar diagram for a hard training day. Include examples of foods that you like for each group.

4 This meal plan shows what a Jamaican sprinter ate on a hard training day.

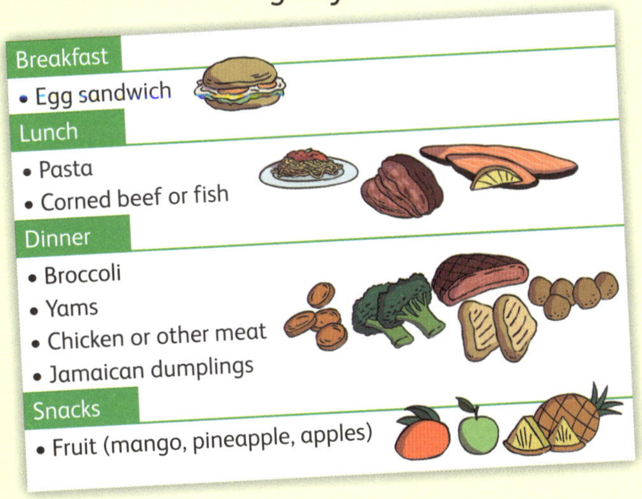

a Does this meal plan match the guidelines? Explain why or why not.

b Draw this athlete's plate for the dinner in the diagram. Show how much of each food should be on the plate.

TERM 2

UNIT 1: NUMBER p133

Solving problems with fractional numbers

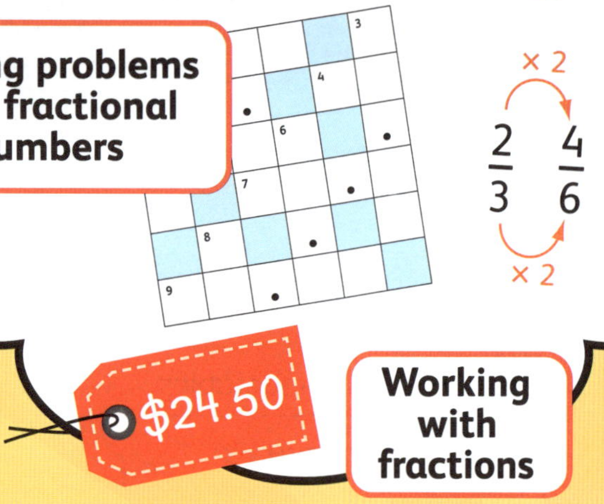

Working with fractions

UNIT 2: MEASUREMENT p175

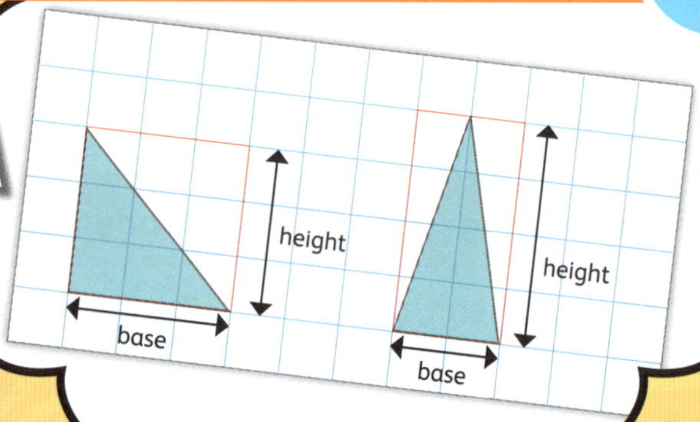

Perimeter and area

UNIT 3: GEOMETRY p195

Properties of triangles and quadrilaterals

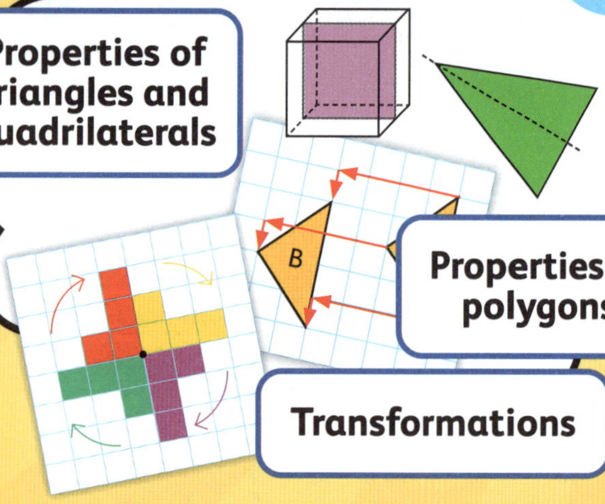

Properties of polygons

Transformations

UNIT 4: ALGEBRA p222

Patterns and rules

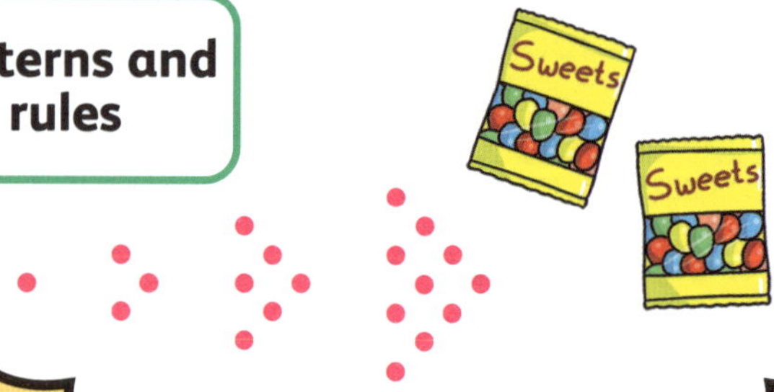

Expressions and variables

PEP TASK p238

Unit 1: Number

Chapter 10 Solving problems with fractional numbers

In this chapter you will explore these focus questions:

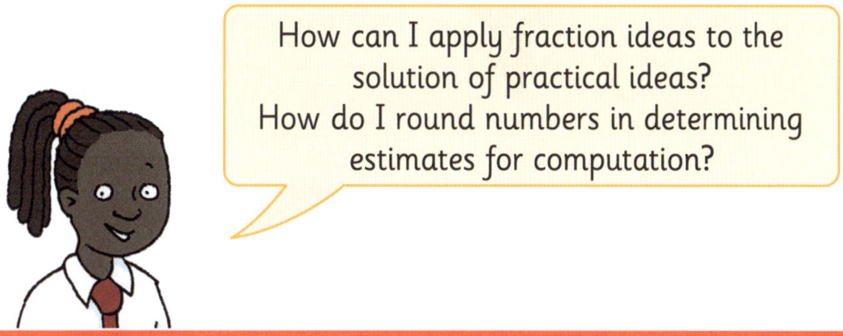

To answer these questions, you will:

- make calculations and solve problems involving decimals and fractions
- work with decimals to three decimal places
- compare, round off and convert fractions in different forms
- solve problems involving decimals and money amounts.

Starting point

Jewel beetles are named because their beautiful metallic colours shimmer like a jewel. Did you know that Jamaica is home to 25 species of jewel beetle and 16 of these are not found anywhere else in the world? However, with 15 000 known species of jewel beetles, there are many other kinds not found in Jamaica!

1 Which of these numbers best describes the fraction of Jamaica's jewel beetle species that are not found anywhere else?

a 0.64 **b** $\frac{1}{2}$ **c** $\frac{2}{3}$

2 Which of these numbers best describes the fraction of the world's jewel beetle species found in Jamaica?

a 0.2 **b** 0.02 **c** 0.002

3 **a** Read these facts. Then match the diagrams below to the facts.

A

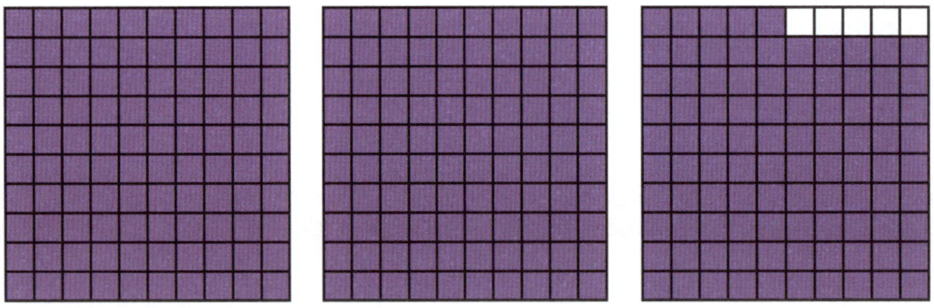

B

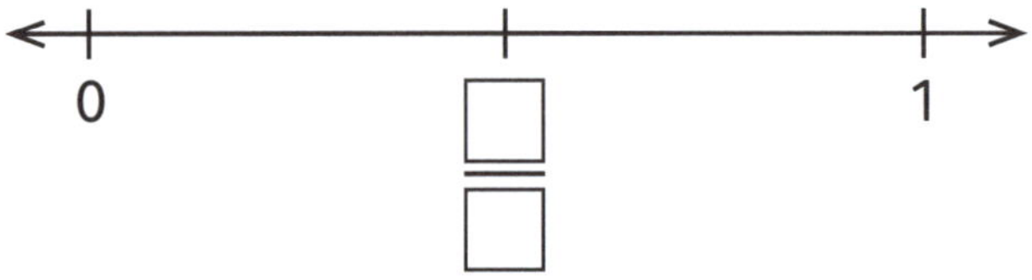

C

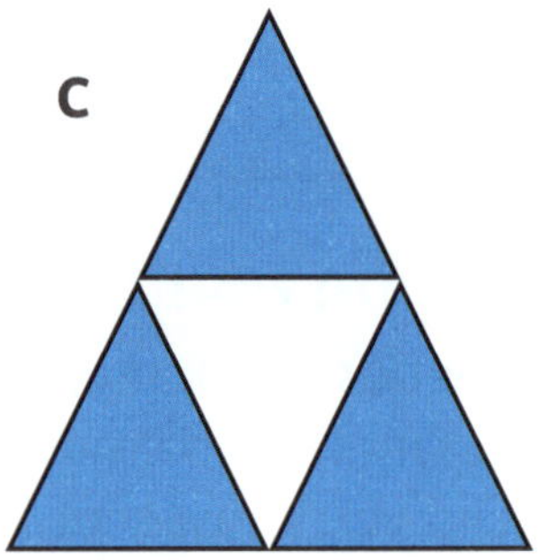

b What other diagrams can you draw to represent the fractions or decimals in each of the facts? Work with a partner. Represent each fact in two different ways.

4 Here is another fact. What will you draw this time?

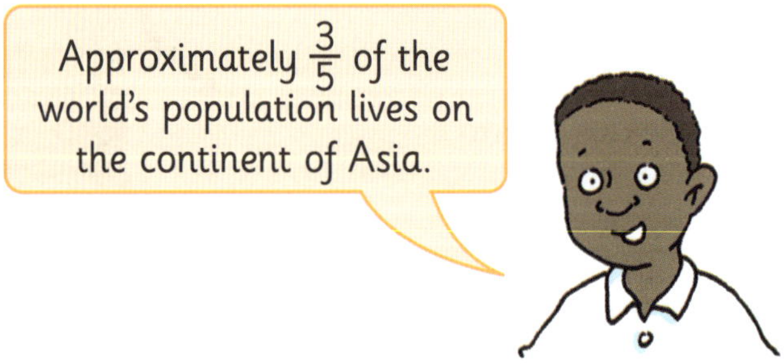

5 Look in a newspaper. Find some information that is given as fractions or decimals. Represent them in your own way.

Decimal place value

Key maths idea

A **decimal fraction** is a type of fraction that has a **denominator** that is a power of ten (such as 10, 100, 1 000 and so on). The **numerator** is expressed using numerals after a **decimal point**. The number of figures after the decimal point tells you how many zeros in the denominator.

For example, 0.12 has a numerator of 12 (the numbers after the decimal point) and a denominator of 100 (the number of figures after the decimal point.)

So $0.12 = \frac{12}{100}$ $\quad 0.015 = \frac{15}{1\,000}$

Decimal numbers have a whole number part and a decimal fraction part.

$57.441 = \frac{57\,441}{1\,000} = 57\,\frac{441}{1\,000}$

Look at these numbers.
What is the same and what is different about them?

0.375 3.75

30.57

0.53 7.053

Talk about

Can you explain which are decimal fractions and which are decimal numbers? How many decimal places does each number have?

Look at the five numbers represented on a place value chart.

Tens	Ones	Tenths	Hundredths	Thousandths
	0	.3	7	5
	3	.7	5	
3	0	.5	7	
	0	.5	3	
	7	.0	5	3

What is the value of the digit 3 in each of the numbers? What is the role of zero each time?

In the decimal fraction 0.375, the digit 3 has the value 3 tenths, $\frac{3}{10}$ or 0.3.

The digit 7 has the value 7 hundredths or $\frac{7}{100}$ or 0.07.

The digit 5 has the value 5 thousandths or $\frac{5}{1\,000}$ or 0.005.

Key words

decimal fraction
denominator
numerator
decimal point
decimal number

Say the value of each digit in the decimal number 7.053.

What happens when you increase the number 7.053 by 7 thousandths? And then by 4 hundredths? Use counters on a place value chart to help you.

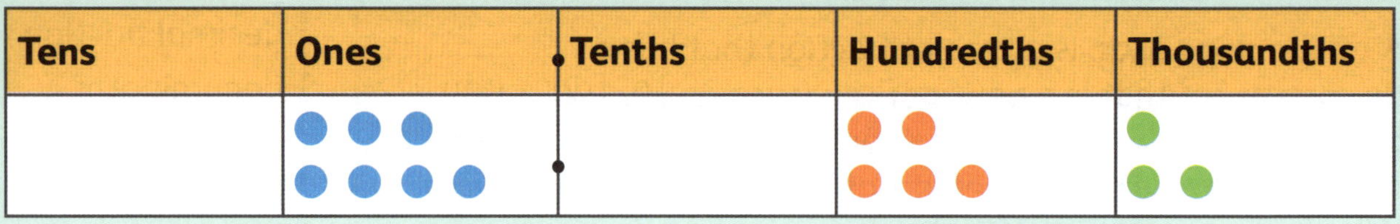

1 Which values are represented here? Say if they are decimal fractions or decimal numbers.

a 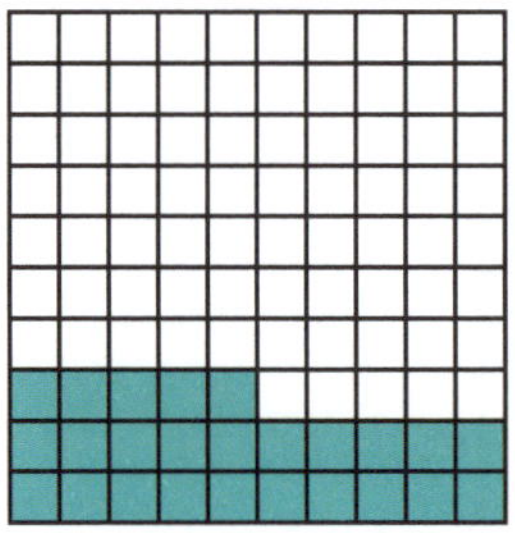

b 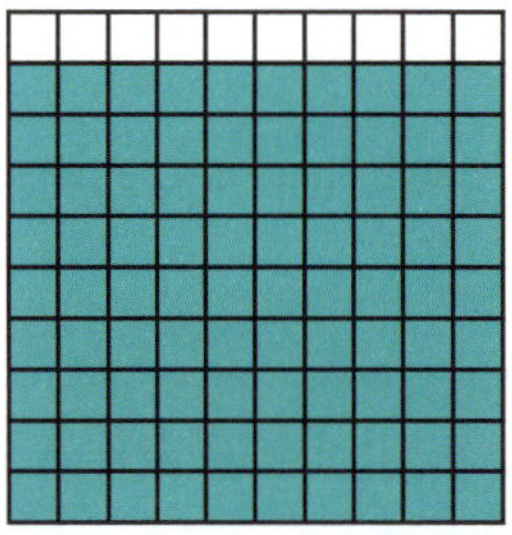

c 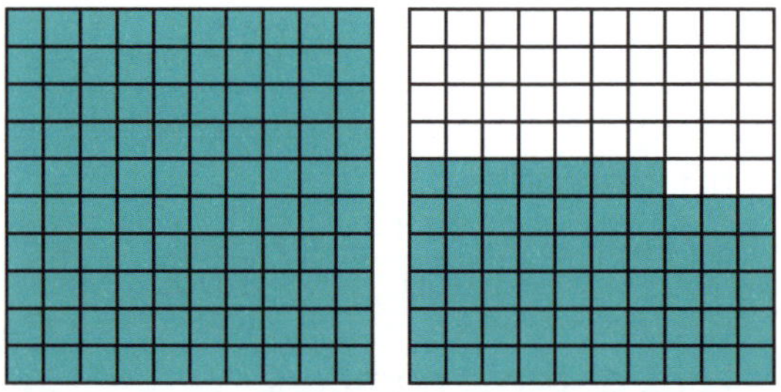

2 Copy and complete these part-whole diagrams.

a 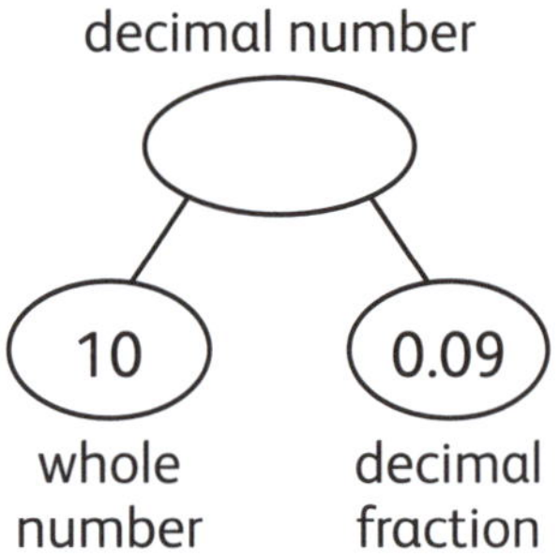

b 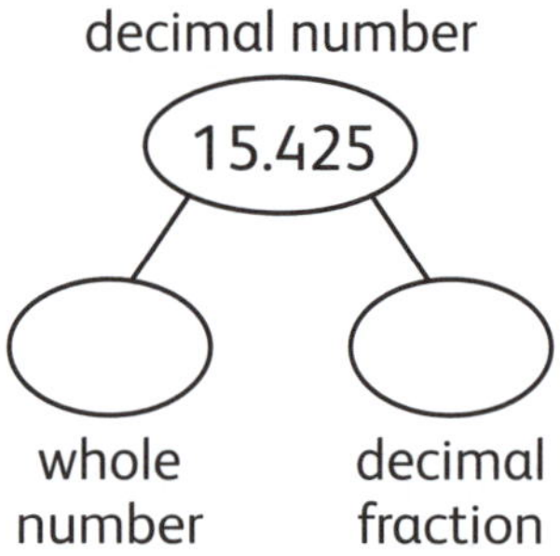

3 Here is a place value chart.

a What pattern do you notice if you go down each column from top to bottom? What is the pattern rule?

b What pattern do you notice if you go along each row from left to right?

This type of chart is called a **Gattegno** chart. It allows you to express numbers in expanded notation. When you expand a number, you express it as the sum of the values in each place value position.

100	200	300	400	500	600	700	800	900
10	20	30	40	50	60	70	80	90
1	2	3	4	5	6	7	8	9
0.1	0.2	0.3	0.4	0.5	0.6	0.7	0.8	0.9
0.01	0.02	0.03	0.04	0.05	0.06	0.07	0.08	0.09
0.001	0.002	0.003	0.004	0.005	0.006	0.007	0.008	0.009

4 Work with a partner. Point to the cells on the chart that you would add to make each number.

a 105.5 **b** 1.08 **c** 12.37 **d** 314.069

5 Use numbers from four cells on the chart to answer each challenge.

a Make a number greater than 0 and less than 1.

b Make a number greater than 30 and less than 31.

c What is the closest number to 1 000 that you can make from this chart?

d Make a decimal number with a fractional part of $\frac{175}{1\,000}$.

e Make a decimal number with a fractional part that is equivalent to $\frac{11}{50}$.

f Make the decimal number that is 1 hundredth more than 145.89.

Expanding decimal numbers

Key maths idea

We can **expand** decimal numbers to show the place value parts.

There are 7 ones, 5 hundredths and 3 thousandths in the decimal number 7.053.

Note that there are no tenths in the tenths position. But did you know that there are 70 tenths altogether in the number 7.053? Can you explain why?

We can show the number in **expanded notation** as:

$$7.053 = 7 + 0.05 + 0.003.$$

Try adding each of the place value parts on a calculator. What is the total? Why?

Key words

expand

expanded notation

1 **a** Here are some digit cards.

2 4 0 5

You can represent each number on a place value chart first.

Use them to make a number with 3 decimal places: ☐.☐☐☐

b Expand the number to show the place value parts.

c Rearrange the four digits to make more numbers. Expand each one to show the place value parts. How many different numbers can you make?

2 Expand these quantities to show their place value parts.

a \$45.36 **b** \$453.60 **c** 15.25 km

d 1.525 km **e** 23.99 kg **f** 2.399 kg

3 True or false? Correct all the statements that are false.

a $4.53 = 4 + 0.5 + 0.03$

b $3.451 = 3 + 0.4 + 0.5 + 0.001$

c $3.702 = 3 + 0.7 + 0.2$

d $37.021 = 30 + 7 + 0.2 + 0.01$

e $370.21 = 300 + 70 + 0.2 + 0.01$

Equivalent decimal fractions

Key maths idea

Two decimal fractions are represented here.

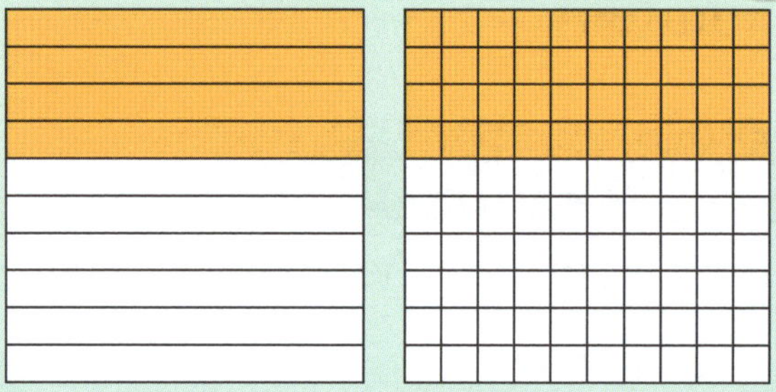

What is the same about them? What is different?

We can use what we know about fractions to write the values in different ways.

4 tenths or $\frac{4}{10}$ 40 hundredths or $\frac{40}{100}$

We can also say that $\frac{4}{10} = \frac{40}{100}$ or that 4 tenths is **equivalent** to 40 hundredths.

We can use what we know about decimal fractions to represent the numbers on a place value chart.

Ones	Tenths	Hundredths	Thousandths
0	.4		
0	.4	0	

Zero acts as a place holder in the ones position. So we know that there are no ones. The zero in the hundredths position is not required as it shows that there are no extra hundredths. Remember that 4 tenths is equivalent to 40 hundredths.

Now look at this representation to show thousandths. What do you notice this time?

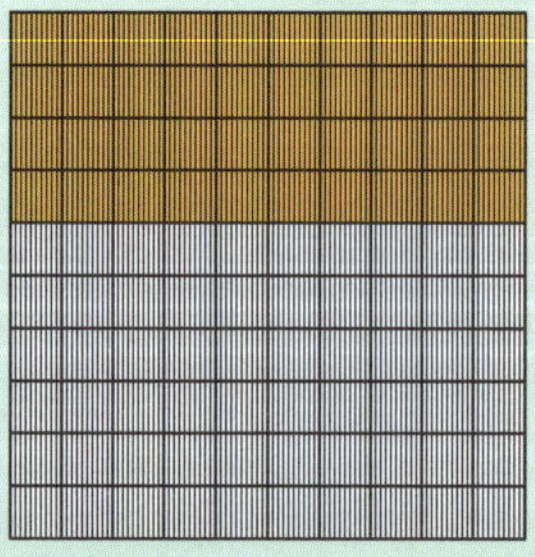

4 tenths = 40 hundredths = 400 thousandths

$\frac{4}{10} = \frac{40}{100} = \frac{400}{1\,000}$

Can you still see the decimal fraction 0.4?

Key word

equivalent

1 Work with a partner. Draw pairs of grids like this.

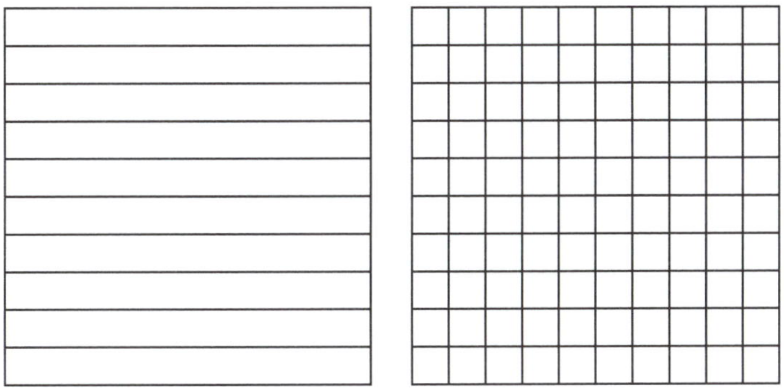

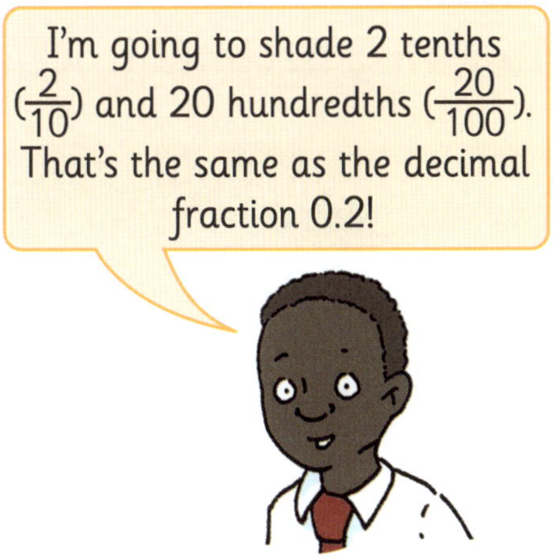

a Shade each grid to show equivalent values. Write the value as fractions and as decimals.

b How many different decimal fractions can you find?

2 Together with your partner, look at the grids you have made. How can you explain how many thousandths are equivalent each time?

Record your findings in the following way each time:

$$\frac{\square}{10} = \frac{\square}{100} = \frac{\square}{1\,000} = 0.\square$$

3 These children are discussing how far they ran yesterday.

Do you agree? Explain your thinking.

4 Work with a partner. Use a computer package to design a new flag. Your flag must have the following fractional parts:

- $\frac{300}{1\,000}$ red
- $\frac{20}{100}$ blue
- $\frac{4}{10}$ yellow
- the remaining part is green.

Write each coloured part as a decimal fraction.

Comparing decimal fractions

Key maths idea

Akiel represents the decimal fractions 0.5 and 0.06 to compare them.

He writes the following statements.

Six hundredths is less than 5 tenths.

0.06 is less than 0.5

0.06 < 0.5

Amy represents the decimal numbers 0.45 and 0.045 to help compare them.

She writes the following statements:

Four tenths and 5 hundredths is greater than 4 hundredths and 5 thousandths.

So 0.45 is greater than 0.045

0.45 > 0.045

1 What should the children write for each of these pairs of decimals? Use grid paper to help represent the decimal fraction part each time.

a 0.09 and 0.9 **b** 0.7 and 0.85 **c** 0.5 and 0.25

d 0.65 and 0.675 **e** 0.4 and 0.375

2 Compare these masses. Which is heavier each time, mass X or mass Y?

	a	b	c	d	e
Mass X	0.4 kg	0.75 kg	0.729 kg	0.54 kg	0.325 kg
Mass Y	0.04 kg	0.8 kg	0.72 kg	0.45 kg	0.33 kg

Ordering decimals

Key maths idea

The decimal numbers on this number line are arranged in **ascending** order.

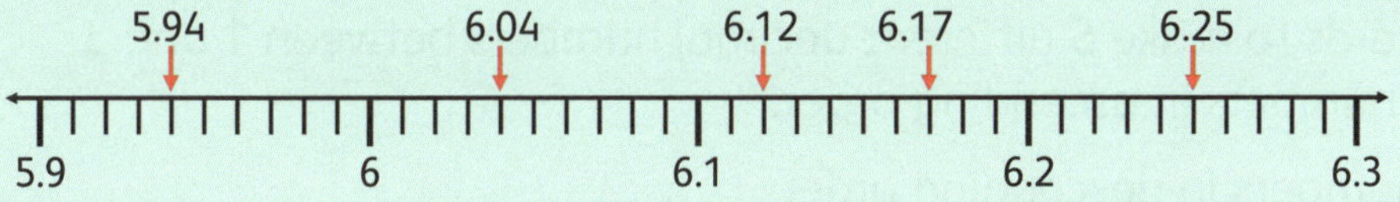

Which numbers have digits with the same ones value?
Which numbers have digits with the same tenths value?
When we are **ordering** decimal numbers, it is important to compare the digits in the same place value positions.

Try putting the numbers 5.39, 5.325 and 5.16 in ascending order.

Step 1 First line up the decimal points: 5.39
5.325
5.16

Step 2 Start on the left. What do you notice about the digits in the ones position?
The digits are the same.
So we move one place to the right.

Step 3 Now check the digits in the tenths position.
The number 5.16 only has 1 tenth in the tenths position, so it is the smallest number.
What do you need to do now?

We write the numbers in ascending order as 5.16, 5.325, 5.39.
Can you write the numbers in **descending** order?

Place value charts are also useful when comparing and ordering decimal numbers.

Ones	Tenths	Hundredths	Thousandths
9	.3	3	
9	.3	2	5

Which number is larger? How do you know?

Key words
ascending
ordering
descending

1 Work in small groups. You will need these cards:

0 1 2 3 .

a Arrange the cards to make 5 different decimal fractions. Write the decimal fractions in ascending order.

b Now arrange the cards to make 5 different decimal numbers between 1 and 3. Write the decimal numbers in ascending order.

> Remember to use a place value chart to help you order these numbers.

2 Arrange the decimal numbers in descending order.

a 5.45 5.49 5.18 5.108 5.99

b 17.23 11.15 13.55 13.2 17.321

3 Arrange the lengths in ascending order.

a 2.2 m 2.02 m 2.002 m 2.112 m 2.222 m

b 13.5 m 13.05 m 13.436 m 13.321 m 13.35 m

Rounding numbers

Key maths idea

Look at the way these numbers have been rounded to the nearest 10, 100 or 1 000.

	Rounded to the nearest 10	Rounded to the nearest 100	Rounded to the nearest 1 000
754	750	800	1 000
2 365	2 370	2 400	2 000

Do you remember the rules for **rounding** whole numbers?

We can use what we know about rounding whole numbers to help us round decimal numbers to the nearest whole number, tenth or hundredth.

Let's think about the number 4.35. It is positioned between the whole numbers 4 and 5 on a number line, but is it closer to 4 or closer to 5?

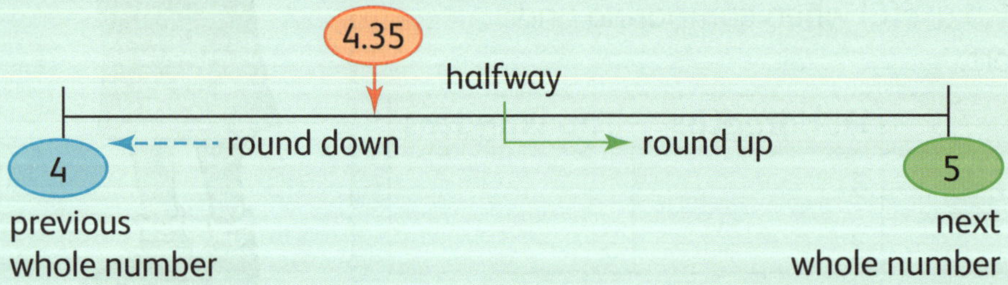

4.35 is closer to 4 because it is less than halfway between the two whole numbers.

Key word
rounding

All decimal numbers with 5 tenths (0.5) or more round up to the nearest whole number. All decimal numbers with fewer than 5 tenths will round down.

The decimal number 4.35 has 3 tenths (0.3), so it rounds down to 4. We can write 4.35 ≈ 4.

We can use similar rules to now round 4.35 to the nearest tenth.

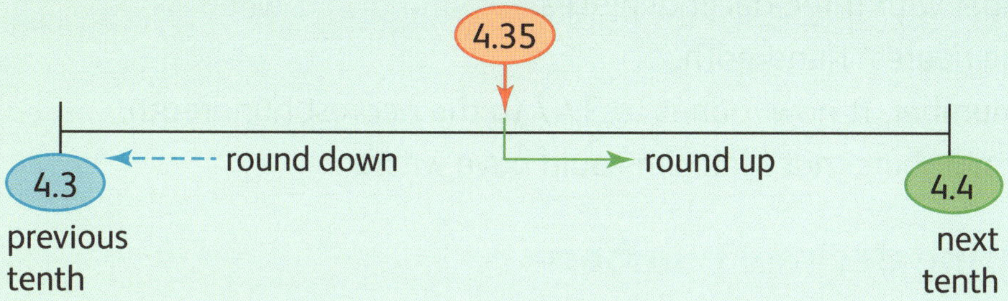

4.35 is halfway between 4.3 and 4.4. All decimal numbers with 5 hundredths (0.05) or more round up to the nearest tenth. The decimal number 4.35 has 5 hundredths so it rounds up to 4.4. We can write 4.35 ≈ 4.4.

The decimal number 4.356 has 3 decimal places. Will it round up or down to the nearest hundredth? Talk to a partner and draw a diagram like the ones above to help you explain.

1 **a** Round 4 505 to the nearest 10, to the nearest 100 and to the nearest 1 000.

b Now round $6 458 to the nearest 10 dollars, 100 dollars and to the nearest 1 000 dollars.

2 Copy and complete this table.

Number	Rounded to the nearest whole number	Rounded to the nearest tenth (0.1)	Rounded to the nearest hundredth (0.01)
0.451			
1.514			
5.14			
4.051			

3 Write five decimal numbers each time that:

a round to 7 to the nearest whole number

b round to 7.1 to the nearest tenth

c round to 6.9 to the nearest tenth

d round to 7.25 to the nearest hundredth

e round to 6.94 to the nearest hundredth.

4 Here are the recorded masses of different penguin species

Adelie	Chinstrap	King	Emperor	Gentoo
4.9 kg	4.49 kg	13.8 kg	22.45 kg	5.921 kg

a Round the mass of each species to the nearest whole kilogram (kg).

b Round each mass to the nearest tenth of a kilogram (0.1 kg).

5 Look at these price tags.

a Round the prices to the nearest dollar.

b Which prices round to $30 to the nearest 10 dollars?

6 Malcolm writes a number with three decimal places.
It rounds to 3.46 to the nearest hundredth.
He adds $\frac{5}{1\,000}$ to his number. It now rounds to 3.47 to the nearest hundredth.
Find the four different numbers that Malcolm could have written.

Adding and subtracting decimal numbers

Key maths idea

We can use mental and written methods of **addition** and **subtraction** to calculate with decimal numbers. Look at the addition and subtraction here. What other methods can you use?

529.3 + 243.1 =
500 + 200 = 700
20 + 40 = 60
9 + 3 = 12
0.3 + 0.1 = 0.4 +
772.4

529.3 − 243.1 =
529.3 − 200 − 40 − 3 − 0.1
= 329.3 − 40 − 3 − 0.1
= 289.3 − 3 − 0.1
= 286.3 − 0.1
= 286.2

The diagrams represent the calculations 0.3 + 0.8 and 1.3 − 0.8.
They help us to think about regrouping decimal fractions.

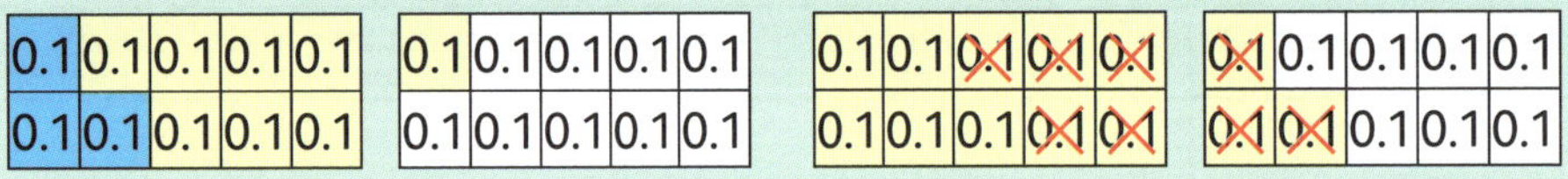

3 tenths + 8 tenths = 11 tenths   13 tenths − 8 tenths = 5 tenths

Can you see how we can use the same facts to help us complete these written calculations?

	100s	10s	1s	.	$\frac{1}{10}$s
	5	2	9	.	3
+	2	4	3	.	8
	7	7	3	.	1
		1	1		

	100s	10s	1s	.	$\frac{1}{10}$s
	4 ~~5~~	1 ~~2~~	8 ~~9~~	.	(13)
−	2	4	3	.	8
	2	8	5	.	5

Key words

addition

subtraction

Let's look at the calculations 5.293 + 2.438 and 5.293 – 2.438.

You can round the decimal numbers to the nearest whole number to help make the estimates 5 + 2 = 7 and 5 − 2 = 3.

How can you use the facts 3 + 8 and 13 − 8 to help you answer these calculations?

What else can help you?

What is different about these calculations?

10 − 1.5   23.65 + 42.3   17.9 − 4.32

I know that 1.5 can be expanded as 1 + 0.5, so I can subtract 1 from 10 first.

I know the answer to 100 − 15, so the answer to 10 − 1.5 must be 10 times as small.

I know that 3 tenths is equal to 30 hundredths, so 42.3 = 42.30.

How do these ideas help you to complete the calculations?

1 Solve these calculations. What method will you use?

- **a** 65.2 + 27.4
- **b** 65.2 – 27.4
- **c** 529.36 + 231.38
- **d** 529.36 – 231.28
- **e** 2.473 + 1.314
- **f** 2.473 – 1.314
- **g** 65.8 + 31.45
- **h** 65.8 – 31.45

2 A class did a project about tadpoles. They recorded the measurements of one of the tadpoles in centimetres (cm). Answer the questions about the tadpoles.

- **a** What was the total length of the tadpole at 3 weeks?
- **b** What was the total length of the tadpole at 12 weeks?
- **c** How much did the tadpole grow between 3 weeks and 12 weeks?
- **d** How much longer was the tadpole at 8 weeks than at 14 weeks?
- **e** How much did the tadpole grow from 3 weeks to 4 weeks?
- **f** At which stage does the tadpole get shorter. Why?

3 Mrs Williams bought four items at the market. She spent these amounts:

$42.75 $28.30 $36.90 $20.80 $39.50

a Estimate whether she can pay using a $200 note.

b How much change would she get from $200?

c Explain in your own words how you worked out the change.

> Check your answers against your estimates and then by using a calculator.

4 Look at the prices of these four items.

$164.75 $185.49 $349.50 $86.73

a Mr Harris buys two items with a total cost that is less than $500. Which two items could he have bought? Find all possibilities.

b Mrs Jackson buys three items with a cost of $500. Which three items could she have bought?

c Mrs Jackson spent 74 cents more in total than Mr Harris. What do you know about the two items that Mr Harris bought now?

> Write your own story problems about adding and subtracting decimal numbers. Swap with other students in your class to solve them.

5 Solve these problems.

a On a hot day, Audrina and Ruth went on a hike. Audrina drank 2.2 litres of water. Ruth drank 3.52 litres. How much more water did Ruth drink than Audrina?

b Dan trimmed the top of his mango tree by 4.14 m. The tree was 7.75 m tall. How tall is it now?

c A parcel has a mass of 8.375 kg. Another parcel has a mass of 2.792 kg. What is the total mass of the two parcels?

Multiplying decimal numbers by 10, 100 or 1 000

Key maths idea

You can use your knowledge of place value to help multiply decimals by 10, 100 and 1 000 mentally.

Let's start with the decimal fraction 0.362.

100	10	1	•	0.1	0.01	0.001
		0	•	3	6	2
		3	•	6	2	
	3	6	•	2		
3	6	2	•			

0.362×10

0.362×100

$0.362 \times 1\,000$

When you multiply a number by 10, the value of each part of the number is made 10 times as large. As a result, the digits all move 1 place to the left.

When you multiply a number by 100, the value of each part of the number is made 100 times as large. As a result, the digits all move 2 places to the left.

Talk about

Talk to your partner about what happens when you multiply a number by 1 000. Remember that the digits move to show the new value. The decimal point does not move.

1 Copy and complete these number chains.

a 63.52 → ×10 → ☐ → ×100 → ☐ → ×1 000 → ☐

b 9.017 → ×10 → ☐ → ×100 → ☐ → ×1 000 → ☐

c ☐ → ×10 → 4.35 → ×100 → ☐ → ×1 000 → ☐

2 Work out the missing numbers. Use a calculator to check your answers.

a 14.46 × ☐ = 144.6

b 17.314 × ☐ = 1 731.4

c 42.75 × ☐ = 427.5

d ☐ × 100 = 7 140

e 73.123 × ☐ = 73 123

f ☐ × 10 = 194.32

g 145.12 × ☐ = 14 512

h ☐ × 100 = 2 300

3 True or false? Correct all the statements that are false.

a 3.45 m is 10 times as far as 3.450 m.

b 1 569 m is 1 000 times as far as 1.569 m.

c 3 600 kg is 100 times as heavy as 3.6 kg.

d 145 000 kg is 1 000 times as heavy as 14.5 kg.

e 75.5 litres is 10 times as much as 75.05 litres.

f $36.00 is 100 times as much as $3.60.

4

A tree is 100 times as tall as a flower.
A bush is 100 times as tall as a blade of grass.
The flower is 10 times as tall as the blade of grass.
The blade of grass is between 0.02 and 0.03 metres tall.

a What could the heights of the blade of grass, flower, bush and tree be? Find at least five solutions.

b Write the calculations you use each time. Record your solutions in a table.

Multiplying whole numbers and decimals

Key maths idea

We use **multiplication** to find the **product** of two or more numbers.

Let's look at multiplying the number 8 in different ways.

$8 \times 1 = 8$ $\quad$ $8 \times 3 = 24$ $\quad$ $8 \times 0.3 = 2.4$

What happens when we multiply 8 by 1?

What happens when we multiply 8 by a number greater than 1?

What happens when we multiply 8 by a number that is less than 1?

When we multiply a whole number by a number that is less than 1, the product will be smaller than the original whole number.

Look at these two arrays. What is the same and what is different about them?

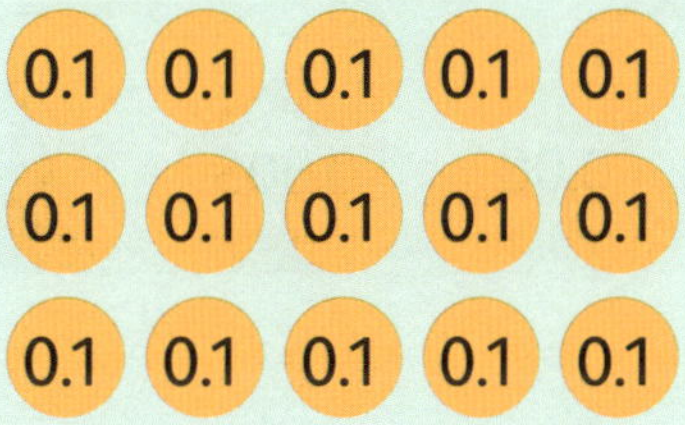

The first array shows that the product of 3×5 is 15.

The second array shows that the product of 3×0.5 is 1.5. The product of 3×5 is 10 times as large as the product of 3×0.5 because 5 is 10 times as large as 0.5.

This array shows that the product of 3×0.05 is 0.15:

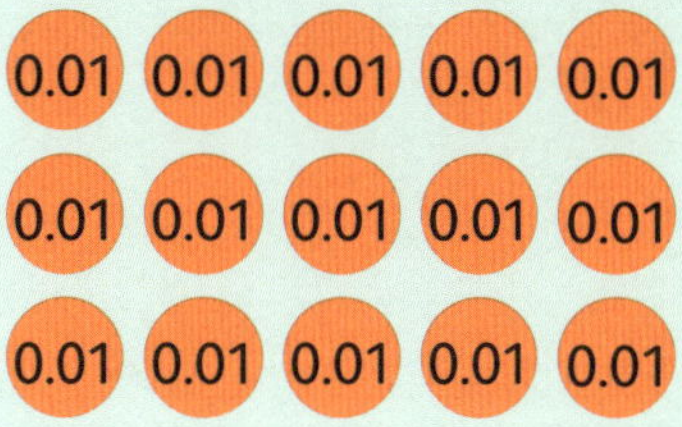

Why is the product 100 times as small as the product for 3×5?

Key words

multiplication

product

1 Use reasoning to sort these multiplications under the following headings. Do not calculate.

a 7 × 0.3 **b** 7 × 1 **c** 12 × 0.2 **d** 12 × 2

e 6 × 1 **f** 9 × 0.04 **g** 6 × 7 **h** 9 × 1

Product is less than the whole number factor	Product is equal to the whole number factor	Product is more than the whole number factor

Now calculate to check if you are correct.

2 Work these out. What patterns do you notice?

a 6 × 7
6 × 0.7
0.7 × 6

b 8 × 5
8 × 0.5
0.8 × 5

c 15 × 2
15 × 0.2
1.5 × 2

3 Answer the questions and write a matching multiplication sentence each time.

What is the total value of:

a the triangles **b** the squares **c** the pentagons?

4 Malika and Amy are discussing how best to calculate 7 × 0.35.

a Try out Malika and Amy's ideas. Check the answer to 7 × 0.35 on a calculator.

b Use what you have found out about this calculating method to now find the following products:

9 × 0.42 9 × 0.425 9 × 1.42	6 × 0.89 6 × 0.893 6 × 2.89	16 × 0.25 16 × 0.125 16 × 3.25

5 Which child spent most in total? How much more did they spend than the child who spent the least?

6 Adam is designing a rectangular flower bed for his garden. Which design will give him the biggest area for his flowers?

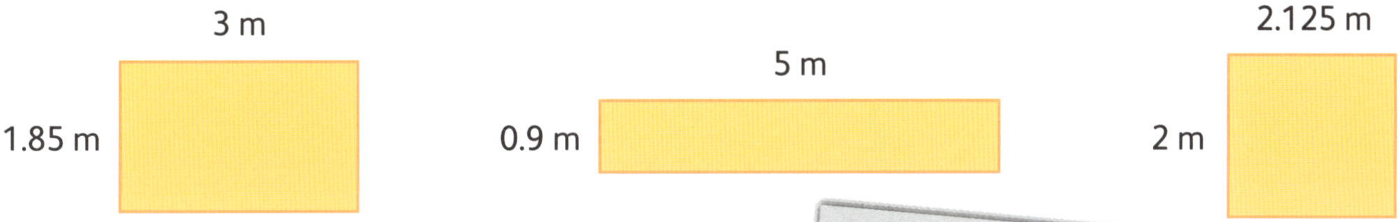

Real-world maths: Electricity charges

Hattie receives the electricity bill for the first three months of the year. How much must she pay in total, including the charges?

Electricity bill

January	$ 18 265.54
February	$ 17 320.65
March	$ 15 048.37
Charges	$ 3 650.75
Total	$

Maths detective: Phenomenal pumpkins

Use your maths detective skills to solve these cases.

Case 1

The school gardening club is holding a giant pumpkin growing competition. They record the six winning pumpkins on the 'Phenomenal Pumpkin' scale.

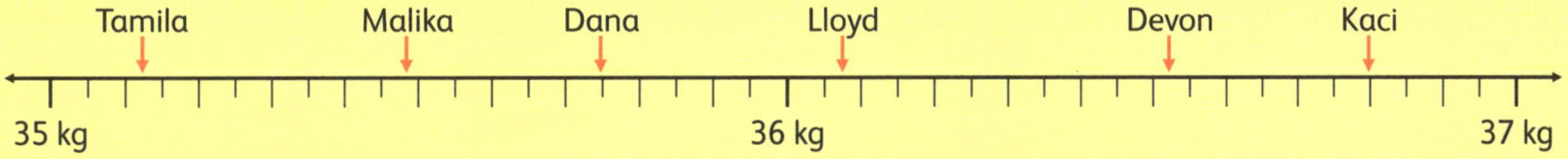

Here are the masses of the 6 winning pumpkins, but which child grew each one?

36.08 kg	35.75 kg	35.13 kg	36.8 kg	35.47 kg	36.525 kg

Name	Tamila	Malika	Dana	Lloyd	Devon	Kaci
Pumpkin mass						

Case 2

The garden club also raffles a pumpkin pie to raise funds. To enter, you have to find at least eight solutions to this riddle.

> You multiply a whole number by a decimal number for your calculation.
> The product has 2 decimal places but rounds to 0.2 to the nearest tenth.
> What could your calculation and the product be?
> Find at least 8 solutions.

Practice questions

1 Expand these numbers to show the place value parts.

a 0.45 b 3.45 c 4.35 d 4.035 e 3.254

2 Find the missing numbers to make each statement true.

a $\frac{4}{10} = \frac{\square}{100}$ b $\frac{700}{1\,000} = \frac{\square}{10}$ c $\frac{50}{100} = \frac{500}{\square}$ d $\frac{3}{10} = \frac{30}{\square}$

3 Copy and complete this table.

Number	Rounded to the nearest whole number	Rounded to the nearest tenth (0.1)	Rounded to the nearest hundredth (0.01)
2.359			
3.451			
4.035			
5.999			

4 Write these values in ascending order.

a 0.04 0.075 0.3 0.325 0.004

b 6.35 5.306 6.035 5.63 6.305

5 The table shows the money two charities raised during four events. Find the missing values a–d.

	Car wash	Sponsored walk	Fair	Exhibition
Charity A	$17 580	$14 768	c	$18 564
Charity B	$19 875	b	$16 305	$23 751
Total	a	$33 599	$27 520	d

6 Look at these distances.

3.45 km 26.675 km 0.999 km

a Make each distance 10 times as far.

b Make each distance 100 times as far.

c Make each distance 1 000 times as far.

7 Copy and complete.

a
4 × 3
0.4 × 3
0.04 × 3
0.004 × 3
4 × 0.03

b
9 × 5
0.9 × 5
9 × 0.5
0.05 × 9
0.005 × 9

What I have learned

Copy and complete the cross-number puzzle. Do not write in the boxes containing the decimal points.

Clues

Across

1. Which is larger, 3.119 or 3.12?
4. Subtract 54.9 from 91.37. Round your answer to the nearest whole number.
5. Subtract 0.29 from 10.
7. Add 26.85 to 26.87. Round your answer to one decimal place.
9. 3.085 × 10

Down

1. 3 × 1.32
2. 5 × 0.35
3. Subtract 13.26 from 100
6. 0.138 × 100
8. Add 38.6, 20.49 and 30.51. Round your answer to the nearest whole number.

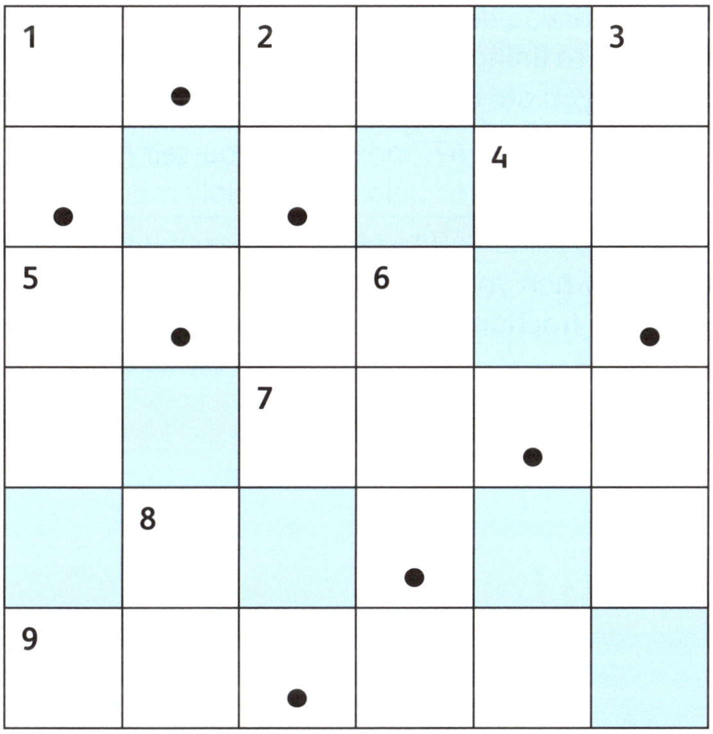

10 Self check

	Check you can do this:	If you have difficulty, try this:
1	Identify decimal fractions.	Go back to page 135 and remind yourself how to write a fraction as a decimal, using the place value of numbers after the decimal point.
2	Expand decimals to show their place value parts.	Remind yourself what you learned last term about place value, face value and true value on page 25. Go back to page 137 and practice expressing a decimal number as the sum of the true value of each place value part.
3	Compare and order decimals.	Go back to pages 140–141, and remind yourself how to put a set of decimals in order from smallest to greatest, or greatest to smallest.
4	Round decimal numbers to the nearest whole number, tenth or hundredth.	Go back to pages 142–143 and remind yourself how to round whole numbers and decimals.
5	Add and subtract decimal numbers.	Go back to pages 144–145 and remind yourself of the work you did in that section. Pay special attention to lining up the place values of the numbers you are adding or subtracting.
6	Multiply decimal numbers by 10, 100 and 1 000.	Go back to page 147 and remind yourself how to use place value to help you multiply mentally.
7	Find the product of a whole number and a decimal.	Go back to page 149 to remind yourself what happens when you multiply a whole number by a decimal fraction.

Chapter 11 Working with fractions

In this chapter you will explore these focus questions:

How can I apply fraction ideas to find solutions for practical ideas?
How do I round numbers to estimate solutions?

To answer this question, you will:

- use number operations (addition, subtraction and multiplication) with fractions
- rename two or more fractional numbers with unlike denominators to show the same denominator
- compare fractional numbers in any form
- round mixed numbers to the nearest whole number
- solve problems involving fractions.

Starting point

1 Look at this number line. Tell your partner:

a what you notice about the numbers above and below the number line

b what you think the numbers should be on the unlabelled part of the number line.

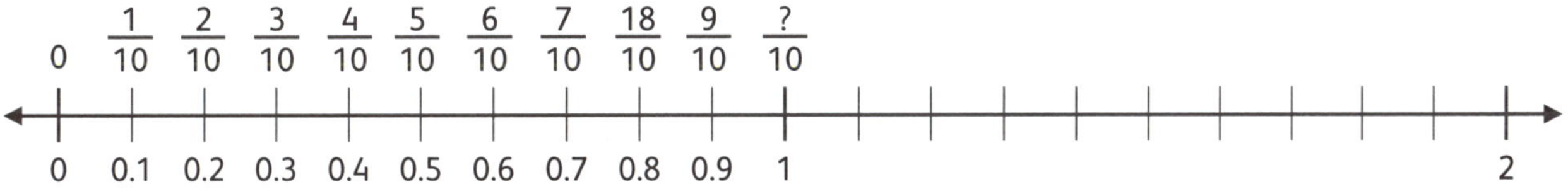

2 Practise counting in tenths, first as fractions and then as decimals. Can you count up to 2?

3 Talk to your partner about the way that you can represent the whole numbers 1 and 2 as fractions with denominator 10. Is there a different denominator you can use?

4 Where will you find the following numbers on this number line:

a the mixed number $1\frac{3}{10}$

b the mixed number $1\frac{1}{2}$

c the improper fraction $\frac{17}{10}$

d the improper fraction $\frac{7}{5}$

5 The values 0.2 and $\frac{2}{10}$ are equivalent. Tell your partner about other fractions that you know that are also equivalent. Explain why.

6 How many ways can you make the total 1 by adding pairs of fractions with denominator 10?

Equivalent fractions

Key maths idea

This fraction wall shows that $\frac{2}{4}$ and $\frac{5}{10}$ are **equivalent**.
The fractions $\frac{2}{4}$ and $\frac{5}{10}$ have the same position on the number line and represent the same proportion of the whole.
Equivalent fractions have the same value but can look different.

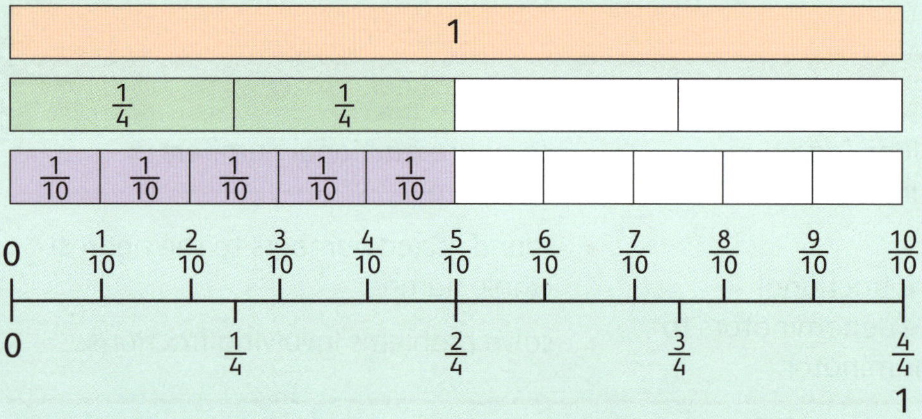

We can use our knowledge of multiplication tables and scaling to help find sequences of equivalent fractions. Look at the multiplication table:

Numerator and denominator both multiplied by 2, and then 3, etc.

×	1	2	3	4	5	6	7	8	9	10
1	1	2	3	4	5	6	7	8	9	10
2	2	4	6	8	10	12	14	16	18	20
3	3	6	9	12	15	18	21	24	27	30
4	4	8	12	16	20	24	28	32	36	40
5	5	10	15	20	25	30	35	40	45	50
6	6	12	18	24	30	36	42	48	54	60
7	7	14	21	28	35	42	49	56	63	70
8	8	16	24	32	40	48	56	64	72	80
9	9	18	27	36	45	54	63	72	81	90
10	10	20	30	40	50	60	70	80	90	100

$\frac{2}{3} \xrightarrow{\times 2} \frac{4}{6}$ (numerator × 2, denominator × 2)

$$\frac{2}{3} = \frac{4}{6} = \frac{6}{9} = \frac{8}{12} \ldots$$

> Did you know that the short line between the numerator and denominator is called the **vinculum**?

When we find equivalent fractions, we must scale the **numerator** and the **denominator** by the same number.
If we do not scale by the same number, the fractions are no longer equivalent.

Key words
equivalent
numerator
denominator
vinculum

1 Work with a partner. Use strips of grid paper like this.

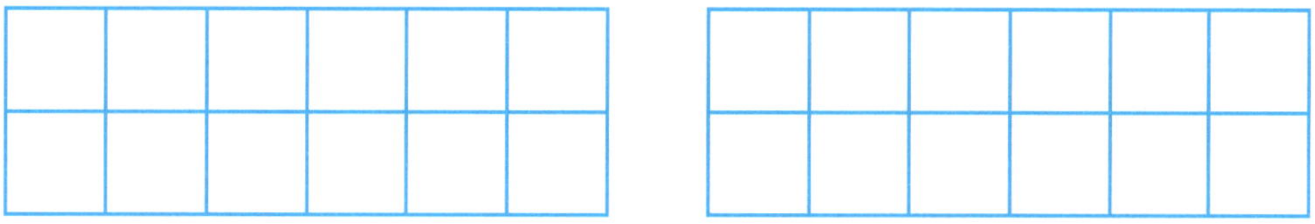

Colour the squares to explore equivalent fractions. Record what you find in words and in numerals, for example, 1 half = 3 sixths and $\frac{1}{2} = \frac{3}{6}$.

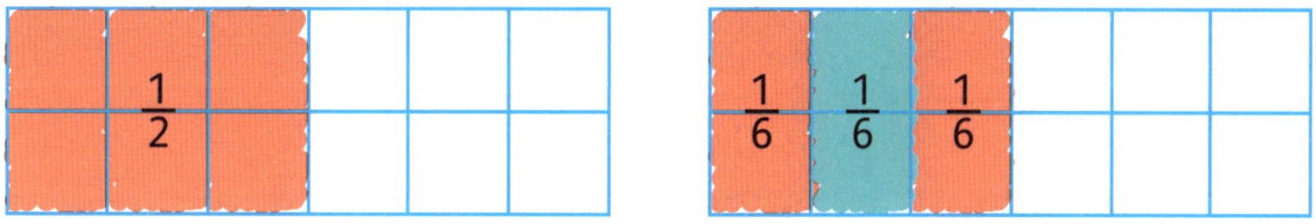

2 Use grid paper and colour pencils or crayons in two different colours. Look at each shape. Draw dots on the grid paper to makes shapes that are equivalent to each fraction. Next to each shape, write the fraction that represents each colour.

a

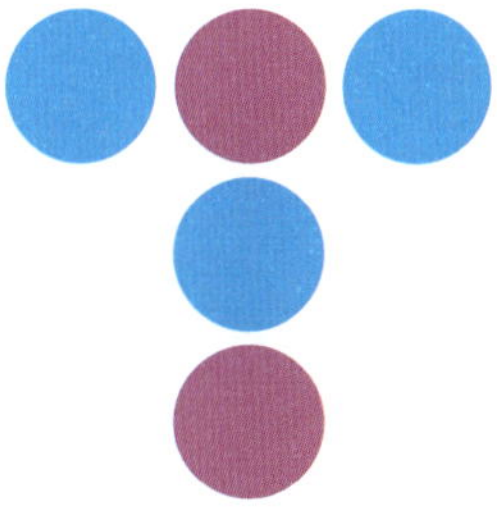

b

c

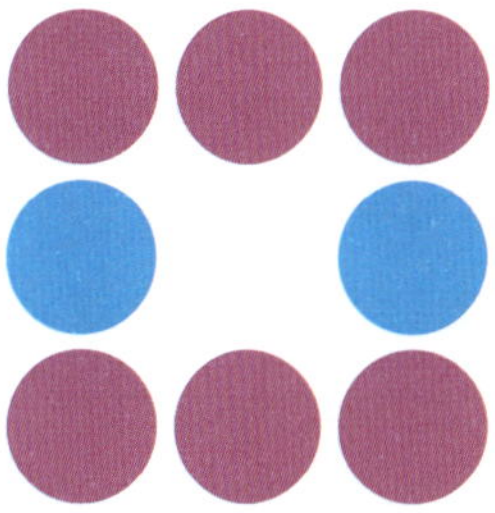

3 Which fraction is the odd one out?

a $\frac{8}{10}, \frac{4}{5}, \frac{16}{25}, \frac{12}{15}$

b $\frac{10}{12}, \frac{20}{24}, \frac{5}{6}, \frac{15}{20}$

c $\frac{7}{8}, \frac{14}{18}, \frac{28}{36}, \frac{21}{27}$

d $\frac{5}{20}, \frac{3}{12}, \frac{6}{24}, \frac{9}{32}$

Use what you know about equivalent fractions to help you.

4 Three children each have a bag of marbles. The fraction of red marbles in each child's bag is shown here.

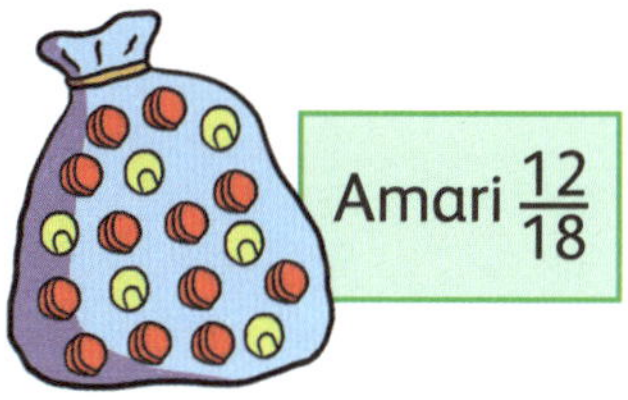

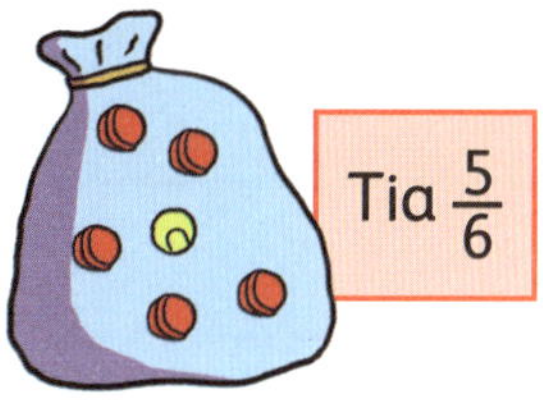

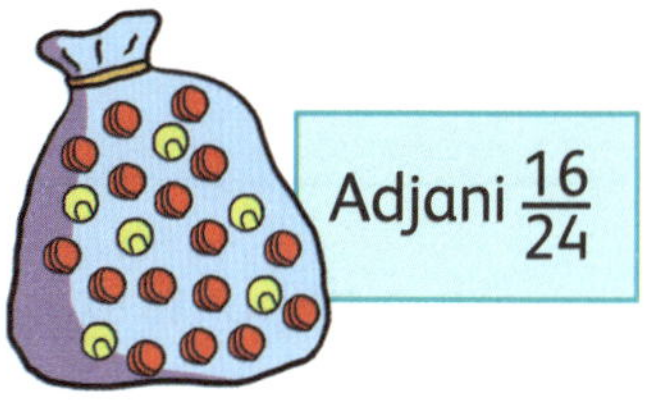

a Which two children have a bag with an equivalent fraction of red marbles?

b Which child has the greatest fraction of red marbles?

Represent the fractions to show how you know this.

More about fraction and decimal equivalence

Key maths idea

You have already worked with equivalent fractions and decimals for tenths, hundredths and thousandths. We use the term **decimal denominators** to describe denominators such as 10, 100 and 1 000.

0.4 is 4 tenths or $\frac{4}{10}$.

0.09 is 9 hundredths or $\frac{9}{100}$.

0.005 is 5 thousandths or $\frac{5}{1\,000}$.

Sometimes we need to find fraction and decimal equivalents for other fractions. We can do this by first converting a fraction to an equivalent fraction with a decimal denominator.

For example:

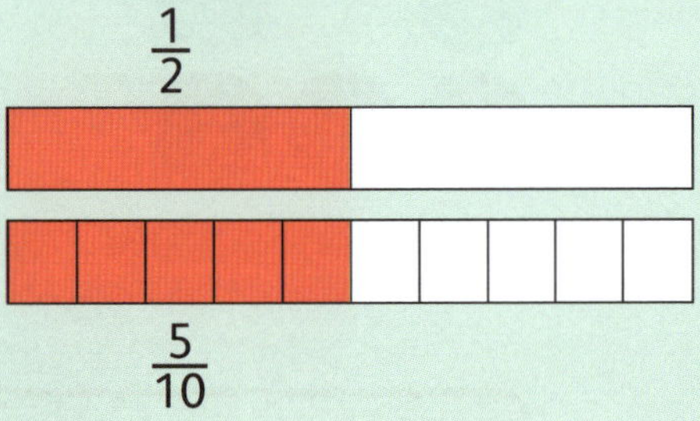

$\frac{1}{2} = \frac{5}{10} = 0.5$

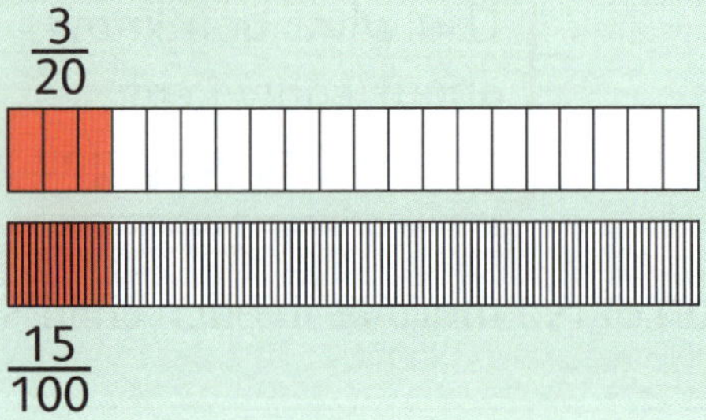

$\frac{3}{20} = \frac{15}{100} = 0.15$

100 is five times as large as 20. So we must make the numerator five times as large to make the fractions equivalent.

Key word

decimal denominator

1 Write these with a denominator of 10.

a $\frac{4}{5}$ b $\frac{1}{2}$ c $\frac{2}{5}$ d $\frac{3}{6}$

2 Write these as decimals.

a $\frac{1}{5}$ b $\frac{3}{5}$ c $2\frac{1}{2}$ d $6\frac{4}{5}$

3 Write these with a denominator of 100, and then as decimals.

a $\frac{3}{10}$ b $\frac{3}{4}$ c $\frac{20}{50}$ d $\frac{1}{4}$

e $\frac{1}{10}$ f $\frac{19}{50}$ g $\frac{3}{25}$ h $\frac{5}{20}$

4 Copy these number sentences. Then fill in the missing values and write the fractions as decimals.

a $\frac{1}{20} = \frac{\square}{100} = $ ______ **b** $\frac{3}{5} = \frac{\square}{10} = $ ______

c $\frac{6}{25} = \frac{\square}{100} = $ ______ **d** $7\frac{4}{5} = 7\frac{\square}{\square} = $ ______

5 Write these decimals as fractions in their simplest form.

a $0.6 = \frac{\square}{10} = \frac{\square}{\square}$ **b** $0.08 = \frac{\square}{100} = \frac{\square}{\square}$ **c** $1.8 = 1\frac{8}{\square} = \frac{\square}{\square}$

Simplifying fractions

Key maths idea

When you **simplify** a fraction, you divide the numerator and the denominator by the same whole number. You can keep doing this as long as both the numerator and the denominator have a common factor.

We can say that the fraction $\frac{3}{4}$ is in its **simplest form**, as we cannot divide the numerator and denominator by a **common factor** other than 1.

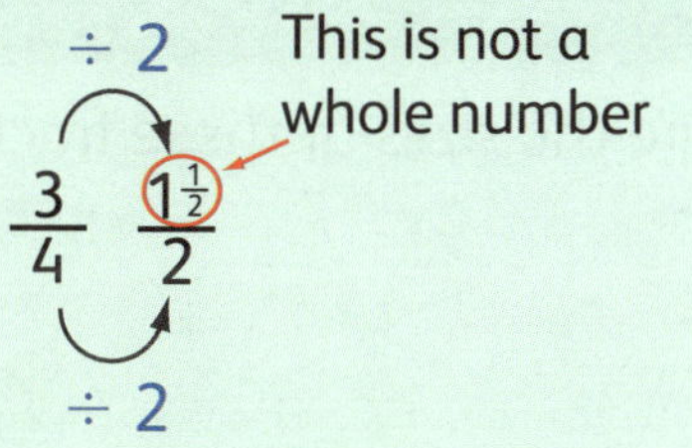

But we can see that the fraction $\frac{3}{6}$ is not in its simplest form because we can divide the numerator and denominator by a common factor.

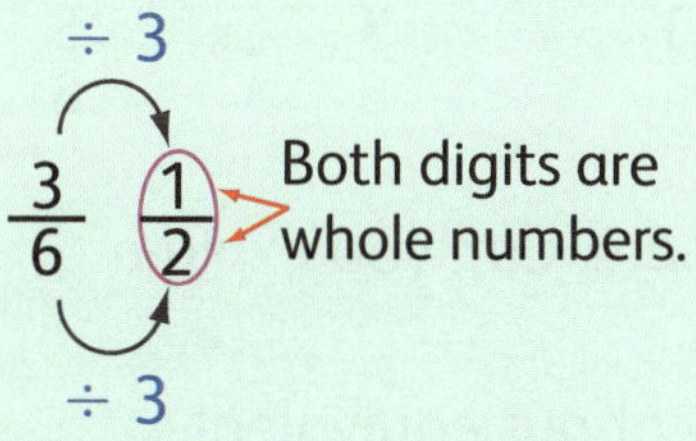

How do you know that you cannot simplify $\frac{1}{2}$ any further? Explain this to a partner.

We look for the **highest common factor (HCF)** when we want to simplify fractions as we then know that the fraction will be in its simplest form.

Let's simplify $\frac{8}{12}$.

Key words

simplify
simplest form
common factor
highest common factor (HCF)

Remember that a factor divides into a given number exactly, with no remainder!

2 is a common factor of 8 and 12. So we can simplify the fraction as $\frac{4}{6}$ but it can still be simplified further.

4 is the highest common factor (HCF) of 8 and 12 so we can simplify the fraction $\frac{8}{12}$ as $\frac{2}{3}$. How do we know that it is now in its simplest form?

1 Which of these fractions are in their simplest form? Explain why you say so.

$\frac{1}{2}$ $\frac{4}{10}$ $\frac{5}{8}$ $\frac{3}{15}$ $\frac{5}{20}$ $\frac{2}{5}$

2 Are these fractions equivalent? Write 'true' or 'false'. Correct any that are false.

a $\frac{5}{10} = \frac{1}{2}$ **b** $\frac{15}{25} = \frac{3}{4}$ **c** $\frac{40}{50} = \frac{4}{10}$

d $\frac{9}{12} = \frac{3}{4}$ **e** $\frac{70}{100} = \frac{3}{5}$

3 Use all the number cards to make four fractions that are written in their simplest form.

2 5 9 4 8 10 3 15

Try to find the highest common factor (HCF) each time so you know the fraction is in its simplest form.

Comparing and ordering fractions

Key maths idea

Here's how we can compare the sizes of these fractions:

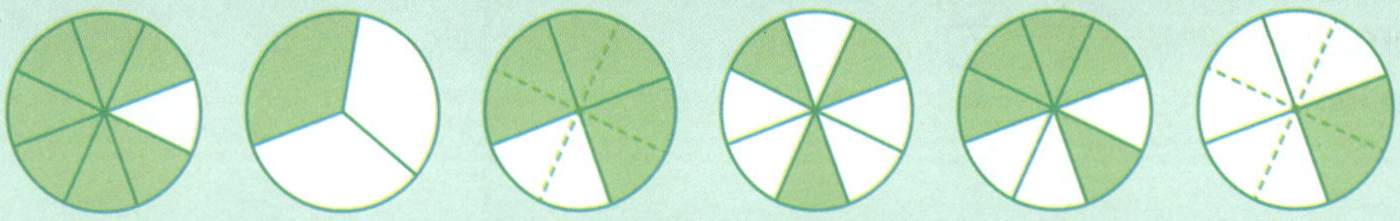

$\frac{7}{8}$ $\frac{1}{3}$ $\frac{3}{4}$ $\frac{3}{8}$ $\frac{5}{8}$ $\frac{1}{4}$

$\frac{7}{8}$ and $\frac{3}{8}$ are easy to compare. Can you explain why?

We can use what we know about equivalent fractions to help us compare $\frac{1}{4}$ and $\frac{3}{8}$.

$\frac{1}{4} \xrightarrow{\times 2} \frac{2}{8}$ (× 2 above and × 2 below)

8 is the **lowest common multiple (LCM)** of 4 and 8, so we can use denominator 8 to rename $\frac{1}{4}$.

$\frac{2}{8} < \frac{3}{8}$ so $\frac{1}{4} < \frac{3}{8}$

Key words

lowest common multiple (LCM)
ascending
descending
ordering

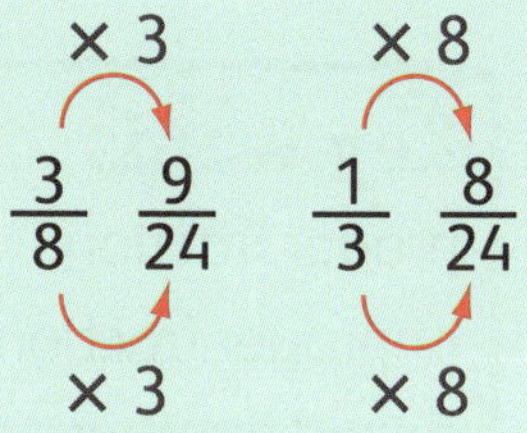

We can also use equivalent fractions to compare $\frac{1}{3}$ and $\frac{3}{8}$.
24 is the lowest common multiple (LCM) of 3 and 8, so we can rename both to fractions with the same denominator, 24.

$\frac{8}{24} < \frac{9}{24}$ so $\frac{1}{3} < \frac{3}{8}$

We can now write all six fractions in **ascending** order:
$\frac{1}{4}, \frac{1}{3}, \frac{3}{8}, \frac{5}{8}, \frac{3}{4}, \frac{7}{8}$
Can you write them in **descending** order now?

1 Compare each pair of fractions. Use <, > or = to show what you find out.

a $\frac{2}{4}$ and $\frac{5}{8}$ b $\frac{2}{4}$ and $\frac{3}{5}$

c $\frac{2}{3}$ and $\frac{5}{8}$ d $\frac{2}{3}$ and $\frac{6}{10}$

e $\frac{4}{5}$ and $\frac{7}{8}$

2 Write these fractions in ascending order.

a $\frac{4}{5}, \frac{2}{5}, \frac{3}{10}$ b $\frac{7}{8}, \frac{3}{4}, \frac{5}{7}, \frac{1}{4}$

c $\frac{2}{3}, \frac{7}{9}, \frac{4}{9}, \frac{1}{3}$ d $\frac{5}{6}, \frac{4}{6}, \frac{1}{2}, \frac{1}{3}$

When comparing or **ordering** fractions with unlike denominators, remember to use what you know about equivalent fractions to help you rename them. Try to use the LCM each time.

3 Benjamin buys $\frac{4}{5}$ kg of bananas. Pia buys $\frac{1}{2}$ kg of star fruit. Leah buys $\frac{7}{10}$ kg of guavas.

a Who bought the greatest mass of fruit?

b Order the masses from lightest to heaviest.

4 Compare the fractions that each colour represents in the diagram.

Proper fractions, improper fractions and mixed numbers

Key maths idea

$\frac{3}{5}$ is a **proper fraction** because it is smaller than 1. The numerator is less than the denominator.

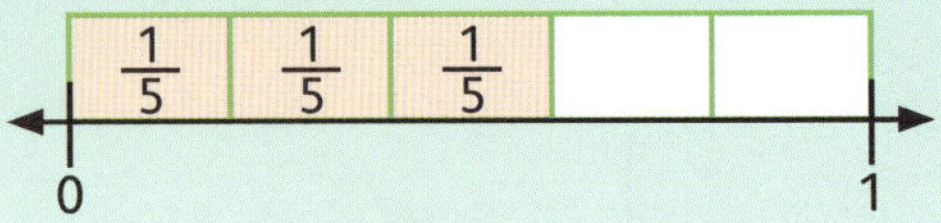

We say that $\frac{8}{5}$ is an **improper fraction** because the numerator is larger than the denominator.

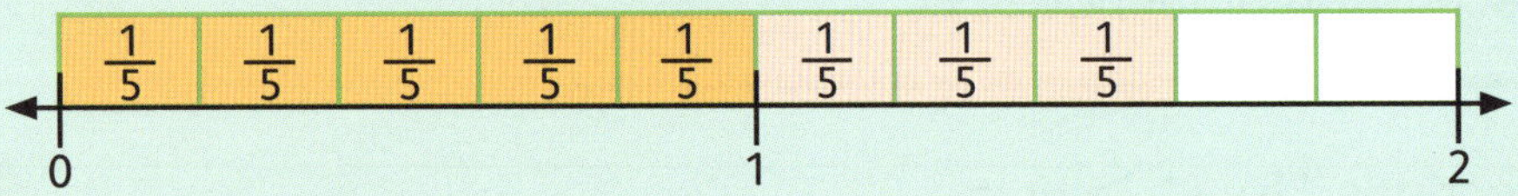

We can say that $\frac{8}{5}$ is equal to $\frac{5}{5} + \frac{3}{5}$ or $1 + \frac{3}{5}$.

We can write this as the **mixed number** $1\frac{3}{5}$.

A mixed number has a whole number part and a proper fraction part.

We can use what we know about the rules for **rounding** to help round mixed numbers to the nearest whole number.

Let's look at $1\frac{3}{5}$. We know that $\frac{3}{5}$ is equivalent to $\frac{6}{10}$ (0.6). This is greater than $\frac{5}{10}$ ($\frac{1}{2}$ or 0.5), so $1\frac{3}{5}$ rounds up to 2 to the nearest whole number.

Explain to a partner why $1\frac{2}{5}$ rounds down to 1 to the nearest whole number.

Key words

- proper fraction
- improper fraction
- mixed number
- rounding

1 Find the odd one out. Which fraction in each group is not a proper fraction? Explain how you know.

a $\frac{1}{4}$ $\frac{4}{5}$ $\frac{5}{5}$ $\frac{1}{2}$

b $\frac{9}{10}$ $\frac{3}{4}$ $\frac{1}{6}$ $\frac{6}{1}$

c $\frac{4}{9}$ $\frac{9}{8}$ $\frac{8}{9}$ $\frac{5}{9}$

> Two of the fractions in Exercise 1 represent whole numbers. Which are they?

2 Convert the improper fractions to mixed numbers. Draw fraction bars to help you work out the answers. For example:

$\frac{5}{2} = \frac{2}{2} + \frac{2}{2} + \frac{1}{2} = 2\frac{1}{2}$

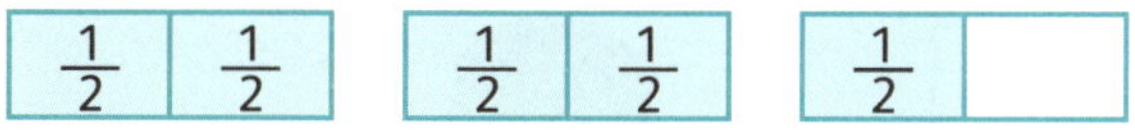

a $\frac{8}{5}$ **b** $\frac{9}{4}$ **c** $\frac{7}{2}$ **d** $\frac{22}{10}$ **e** $\frac{17}{5}$

3 Convert the mixed numbers to improper fractions.

For example: $2\frac{2}{4} = \frac{4}{4} + \frac{4}{4} + \frac{2}{4} = \frac{10}{4}$

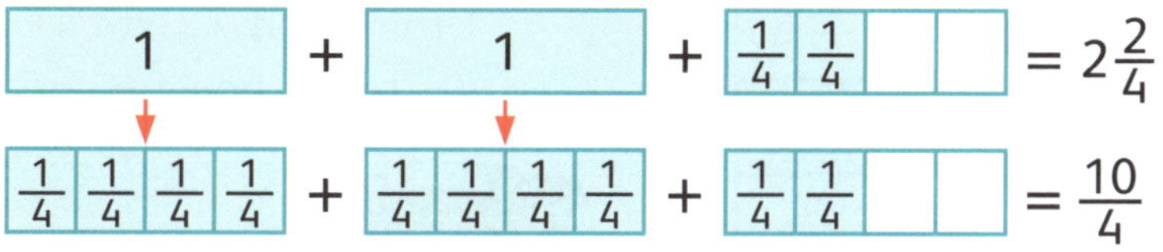

a $5\frac{1}{2}$ **b** $2\frac{3}{4}$ **c** $3\frac{4}{5}$ **d** $5\frac{3}{10}$

4 Match these mixed numbers to the letters on the number line.

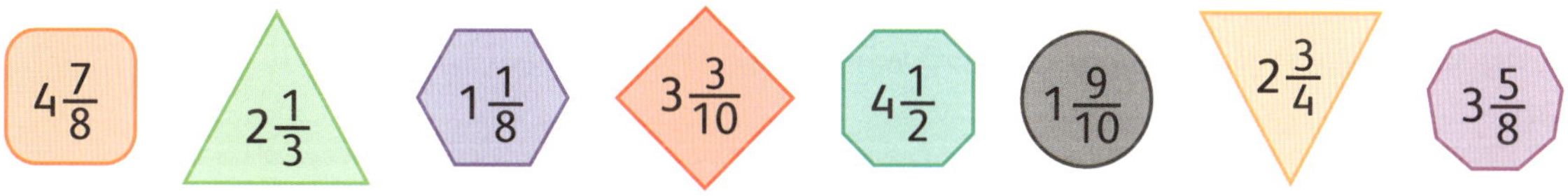

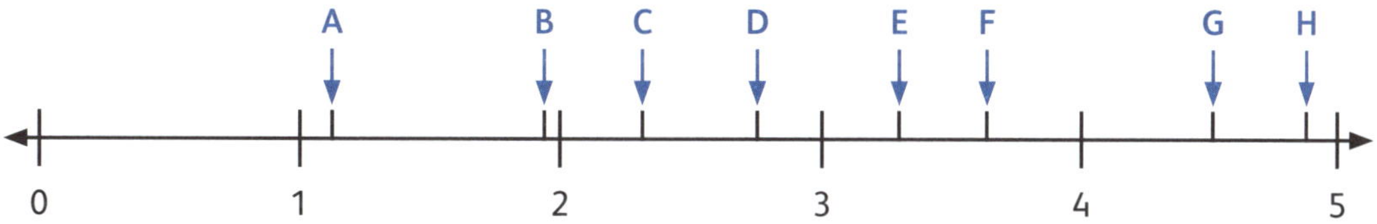

5 Round each of the mixed numbers in Exercise 4 to the nearest whole number.

a How can you use the number line to help make decisions?

b Record your answers using the symbol '≈'.

6 Malika is rounding some mixed numbers to the nearest whole number.

$4\frac{1}{4}$ $4\frac{2}{5}$ $3\frac{3}{4}$ $3\frac{2}{3}$ $4\frac{1}{6}$ $3\frac{3}{8}$

a Malika thinks that each of these mixed numbers rounds to 4. Do you agree? Explain how you know.

b Write down another mixed number that Malika can use.

> $4\frac{1}{4}$ must round down to 4 because $\frac{2}{4}$ is equivalent to $\frac{1}{2}$.

Fractions that total 1

Key maths idea

Look at the fraction bars in this fraction wall.
We can see that $\frac{1}{2} + \frac{1}{2} = 1$ and that $\frac{1}{3} + \frac{1}{3} + \frac{1}{3} = 1$.
Why can we also say that $\frac{2}{3} + \frac{1}{3} = 1$?
What other fractions total 1?

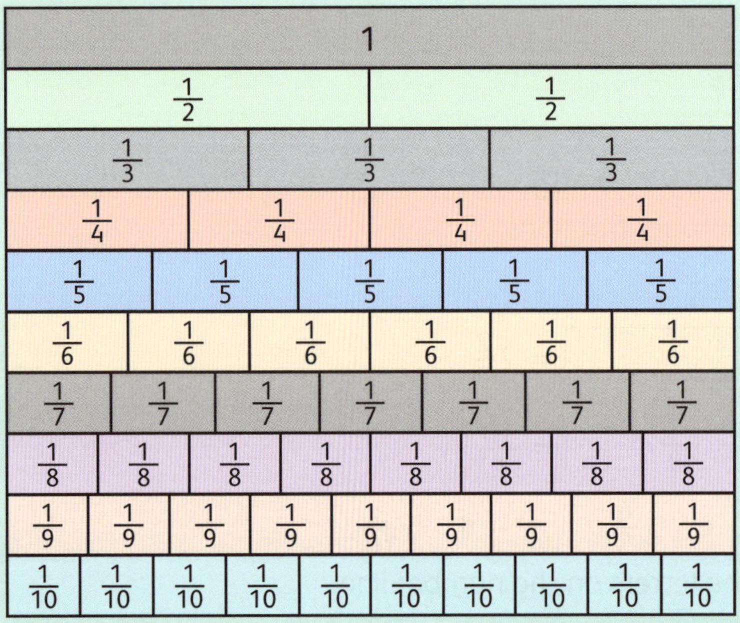

Do you agree with these students? Use what you know about equivalent fractions to help you.

1 Use your own set of fraction bars or the ones above to help you find the missing fractions in these additions.

a $\frac{\square}{4} + \frac{1}{2} = 1$ **b** $\frac{2}{3} + \frac{3}{\square} = 1$ **c** $\frac{2}{5} + \frac{6}{\square} = 1$

d $\frac{1}{5} + \frac{2}{10} + \frac{\square}{5} = 1$ **e** $\frac{1}{6} + \frac{3}{9} + \frac{1}{6} + \frac{\square}{\square} = 1$

2 Use fraction bars to find your own additions that total 1. You may need to make bars for twelfths and sixteenths.

a halves, quarters and eighths
b thirds and ninths
c quarters, eights and sixteenths
d fifths, sixths and tenths
e thirds, sixths and twelfths

3 Solve these problems.

a Malcolm has a bag of fruit. One-third of the bag is apples, five-twelfths is guavas. The rest are bananas. What fraction of the bag is made up of bananas?

b At a playpark, two-fifths of the children are on the swings, and eight-twentieths of the children are playing on the climbing frame. The rest are in the sandpit. What fraction of the children are in the sandpit?

Adding and subtracting fractions

Key maths idea

Let's think about addition and subtraction with fractions.

When fractions share the same denominator, we can add or subtract them easily.

$\frac{3}{5} + \frac{4}{5} = \frac{7}{5} = 1\frac{2}{5}$

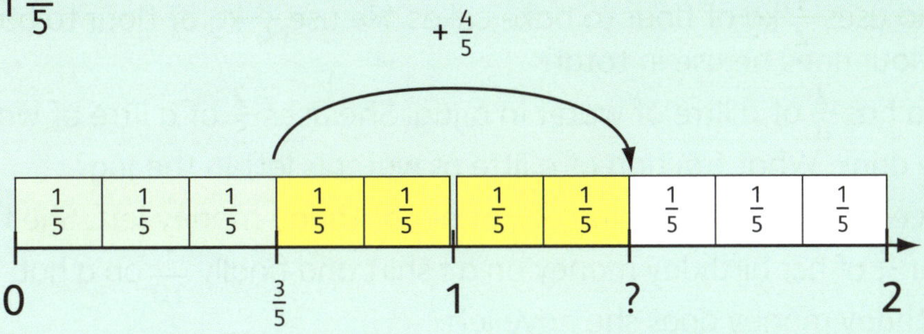

When fractions do not share the same denominator, we use our knowledge of equivalent fractions to help us.

For example:

For the subtraction $2\frac{3}{4} - \frac{2}{3}$ we find a common denominator.

Both 4 and 3 are factors of 12, so we can rename the fractions with the denominator 12.

$\frac{3}{4} = \frac{9}{12}$ (× 3) $\frac{2}{3} = \frac{8}{12}$ (× 4)

Now we can complete the subtraction:

$2\frac{3}{4} - \frac{2}{3}$

$= 2\frac{9}{12} - \frac{8}{12} = 2\frac{1}{12}$

1 Add.

a $\frac{3}{5} + \frac{4}{5}$ **b** $\frac{5}{6} + \frac{4}{6}$ **c** $\frac{7}{8} + \frac{3}{8}$ **d** $\frac{2}{9} + \frac{7}{9} + \frac{2}{9}$

e $1\frac{3}{5} + \frac{4}{5}$ **f** $1\frac{5}{6} + 2\frac{4}{6}$ **g** $3\frac{7}{8} + 3\frac{3}{8}$ **h** $\frac{11}{9} + \frac{14}{9} + \frac{2}{9}$

2 Subtract.

a $\frac{9}{10} - \frac{4}{10}$ **b** $\frac{5}{8} - \frac{4}{8}$ **c** $\frac{6}{6} - \frac{4}{6}$ **d** $\frac{8}{9} - \frac{3}{9} - \frac{1}{9}$

e $1\frac{9}{10} - \frac{4}{10}$ **f** $1\frac{5}{8} - 1\frac{4}{8}$ **g** $\frac{9}{6} - \frac{4}{6}$ **h** $\frac{14}{9} - \frac{3}{9} - \frac{1}{9}$

3 Work out these sums and give the answers in their simplest form.

a $\frac{1}{4} + \frac{1}{2}$ **b** $\frac{5}{6} + \frac{5}{12}$ **c** $1\frac{5}{9} + \frac{1}{3}$ **d** $\frac{9}{5} + \frac{3}{10} + \frac{2}{5}$

e $\frac{2}{7} + \frac{1}{2}$ **f** $\frac{2}{3} + \frac{2}{5}$ **g** $2\frac{4}{5} + 1\frac{1}{3}$ **h** $\frac{11}{8} + \frac{4}{5} + \frac{1}{2}$

4 Subtract. Give the answers in their simplest form.

a $\frac{7}{8} - \frac{3}{4}$ **b** $\frac{5}{6} - \frac{5}{12}$ **c** $1\frac{5}{9} - \frac{1}{3}$ **d** $\frac{9}{5} - \frac{3}{10} - \frac{1}{5}$

e $\frac{5}{6} - \frac{4}{9}$ **f** $\frac{3}{4} - \frac{2}{3}$ **g** $2\frac{4}{5} - 1\frac{1}{3}$ **h** $\frac{15}{10} - \frac{1}{4} - \frac{1}{5}$

5 The answer to a question is $\frac{11}{12}$. Make up a possible question that uses:

- **a** addition of fractions with like denominators
- **b** subtraction of fractions with like denominators
- **c** addition of fractions with unlike denominators
- **d** subtraction of fractions with unlike denominators.

6 Solve these problems. Write down all your calculations.

- **a** Romano uses $\frac{1}{2}$ kg of flour to bake cakes. He use $\frac{3}{4}$ kg of flour to bake cookies. How much flour does he use in total?
- **b** Melisha has $\frac{7}{8}$ of a litre of water in a jug. She uses $\frac{2}{5}$ of a litre of water to make orange drink. What fraction of a litre of water is left in the jug?
- **c** Candace buys a comic and has $\frac{9}{10}$ of her birthday money left. She then spends another $\frac{2}{3}$ of her birthday money on a t-shirt and finally $\frac{1}{10}$ on a hat. What fraction of her birthday money does she have left?

Make up some word problems about adding and subtracting fractions or mixed numbers. Ask others to solve them.

Subtracting fractions from whole numbers

Key maths idea

Sometimes we need to subtract a fraction from a whole number.
There are different ways to solve subtraction.
Let's find two ways to calculate $2 - \frac{1}{3} = \square$.

Method 1:

$2 - \frac{1}{3}$
$= 1 + 1 - \frac{1}{3}$
$= 1 + \frac{2}{3}$
$= 1\frac{2}{3}$

Method 2:

We can see that 2 ones = 6 thirds or $\frac{6}{3}$
$\frac{6}{3} - \frac{1}{3}$
$= \frac{5}{3}$
$= 1\frac{2}{3}$

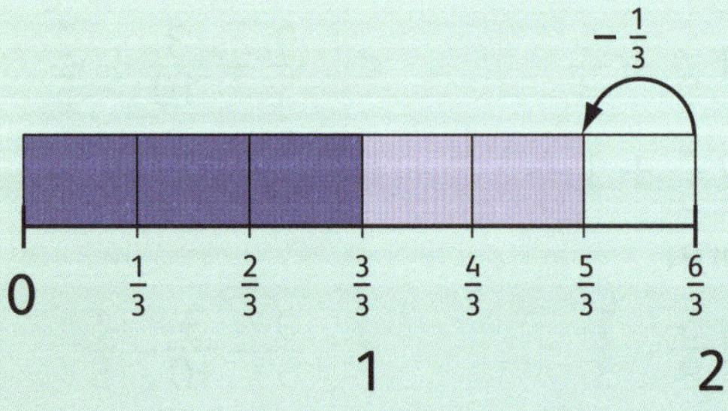

1 Use both methods above to do these subtractions. Copy and shade the diagrams to show your answers.

a $2 - \frac{5}{6} = \square$

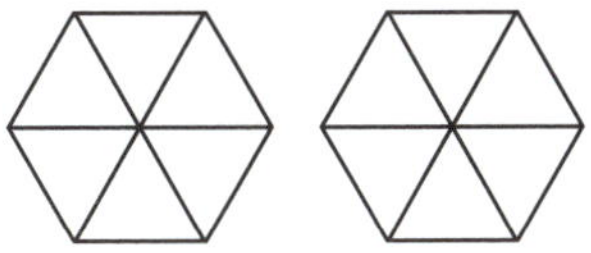

b $3 - 2\frac{1}{7} = \square$

2 Now draw your own diagrams or number lines to complete these subtractions.

a $5 - \frac{3}{4}$ **b** $6 - \frac{5}{8}$ **c** $1 - \frac{5}{10}$

d $7 - 2\frac{1}{3}$ **e** $3 - 2\frac{7}{12}$ **f** $4 - 2\frac{4}{5}$

3 Write your own number sentences and solve these word problems. Give your answers as mixed numbers.

a A group of children ordered 4 pizzas to share. They ate $2\frac{2}{3}$ pizzas. How many pizzas were left?

b Jalisia had 10 bottles of juice at her birthday party. The guests drank $7\frac{1}{3}$ bottles of juice at the party. How much juice was left?

c Sara has 2 kg of rice. She uses $1\frac{1}{4}$ kg of rice. How much rice does she have left?

d Mark brought 4 bags of mangoes to share with his classmates and teacher. He gave his teacher $1\frac{3}{6}$ bags of mangoes. How many bags of mangoes are left?

e Your grandmother has 5 metres of ribbon. She uses $2\frac{3}{4}$ metre of ribbon to wrap around a present. How much does she have left?

Finding fractions of quantities

Key maths idea

Division can help you find fractions of quantities. Look at these bar models:

First you divide the whole quantity by the denominator to find the value of one equal part.

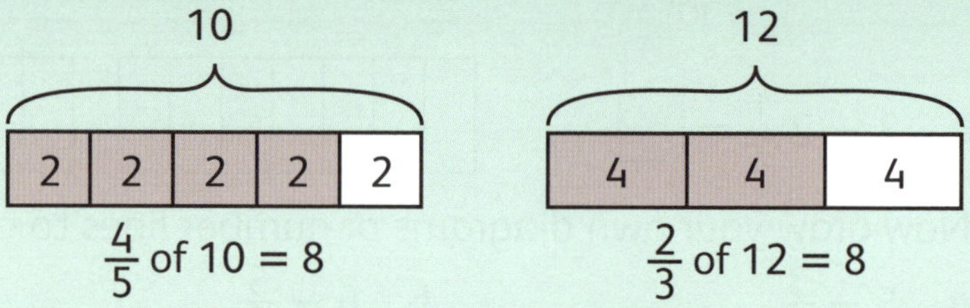

Then you multiply the value of one equal part to find the fraction of the whole you need.

This diagram shows the cost of different fractions of a kilogram of apples.

Work with a partner to find the cost of $\frac{3}{8}$ kg of apples, and then $\frac{7}{8}$ kg of apples.

1 kg apples = $ 480							
$\frac{1}{8}$	$\frac{1}{8}$	$\frac{1}{8}$	$\frac{1}{8}$	$\frac{1}{8}$	$\frac{1}{8}$	$\frac{1}{8}$	$\frac{1}{8}$
$60	$60	$60	$60	$60	$60	$60	$60

1 From each set of four quantities, write down the two that have the same value. Show them as two bars.

a $\frac{1}{4}$ of 16 $\frac{1}{3}$ of 12 $\frac{1}{5}$ of 10 $\frac{1}{2}$ of 30

b $\frac{1}{7}$ of \$35 $\frac{1}{3}$ of \$21 $\frac{1}{6}$ of \$60 $\frac{1}{8}$ of \$40

c $\frac{3}{8}$ of 24 hours $\frac{1}{6}$ of 54 hours $\frac{7}{12}$ of 24 hours $\frac{5}{9}$ of 18 hours

2 The table shows the fraction of customers who bought different fruits from the market on Saturday. There were a total of 360 customers that day.

Apples	Guavas	Sweetsop	Mango	Lemons	Papaya
$\frac{3}{4}$	$\frac{1}{3}$	$\frac{1}{6}$	$\frac{2}{9}$	$\frac{3}{10}$	$\frac{2}{5}$

How many customers bought each type of fruit?

3 A jug contained 720 ml of orange juice. Shawna poured out $\frac{1}{4}$ of the juice, Keeno poured out $\frac{3}{10}$ and Jonathan poured $\frac{3}{8}$.

a How much did they each pour? Give your answer in millilitres.

b How many millilitres are left in the jug?

4 The pie chart shows the number of children, teenagers and adults who attended a sporting event.

a How many children attended the event?

b How many more teenagers than adults attended?

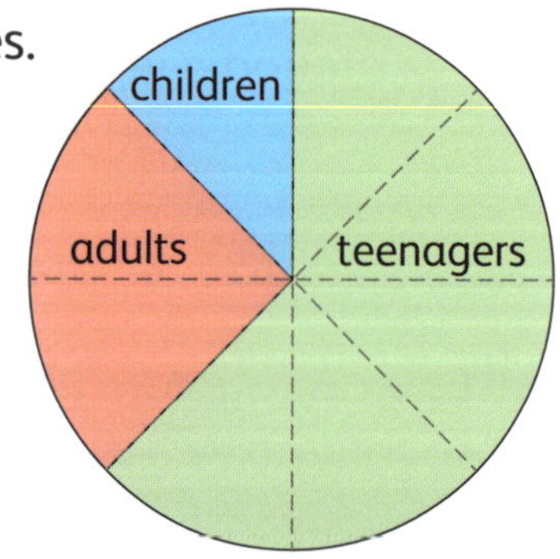

640 people

In a pie chart, each segment represents a fraction of the whole.

Multiplying proper fractions

Key maths idea

Think about what we mean when we multiply fractions. Remember, we can think of multiplication as repeated addition.

I shade two pieces of card. Each shaded piece is $\frac{1}{3}$ of the whole strip.

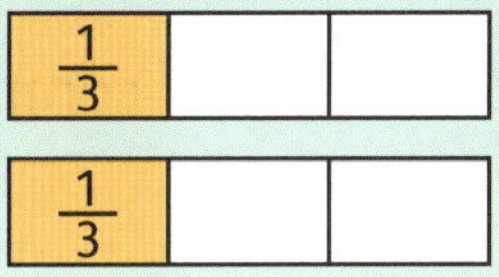

There are two parts of $\frac{1}{3}$.

So $\quad 2 \times \frac{1}{3} = \frac{2}{3}$

or $\quad \frac{1}{3} \times 2 = \frac{2}{3}$

Another way to think of this is:

I shade $\frac{1}{3}$ of 2 strips of card.

$\frac{1}{3}$ of $2 = \frac{1}{3} \times \frac{2}{1}$

$= \frac{2}{3}$

When we multiply, the answer is called the **product** of the numbers that we multiplied.

Sometimes we use multiplication to find a fractional part of a whole number. For example, imagine you have 30 counters. What is $\frac{2}{5}$ of the whole group?

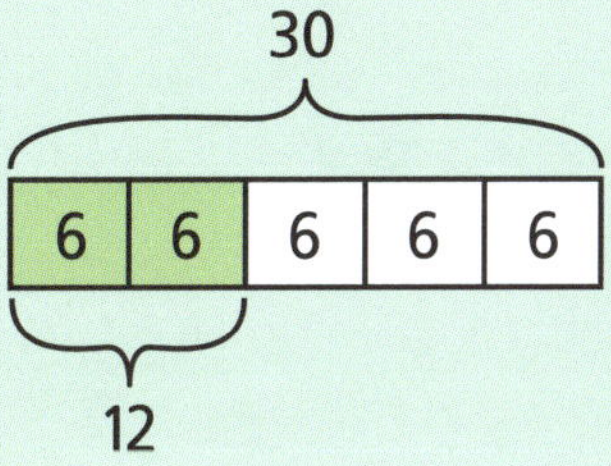

$\frac{2}{5}$ of 30:

$\frac{1}{5}$ of $30 = 6$ (Think: Divide 30 into five equal parts. $30 \div 5 = 6$)

So $\frac{2}{5}$ of $30 = 12$

We can also write:

$\frac{2}{5} \times 30 = \frac{2}{5} \times \frac{30}{1}$

$= \frac{60}{5}$

$= 12$

Look how the whole number 30 has been written as a fraction.

We can also multiply a pair of proper fractions.

$\frac{1}{2} \times \frac{2}{3}$ (Think: What is half of $\frac{2}{3}$?)

$= \frac{1}{2}$ of $\frac{2}{3}$

$\frac{1}{2}$ of $\frac{2}{3}$ (or $\frac{1}{2} \times \frac{2}{3}$)

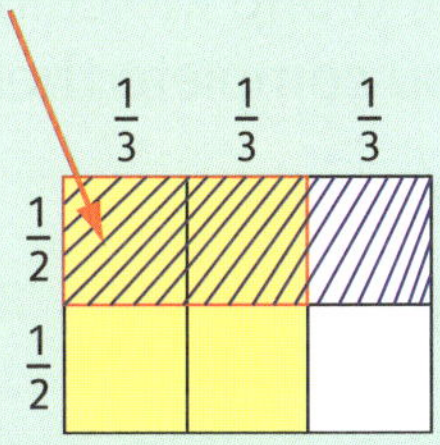

$\frac{1}{2} \times \frac{2}{3} = \frac{2}{6} = \frac{1}{3}$

Look at the way that $\frac{2}{6}$ has been written in its simplest form.

1 Write a multiplication sentence to match each diagram. Find the answers.

a

b

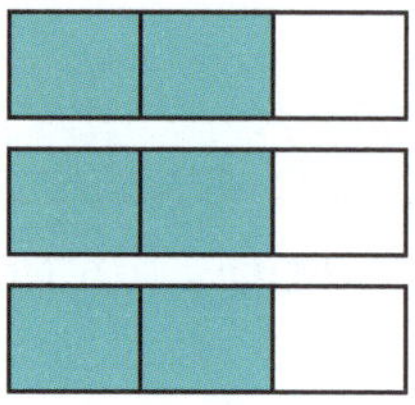

c

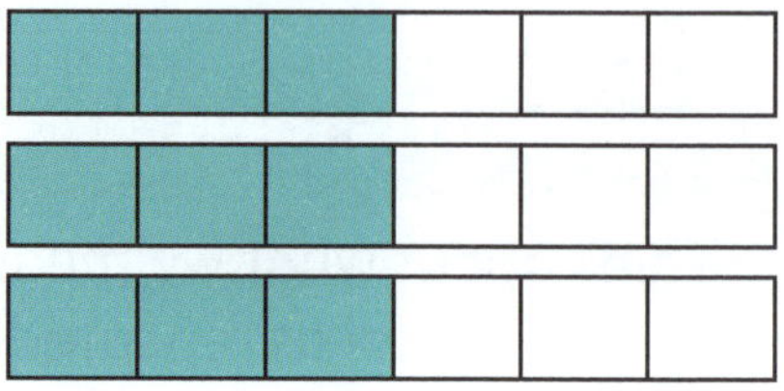

2 Copy and complete. Use diagrams like the ones in the key maths section to help you.

a $3 \times \frac{1}{4}$ **b** $3 \times \frac{3}{4}$ **c** $4 \times \frac{1}{5}$ **d** $4 \times \frac{3}{5}$

e $\frac{3}{4} \times 20$ **f** $\frac{2}{5} \times 15$ **g** $\frac{2}{3} \times 24$ **h** $\frac{5}{8} \times 24$

i $\frac{3}{4} \times \frac{1}{2}$ **j** $\frac{1}{3} \times \frac{2}{5}$ **k** $\frac{2}{3} \times 1/6$ **l** $\frac{3}{8} \times \frac{1}{2}$

3 Uncle Max has four bottles. Each holds $\frac{5}{8}$ litre of plant feed for flowers. He pours all the plant feed into one larger container. How many litres of plant feed are in the container?

4 Dana and Malcolm are discussing ways to simplify the fractions first before multiplying. Look at the methods and what the children say about them.

$$\frac{9}{12} \times \frac{1}{5}$$

$$\frac{{}^{3}\cancel{9}}{{}_{4}\cancel{12}} \times \frac{1}{5} = \frac{3}{20}$$

$$\frac{4}{5} \times \frac{3}{8}$$

$$\frac{{}^{1}\cancel{4}}{5} \times \frac{3}{\cancel{8}_{2}} = \frac{3}{10}$$

I've noticed that the numerator and denominator in the fraction $\frac{9}{12}$ have a common factor, so I can simplify the fraction to $\frac{3}{4}$ before calculating.

I've noticed that the numerator in the fraction $\frac{4}{5}$ and the denominator in the fraction $\frac{3}{8}$ also have a common factor. I can use cancelling to simplify the multiplication to $\frac{1}{5} \times \frac{3}{2}$.

Use Dana and Malcolm's methods to help you complete these multiplications:

a $\frac{3}{4} \times \frac{1}{6}$ **b** $\frac{1}{4} \times \frac{6}{9}$

c $\frac{3}{5} \times \frac{6}{10}$ **d** $\frac{7}{10} \times \frac{5}{6}$

e $\frac{4}{9} \times \frac{3}{8}$ **f** $\frac{8}{11} \times \frac{1}{8}$

Remember that you can only cancel numerators with denominators. You cannot cancel numerators with numerators or denominators with denominators!

Real-world maths: Sleepy creatures

A giant armadillo spends approximately 18 hours of the day asleep.

1 What fraction of the day is this? Represent the fraction in its simplest form.

2 Think about the number of hours you sleep each day. What fraction of your day is this?

3 Make a time line of the other activities you do each day. Work out the fraction of the day you spend doing each activity.

Maths detective: Fraction puzzles

Use the clues and your detective skills to solve these fraction puzzles.

Case 1

Grandma, Grandpa, Uncle Julian and Aunt Tammy are enjoying a lovely cup of tea and some of their favourite fruit pies.

1 Find out below who eats each flavour of pie and what fraction of the pie they eat.

- Grandma bakes three pies: apple, berry and apricot.
- Uncle Julian eats the largest fraction of pie. His piece is equivalent to $\frac{2}{9}$ of the 3 whole pies.
- Aunt Tammy doesn't like apples. She cuts $\frac{5}{8}$ of her favourite pie but only eats $\frac{2}{3}$ of the slice.
- Grandpa chooses the berry pie. He cuts it into three equal slices and eats one of them. It is so delicious that he cuts $\frac{3}{5}$ from another slice and eats that too!

2 Draw a table to organise your solution. What fraction of each pie is left for Grandma?

Case 2

Grandma, Uncle Julian and Aunt Tammy also love quizzes. They talk about the different ones they took part in last week. None of the quizzes had the same number of questions, but strangely they all got an equivalent fraction of their quiz questions correct! Is this possible? Or have they made a mistake? Read what Grandma and Uncle Julian say about the quizzes to help you decide.

> I got more questions correct than Uncle Julian but fewer than Aunt Tammy.

> No quizzes had more than 20 questions in total.

How many questions could have been in each person's quiz? How many could they each have got correct? Draw a table to organise your possible solutions.

What I have learned

Copy and complete each diagram.

Fill in five fractional numbers (fractions or decimals) that are equivalent to the fractional number in the middle.

a

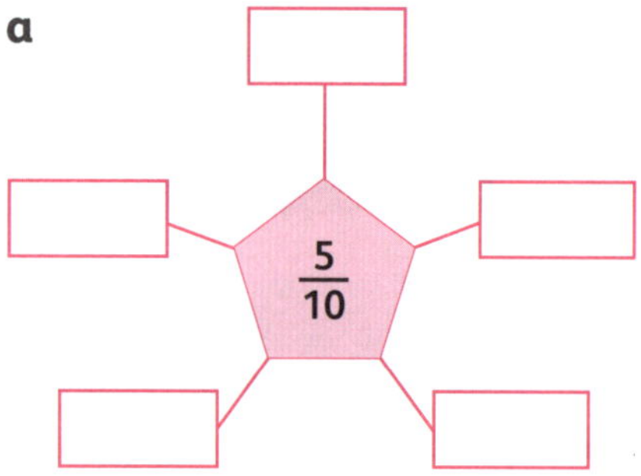

b

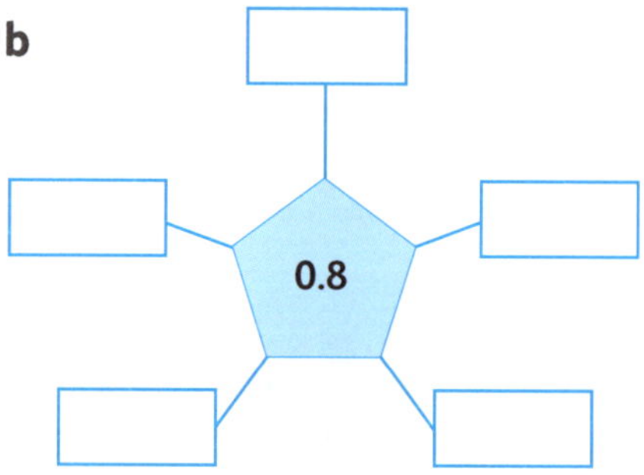

c

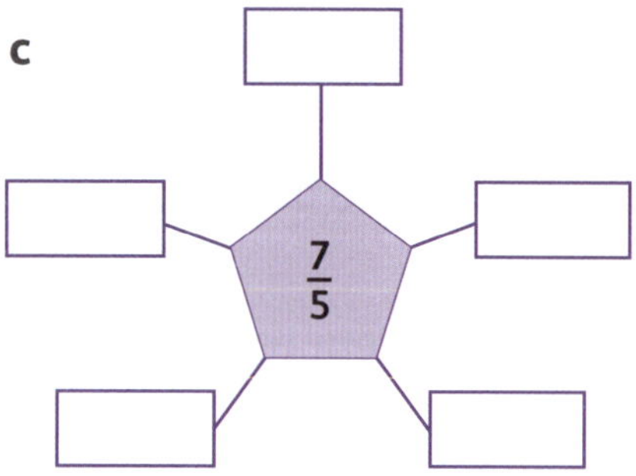

d

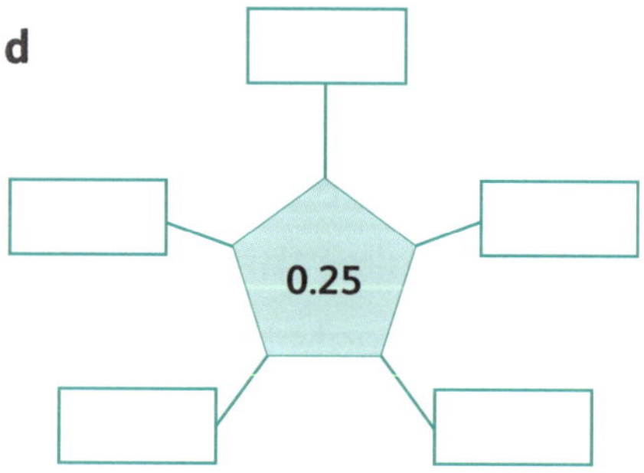

Practice questions

1 Fill in the missing numbers to make each equivalent statement true.

a $\frac{\square}{5} = \frac{8}{10}$ **b** $\frac{2}{3} = \frac{8}{\square}$ **c** $\frac{6}{\square} = 0.6$

d $\frac{\square}{50} = \frac{\square}{100} = 0.06$ **e** $\frac{3}{20} = \frac{\square}{100} = 0.\square$

2 Look at these fractions.

$\frac{6}{10}$ $\frac{1}{3}$ $\frac{3}{15}$ $\frac{2}{7}$ $\frac{6}{12}$

a Which fractions are written in their simplest form?

b Now simplify the other fractions to find their simplest forms.

3 Compare these fractions. Use $<$, $>$ or $=$ to show what you found.

a $\frac{4}{5}$ and $\frac{2}{3}$ **b** $\frac{3}{4}$ and $\frac{5}{6}$ **c** $\frac{5}{9}$ and $\frac{3}{7}$ **d** $\frac{3}{4}$ and $\frac{3}{8}$

4 Write these values in ascending order.

$\frac{3}{8}$ $\frac{2}{3}$ $\frac{3}{4}$ $\frac{5}{6}$ $\frac{1}{2}$

5 Take one fraction from the triangle and another from the rectangle.

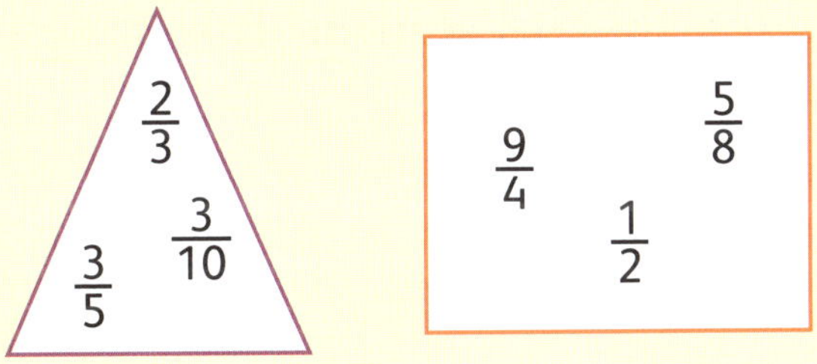

a Make and solve three addition calculations.

b Make and solve three subtraction calculations.

6 Copy and complete.

a $3 \times \frac{5}{6}$ **b** $5 \times \frac{3}{6}$ **c** $\frac{2}{3} \times 24$ **d** $\frac{3}{8} \times 24$

e $\frac{2}{3} \times \frac{9}{15}$ **f** $\frac{3}{4} \times \frac{4}{5}$ **g** $\frac{7}{10} \times \frac{5}{8}$

11 Self check

Check you can do this:	If you have difficulty, try this:
1 Find equivalent fractions.	Go back to page 156. Remember that if you multiply the numerator and denominator of a fraction by the same number, you get an equivalent fraction.
2 Use common factors to simplify fractions and find their simplest form.	Go back to page 159. Remind yourself that simplifying is the inverse of finding equivalent fractions. You divide the numerator and denominator by their common factors.
3 Rename fractions to help compare and order them.	Go back to pages 160–161. Remind yourself how to compare fractions that have unlike denominators. Use what you have learned about equivalent fractions.
4 Convert improper fractions to mixed numbers and vice versa.	Go back to page 162, and remind yourself how to convert between the different types of fraction.
5 Round mixed numbers to the nearest whole number.	Go back to page 162, and remind yourself how to round a mixed number (such as $1\frac{3}{4}$) to the nearest whole number.
6 Add fractions or mixed numbers, with or without renaming.	Go back to page 165.
7 Subtract fractions or mixed numbers, with or without renaming.	Go back to page 165, as well as page 166.
8 Multiply proper fractions.	Revise pages 169–170.

Unit 2: Measurement

Chapter 12 Perimeter and area

In this chapter you will explore these focus questions:

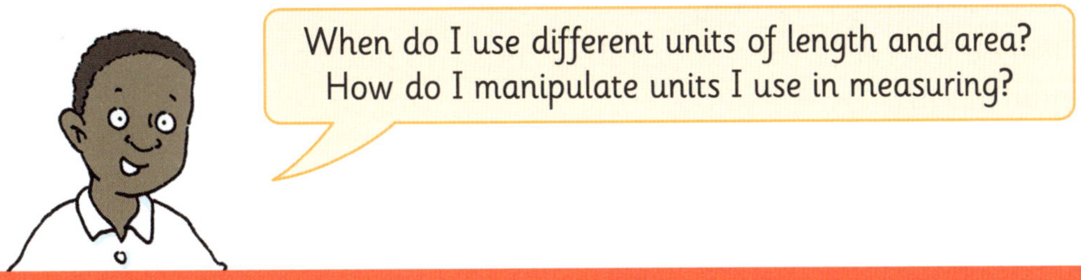

To answer these questions, you will:

- develop relationship(s) between units of length and units of area
- estimate, measure and record distances and perimeters in cm and m
- solve problems involving length, perimeter and area
- identify when to use square centimetres, square metres or hectares.

Starting point

1 This is what a football field looks like from above. It is covered with grass and the white lines are painted on the grass. The garden store sells the grass in patches that are 1 metre long and 1 metre wide.

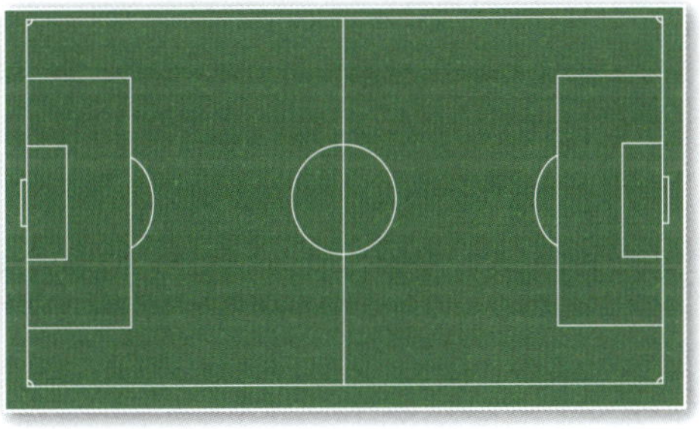

What information would you need to work out how many patches of grass were needed to cover the whole field?

2 These children are talking about the different ways they can give the measurements of this square.

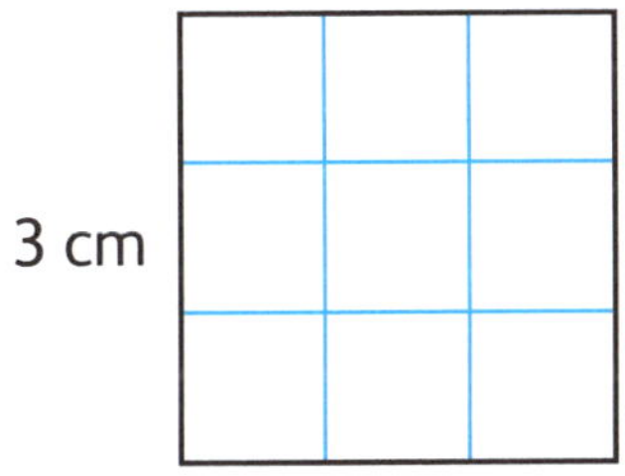

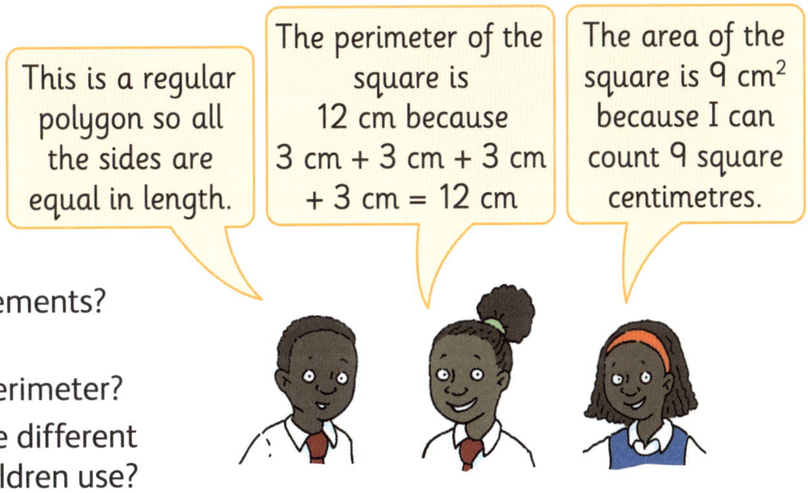

Do you agree with these statements?

What is different about the measurements of area and perimeter?

What do you notice about the different units of measurement the children use?

Explain why.

Working with perimeter

Key maths idea

Perimeter is the distance or path around a two-dimensional (2-D) shape.

You find the perimeter of a **polygon** by measuring the length of each side and adding them together.

Let's find the perimeter of this **rectangle**. It has two pairs of sides of equal length.

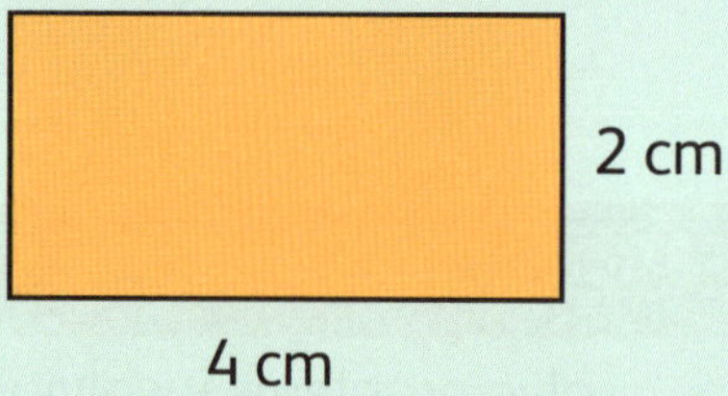

4 cm + 4 cm + 2 cm + 2 cm = 12 cm

So the perimeter of the rectangle is 12 cm.

If the lengths of the sides (or dimensions) are given, just add them up. For curved sides, you can lay a piece of string along the lines, and then measure the string.

> Did you know that the perimeter of a circle has its own special name? It is called the **circumference**. A circle does not have straight sides, so its perimeter is calculated in a different way.

Key words

perimeter
polygon
rectangle
circumference

1 Measure the sides of the shapes. Calculate the perimeter of each one.

a

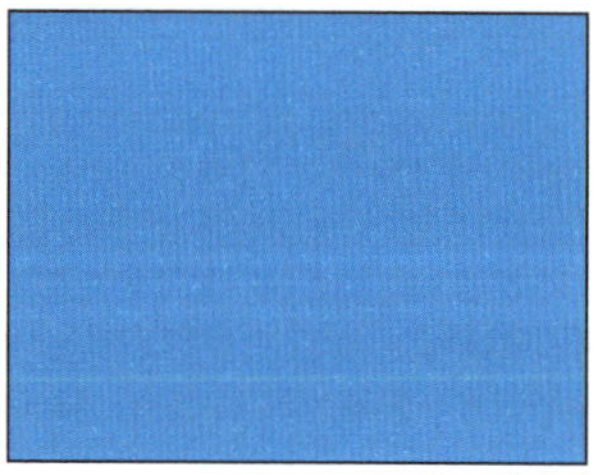

b

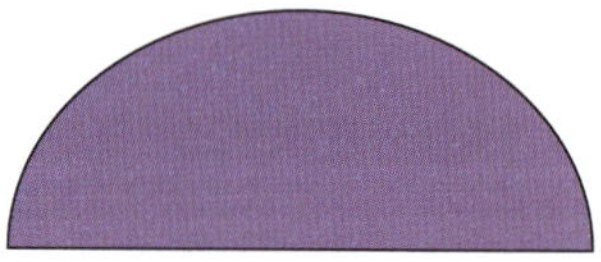

c

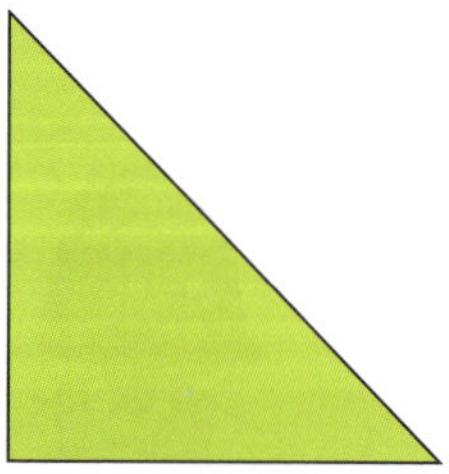

d

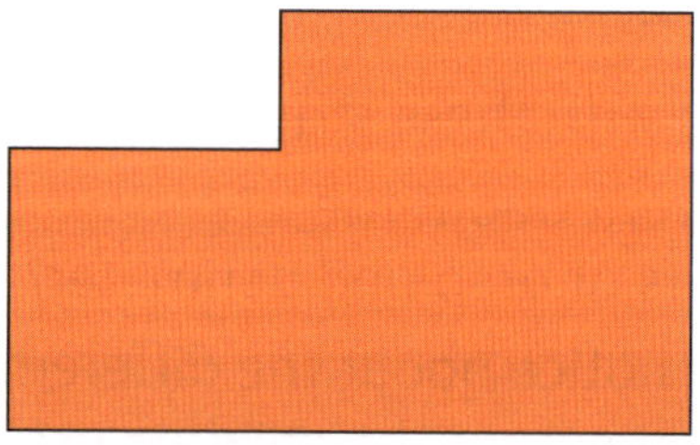

2 Use 1 cm squared grid paper and a ruler to draw the shapes described below. Then calculate the perimeters. Write the calculations you use.

- **a** A square with 5 cm length sides.
- **b** A rectangle with 75 mm longer sides and 35 mm shorter sides.
- **c** A right angled triangle with a 30 mm base and a 40 mm height.

3 Find the perimeter of each shape. Do not measure the sides.

a

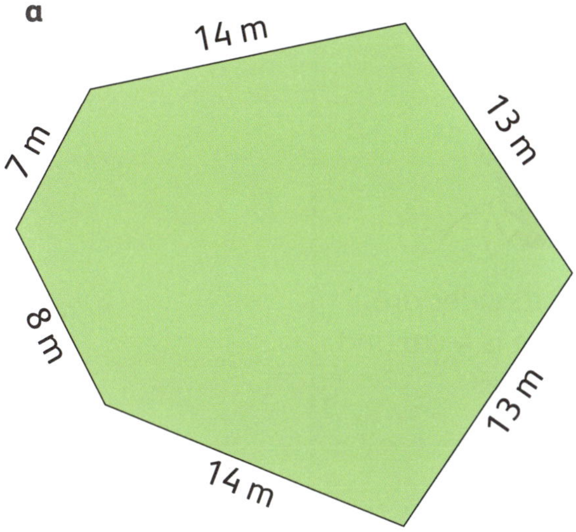

b

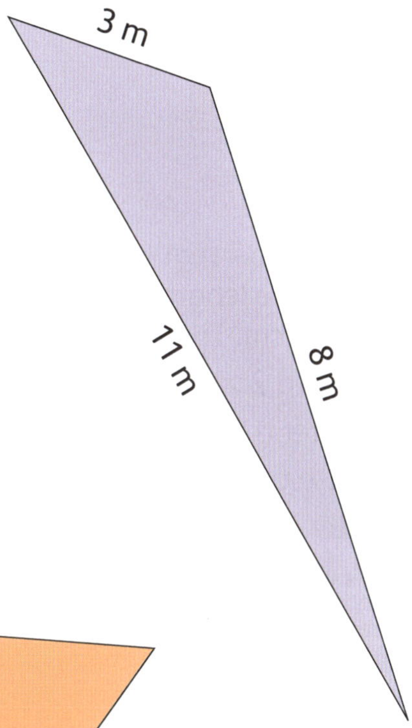

c

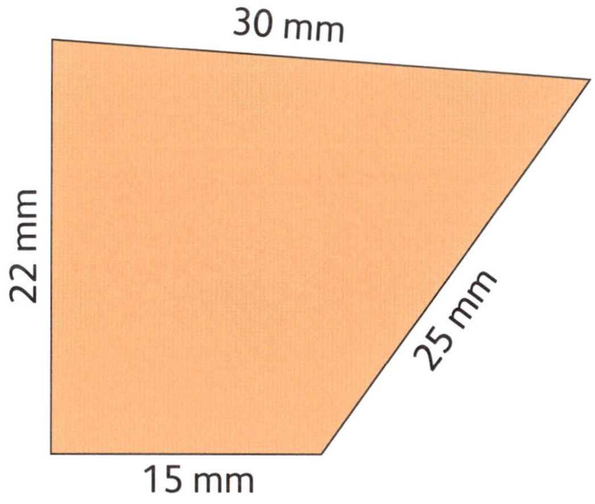

(not to scale)

4 Work in pairs or groups. Your teacher will assign each group a different place in your school, such as a classroom, a field, a quadrangle or a yard. Estimate and then measure the perimeter of each place. What unit of measure will you use?

Finding the perimeter of rectangles

Key maths idea

We can calculate the perimeter of a rectangle in different ways.

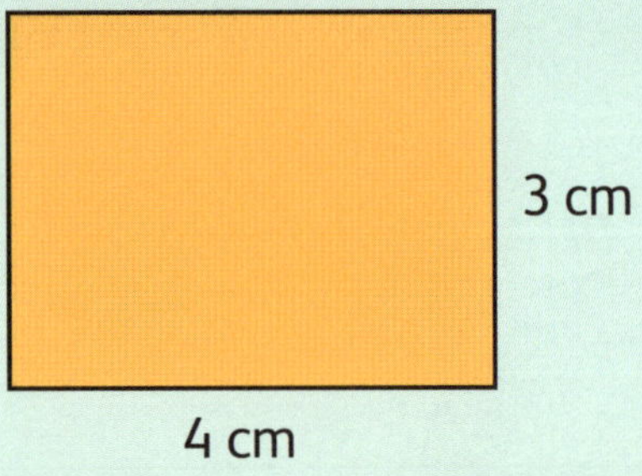

A rectangle is an **irregular polygon**. That means its sides are not all the same length. The length of the rectangle is 4 cm and the width is 3 cm. There are two of each of these sides.

So we can find 2 × 4 cm = 8 cm and 2 × 3 cm = 6 cm.

The perimeter is 8 cm + 6 cm = 14 cm.

We can write this using the formula

perimeter of rectangle = (2 × length) + (2 × width)

So (2 × 4 cm) + (2 × 3 cm) = 8 cm + 6 cm = 14 cm

Talk about

The perimeter of this rectangle can also be found as 2 × 7 cm = 14 cm.
Try to think why. Talk to a partner about your idea.

Key word
irregular polygon

1 **a** Look at these rectangles. Which do you think will have the longest perimeter? Explain how you know.

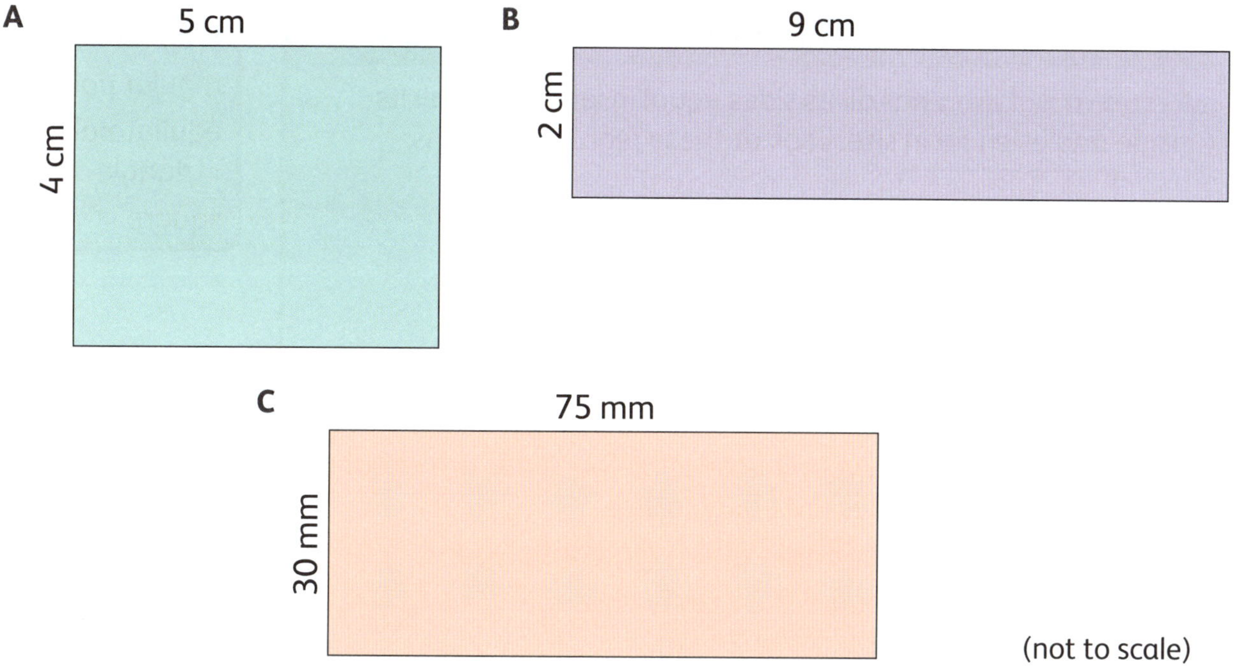

(not to scale)

b Calculate the perimeter of the rectangles. Write the calculations you use.

2 Work with a partner. How many different rectangles can you draw with a perimeter of 60 cm?

3 A rectangle has one side of 11 m and a perimeter of 38 m. What are the lengths of the other three sides? Make a rough sketch and fill in the answers.

4 Solve these problems:

a Jevaughn has made a painting with a length of 90 cm and width of 65 cm. He needs enough wood to make a border all the way along the edge of the poster. Calculate the perimeter of the painting to work out how much wood he needs.

b This is a diagram of a rectangular field.

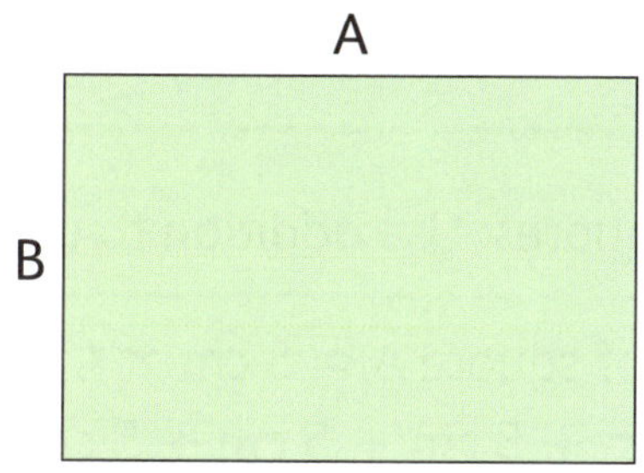

Side A is 30 metres longer than side B. One of the sides of the garden is 105 metres long. What are the two possible perimeters of the garden?

Finding the perimeter of regular polygons

Key maths idea

A **regular polygon** has all its sides equal in length and all its inside angles equal in size. Look at these regular polygons.

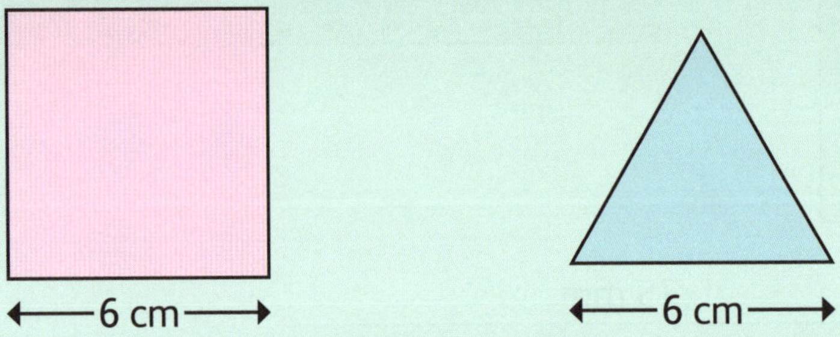

A square is a regular polygon because all four sides are equal in length and all four angles are equal in size.

Why is an **equilateral triangle** also a regular polygon?

The perimeter of the square is 24 cm because
6 cm + 6 cm + 6 cm + 6 cm = 24 cm.

We can also say that 6 cm × 4 = 24 cm or 4 × 6 cm = 24 cm.

What do you notice about the calculations and the number of equal sides of the **square**?

Let's think about the perimeter of the equilateral triangle.

The length of one side is also 6 cm, but this time there are only 3 equal sides.

The perimeter can be calculated as 6 cm + 6 cm + 6 cm or as 6 cm × 3 or 3 × 6 cm.

What is the perimeter of the equilateral triangle?

Key words

regular polygon
equilateral triangle
square

1 Find the perimeter of these squares. Use addition to check your answers. For example:

Perimeter of square A = 5 cm × 4 = 20 cm
Check: 5 cm + 5 cm + 5 cm + 5 cm = 20 cm

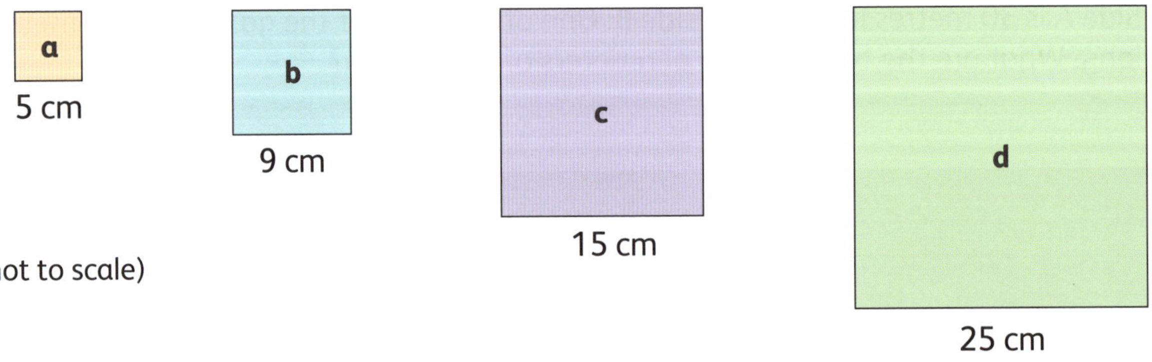

(not to scale)

2 Write the calculations you have to do to find the perimeter of shapes A to D.

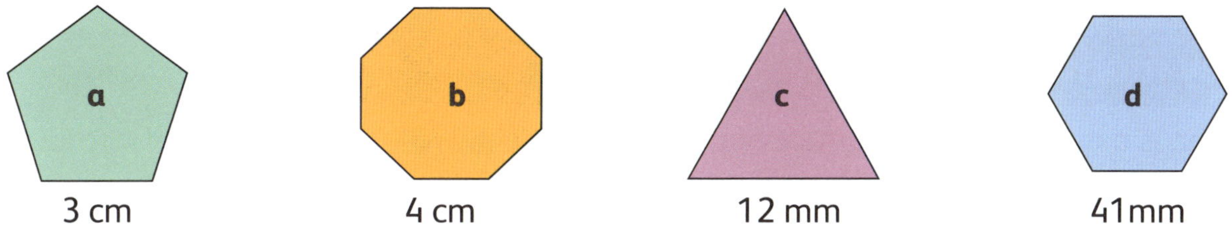

(not to scale)

What do you notice about your calculations?

Write a rule to help you find the perimeter of any regular polygon.

3 Write the length of one side of each of these shapes:

a a regular pentagon with a perimeter of 25 cm

b a square with perimeter equal to 20 m

c a regular octagon with perimeter equal to 64 cm

d a regular hexagon with perimeter of 42 cm

e an equilateral triangle with perimeter of 99 mm.

4 An aerodrome is marking out target areas for parachutists to land. They decide to use regular polygons for these target areas. Use the information to work out the regular shape they use each time.

5 Make up another clue of your own for another regular shaped target. Swap with a partner.

Measuring area

Key maths idea

Area is a measure of the surface of a shape or object. We measure area in square units such as **square centimetres (cm^2)** or **square metres (m^2)**. You can find the area of these shapes by counting the squares they cover. Each square here is 1 cm^2.

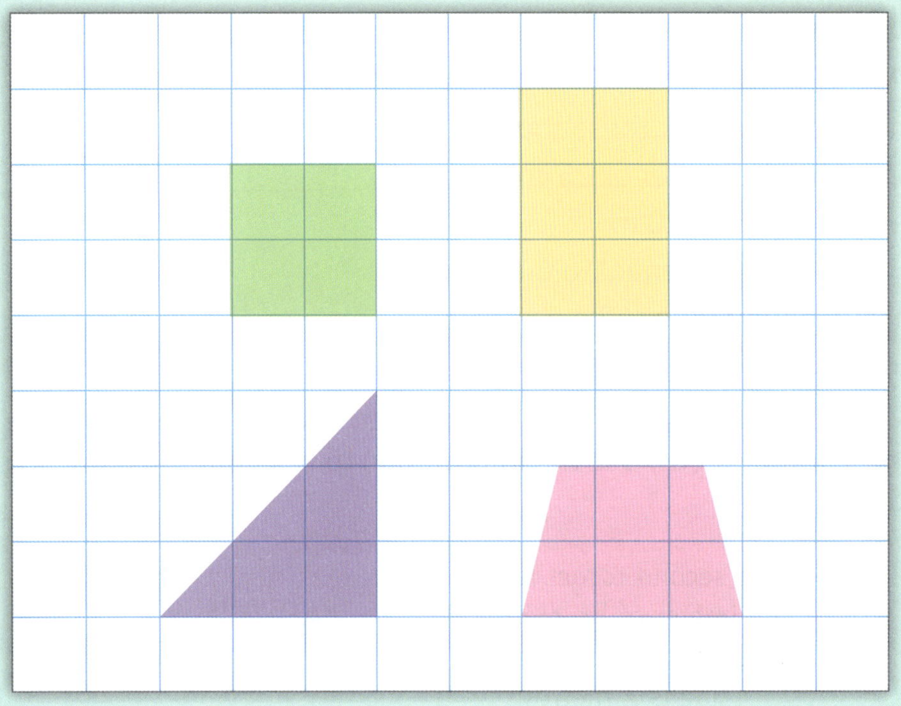

Which shapes have areas that are easier to find by counting squares?

Which shapes have areas that are more difficult to find by counting squares? Why?

Look at this shape. Is it possible to estimate its area in square units?

How can you work it out?

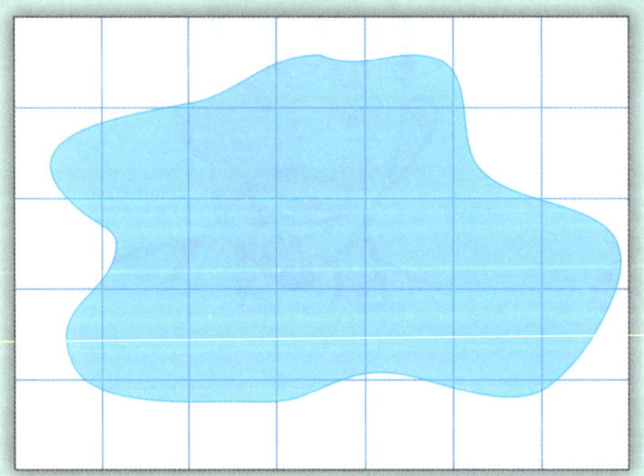

Why do you think we measure area in square units such as square centimetres (cm^2), but perimeter in units such as centimetres (cm)?

Key words

area

square centimetre (cm^2)

square metre (m^2)

1 Count the squares to find the areas of these polygons.

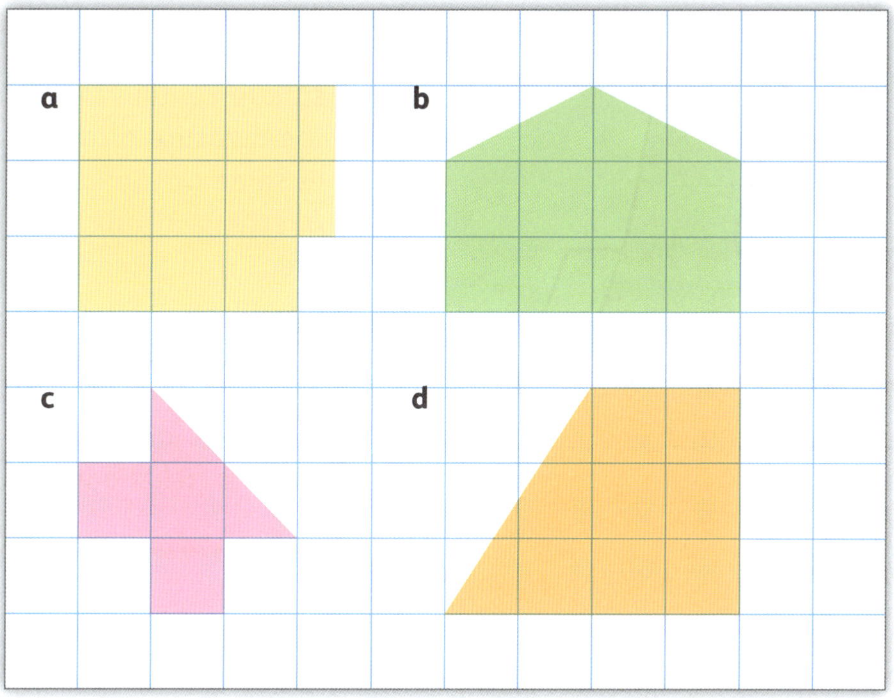

2 **a** Estimate the area of each of these shapes in square units.

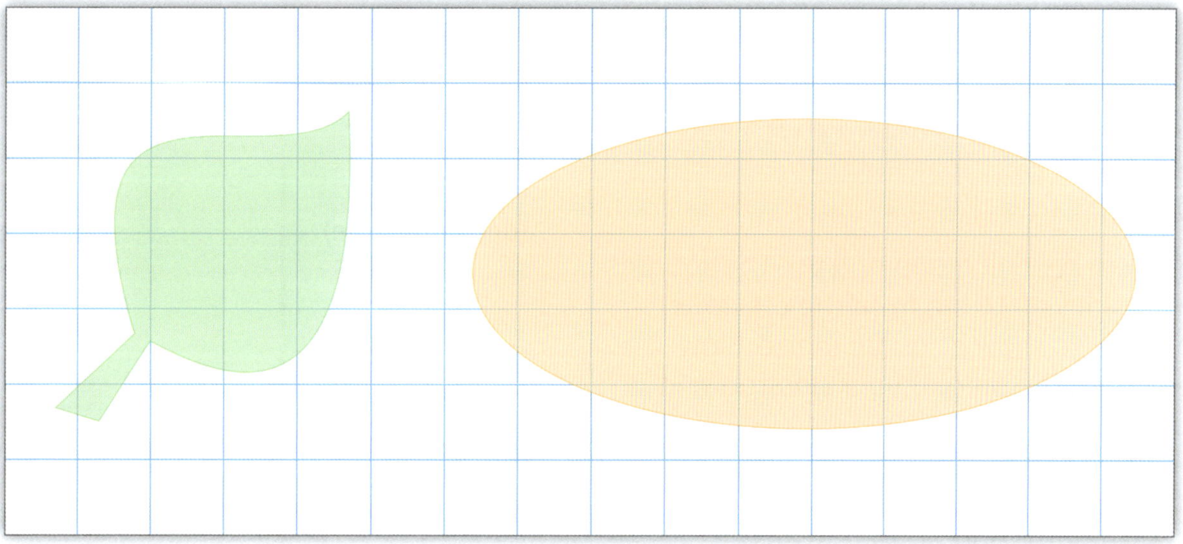

b Why can we not give an accurate answer for the areas of these shapes?

3 Cut a square piece of paper with sides of 10 cm. This is one square. How many of these squares would you need to cover the area of:

- your desk
- the whiteboard or blackboard
- the window in your classroom?

Keep your square so you can use it for another activity.

4 Work with a partner. Trace the outline of your foot on 1 cm squared grid paper.

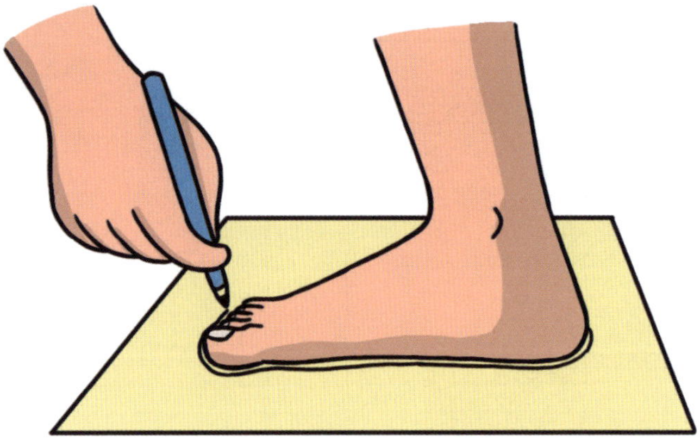

Estimate the area of your foot. Compare with a partner and decide whose foot has the larger area. Count the squares to find out.

> You have been using square centimetres as the measure of area in some of the activity questions. Think about areas that would be measured in square metres. Research the measure of area used for much larger areas, such as fields and forests.

Square units

Key maths idea

This is 1 square centimetre. We write this as 1 cm^2. It means a square with each side equal to 1 cm.

1 cm
1 cm
(drawn to scale)

We write 1 square metre as 1 m^2.

Work in a small group and use the squares with sides of 10 cm from the previous activity to show a square metre. Now estimate the area of your classroom in square metres.

A **hectare** is a square unit that measures 10 000 m^2. It is used to measure much larger areas.

100 m
1 hectare
(not to scale)
100 m

> The word 'hectare' comes from the Greek word 'hekaton', which means a hundred, and 'are', which is a square unit that measures 100 square metres.

Key word
hectare

1 Say whether you would measure the area of each of these in cm^2, m^2 or hectares.

a a large sugar farm

b a cinema

c a notepad

d a large island

e a painting

f a swimming pool

2 A square piece of land measures 7 hectares. What is its area in square metres?

Talk about

Work with a partner. Use books, your own drawing and images or maps online. Discuss and compare small regions (for example, in diagrams) with large regions in the real world. Talk about the square units you would use to measure or show the different areas. Make a PowerPoint presentation of your findings to share with the class.

Areas of rectangles and squares

Key maths idea

Look at the rectangle and square here. Count squares to find the area.

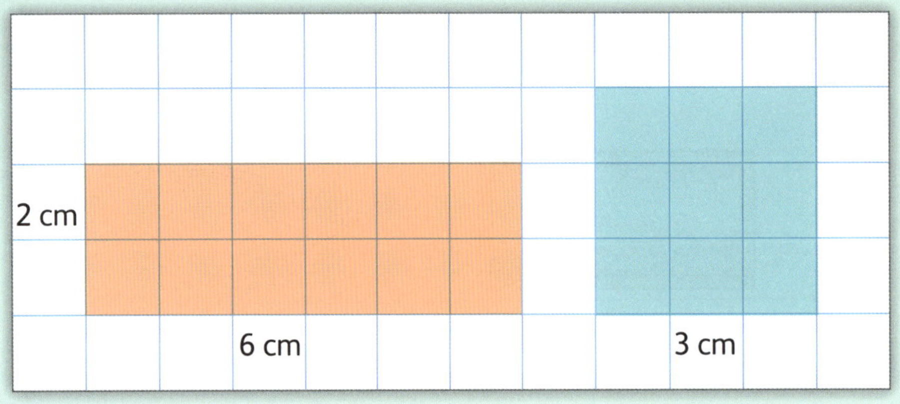

Now think about the shapes as arrays. What do you notice about the number of rows and columns, and the total area they cover?

The rectangle is a two-by-six array covering 12 squares.

The square is a three-by-three array covering 9 squares.

You can use the formula **area = length × width** to help calculate the area of rectangles and squares easily:

area of rectangle = 6 cm × 2 cm = 12 cm²

area of square = 3 cm × 3 cm = 9 cm²

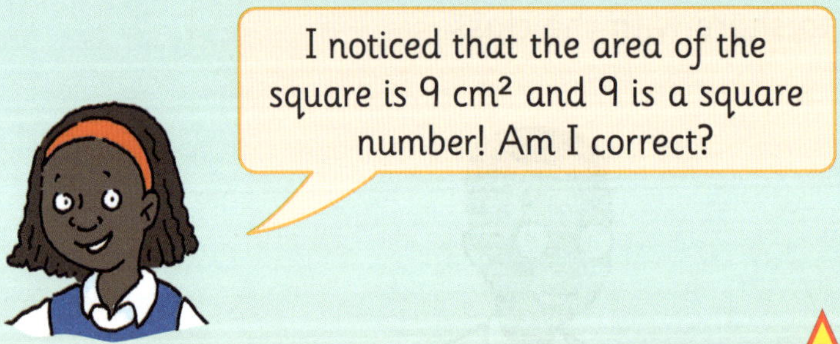

Tiles and other flooring materials can be used as **floor covering** for a room. If you have access to computer software, design a floor plan for a simple house. Draw it, and make sure you use appropriate units. Calculate the floor areas for the **tiling** that you need.

Key words

floor covering

tiling

1 Use the formula for rectangles and squares to help you find the area of each shape.

a

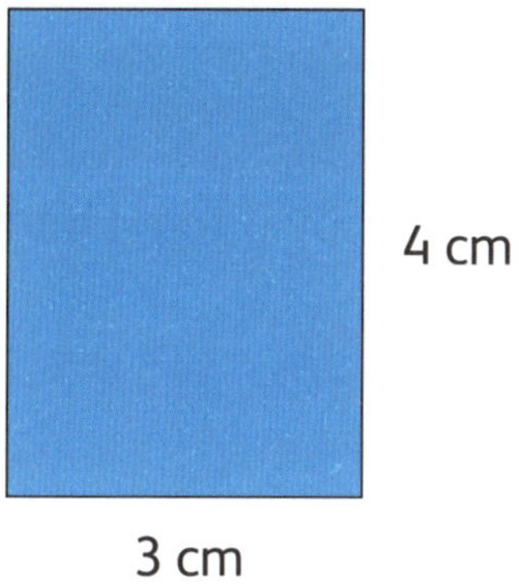

b

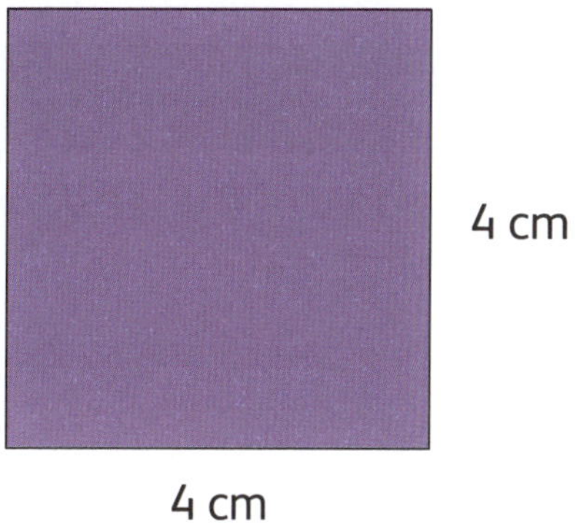

c

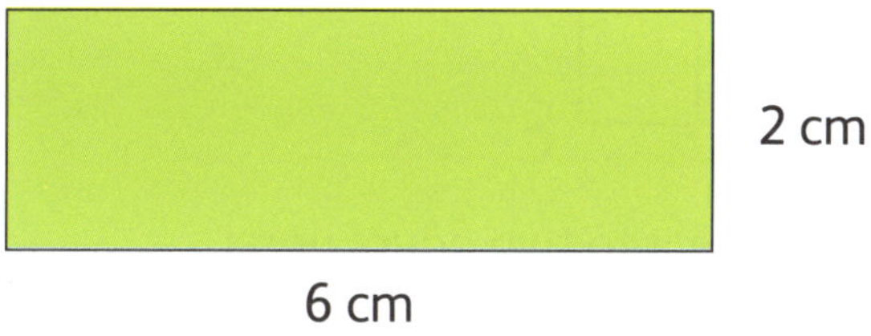

d

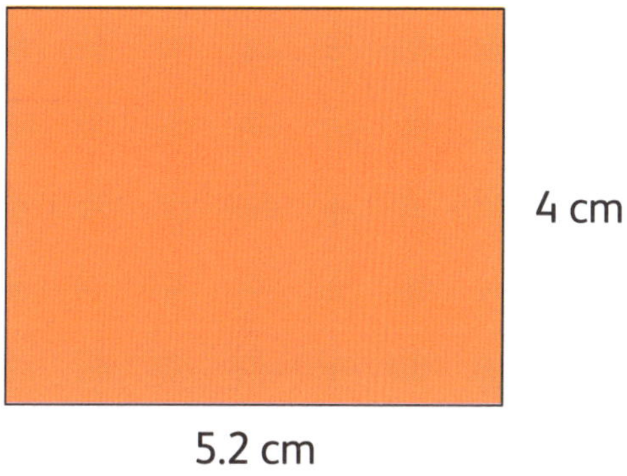

2 Akiel draws a rectangle with an area of 24 cm^2. What could the length and the width of the rectangle be? Find at least 3 solutions.

3 **a** A rectangular classroom has an area of 80 m^2. Its length is 10 m. How wide is the classroom?

b A square classroom has an area of 81 m^2. What are the dimension of the classroom?

> You recently estimated the area of your classroom in square metres. How could you calculate the actual area?

4 Devon's workshop is 10 m wide and 15 m long. He wants to cover the whole area with non-slip flooring. The floor covering comes in pieces that measure 5 m^2.

How many pieces must Devon buy to cover the floor?

5 Mr Williams tends to a field with a length of 900 m and a width of 500 m.

a Express the area in square metres.

b How many hectares of sugar cane can Mr Williams plant on this field?

Areas of triangles

Key maths idea

We can find the area of the rectangle ABCD by counting squares or by using a formula.
Area of ABCD = 15 cm^2.

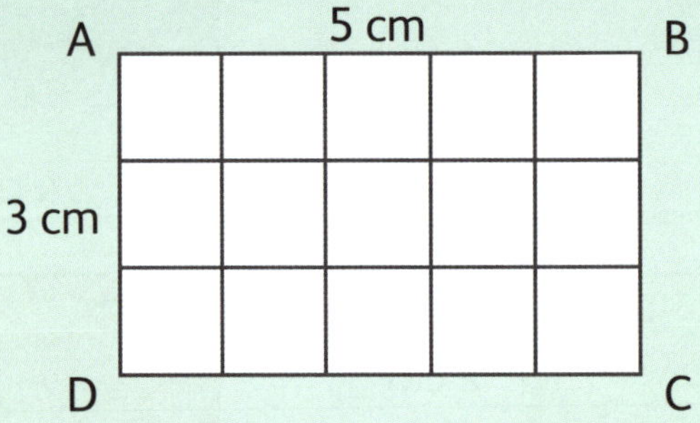

Look at what happens when we divide the rectangle into two equal parts.

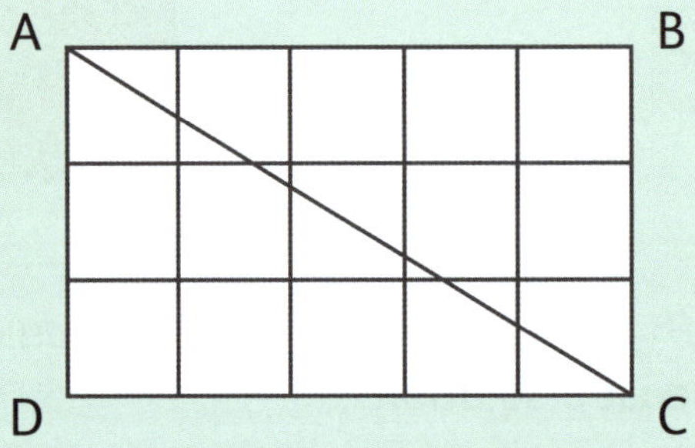

Each part covers exactly half of the area of the whole shape.
Name the two smaller shapes that you can see.
We can calculate the area of a triangle by finding half of the area of the rectangle with the same height and width.

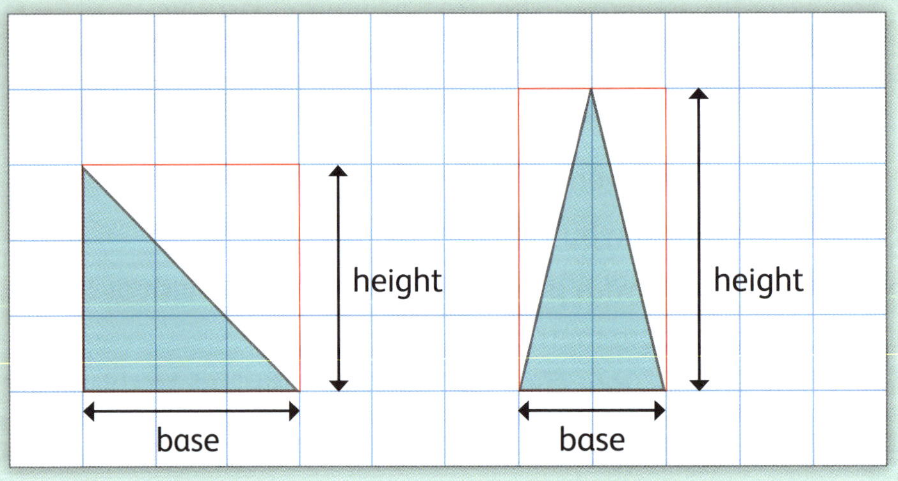

So, we find the area of the rectangle as length × width and then divide by 2.
Area of triangle = (base × height) ÷ 2 or $\frac{1}{2}$ (base × height)
Work with a partner to calculate the area of the two triangles above.

1 Calculate the area of shape QRST.

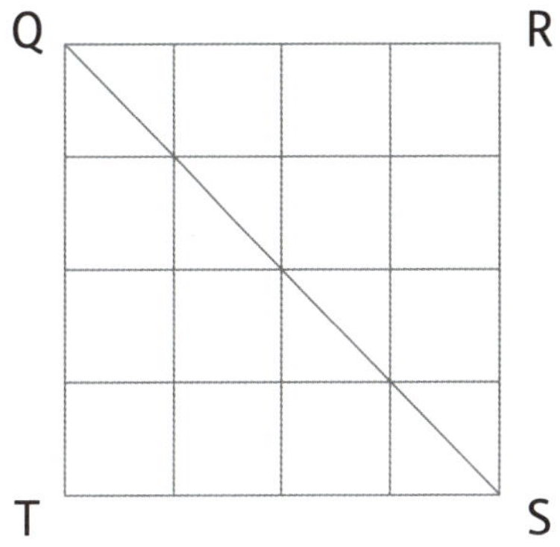

2 **a** Make a paper cut-out of shape QRST on 1 cm squared grid paper.

b Fold the square into two triangles.

c Are the two triangles the same size?

d What is the area of each triangle?

3 Find the area of each of the shaded triangles.

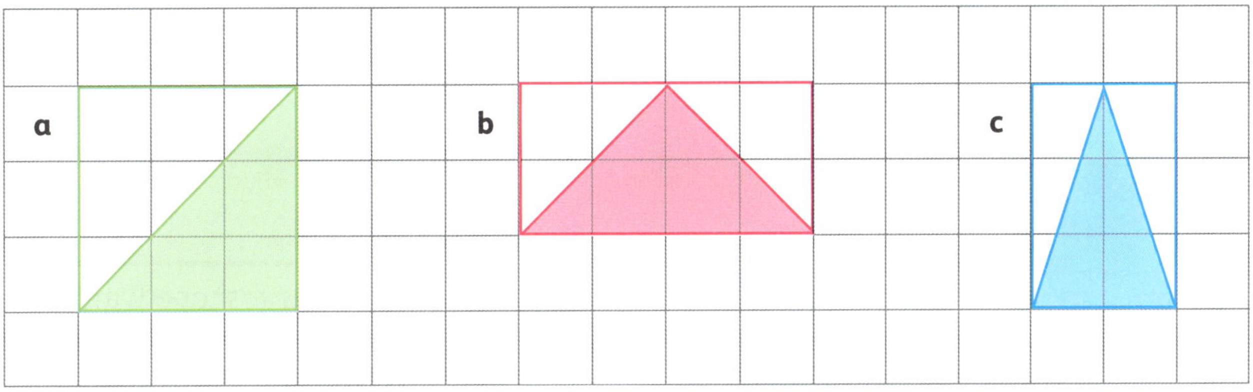

4 The area of a triangle is 18 cm². Use what you know about finding the area of a triangle to work out some possible dimensions. You could sketch the triangles on paper or use drawing tools in computer drawing software.

Reasoning about perimeter and area

Key maths idea

Which of these two shapes has the larger area?
Which of these two shapes has the longer perimeter?

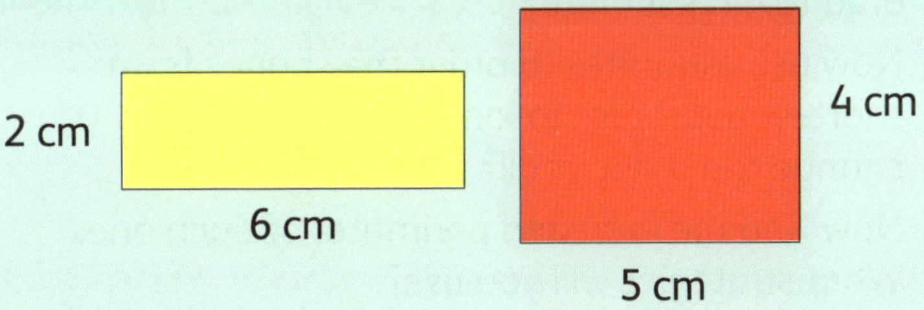

A **compound shape** is made up of two or more shapes joined together. In this example, the yellow and the red rectangles have been joined to make a compound shape.

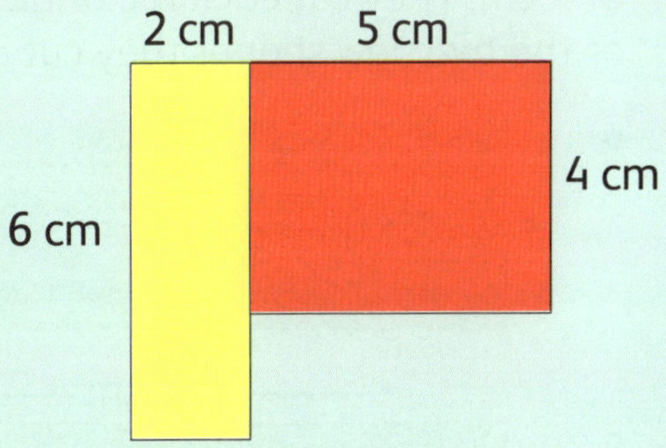

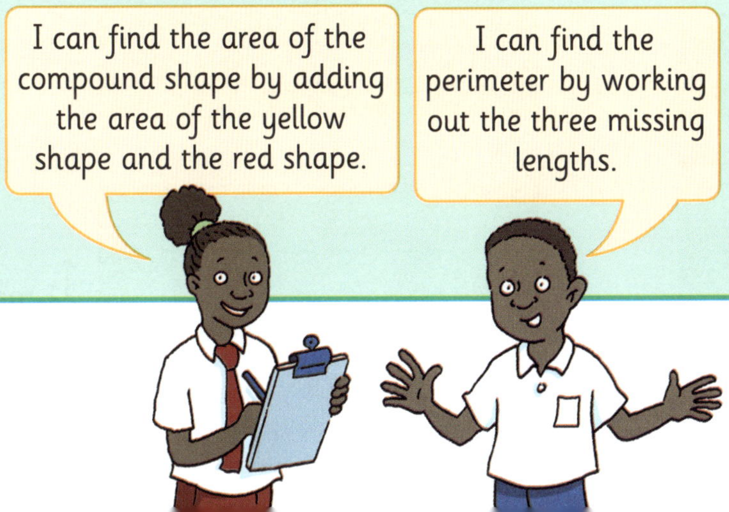

Try out these ideas.

1 Measure the dimensions of this shape. Work out the area and the perimeter.

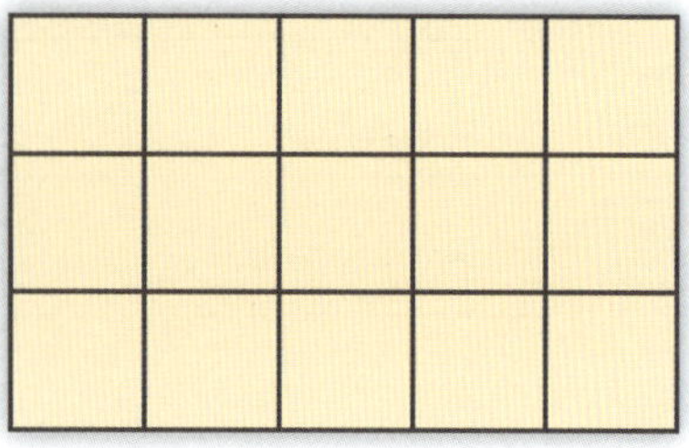

2 Your teacher will give you some 5 cm by 3 cm cards. Use your ruler to draw horizontal and vertical lines to make centimetre squares, like the rectangle in Exercise 1. Explore the different areas and perimeters you can find by placing 2 or 3 shapes in different positions, for example:

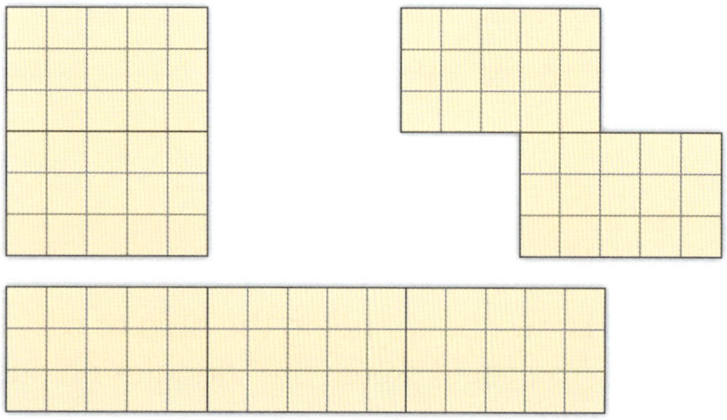

> Does a shape with the largest area always have the largest perimeter?

What strategies will you use to help you calculate? Talk to your partner about them.

Record and compare your findings. Discuss any patterns you notice with the class.

3 You need some polygons from the maths area or find some pictures to use.

- **a** Use estimates to order the shapes from smallest area to largest area.
- **b** Now use estimates to order the shapes from shortest perimeter to longest perimeter. Did you change the order at all? Discuss why.
- **c** Now find the area and perimeter of each one. What strategies will use?

> Use 1 cm squared grid paper and tracing paper to help you find the area and perimeter of the polygons in Exercise 3.

4 Devon and Candace have four rectangles that each have a length of 12 cm and width of 4 cm. They cut each rectangle into two new shapes. Work out the perimeter of each of the two new shapes they cut out.

a

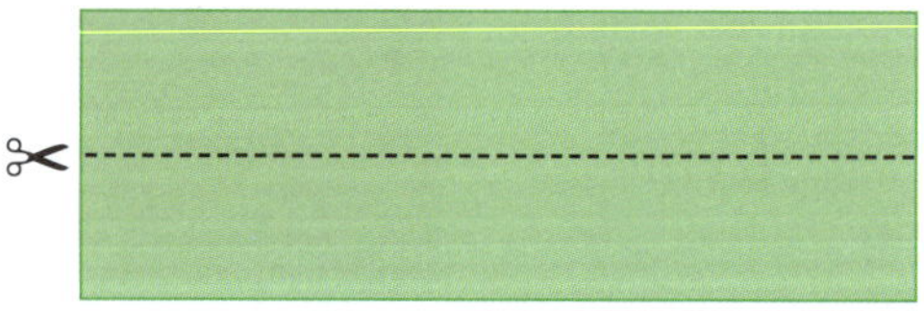

b

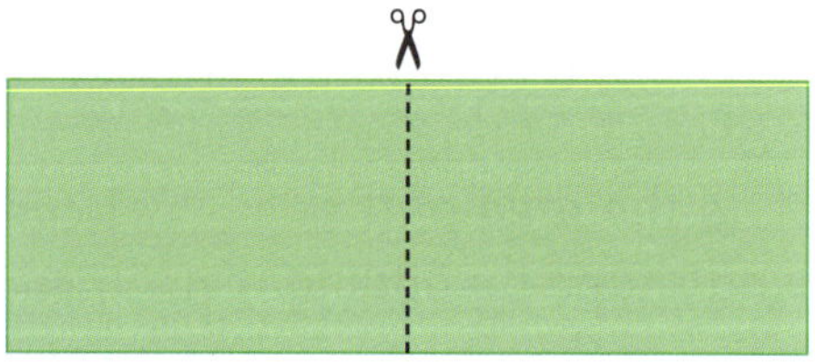

c

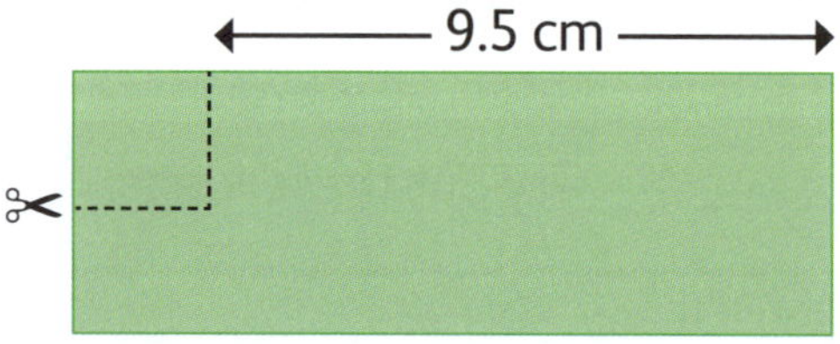

d

(not to scale)

Real-world maths: Let's go camping

1 This is a tent for two adults. It has a rectangular base. What would be a good estimate for the area of ground it covers? Work with a partner to decide how you can work it out. What measurements will you need to make? Use large paper to represent the area or use chalk to mark it on the playground.

2 A camping ground has a square field with an area of 1 hectare. The field is divided into rectangular camping pitches that each measure 5 metres by 10 metres.

   a How many camping pitches of this size are there?

   b What is the perimeter of the square field?

Maths detective: Rectangles and grids

You have been thinking a lot about the area and perimeter of shapes. Now use your maths detective skills in these challenges.

Case 1

You can use 10 matchsticks to make this shape on large grid paper. Each stick is 1 unit long.

Explore how to solve scenarios 1 to 4. Start with the original shape each time.

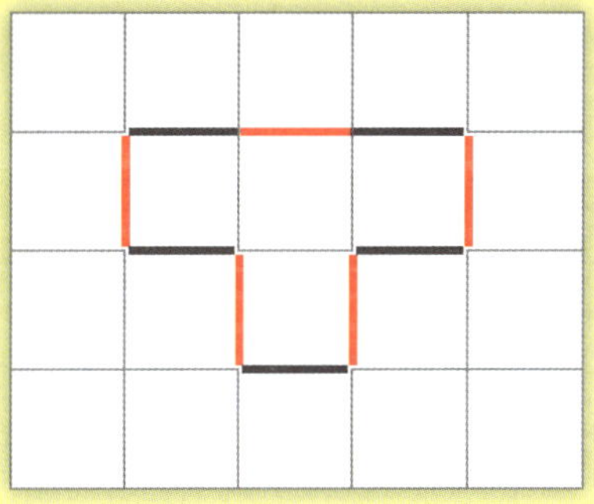

1 You move 2 sticks so the area of the whole shape is larger but the perimeter is the same. Give the area of your new shape in square units and its perimeter in units.

2 You move 4 sticks so that your whole shape is now a rectangle. Calculate the area and perimeter of your rectangle.

3 You move 3 sticks to make another rectangle with a smaller area and a shorter perimeter than the original shape.

4 Add 2 more sticks and move 4 sticks so your shape has the greatest possible area and greatest possible perimeter.

Case 2

Make a rectangle with the same dimensions as the one shown here. Then cut the rectangle into two shapes.

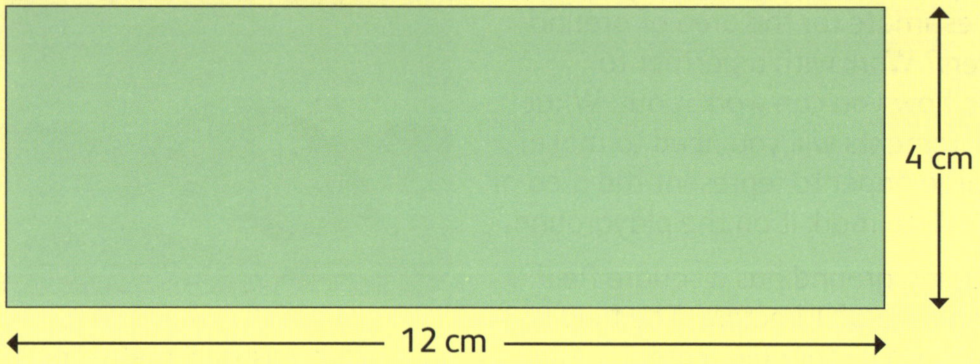

1 Find a solution to cut the rectangle so that the area of one of the new shapes is double the area of the other.

2 Find a solution to cut the rectangle so that one of the areas is triple the area of the other.

3 How can you cut your rectangle so that one of the new shapes has half the perimeter of the other new shape?

What I have learned

1 Do these calculations involve perimeter or area? Copy and complete the table.

Calculation	Area or perimeter?
a Working out how much carpet you need to cover a floor.	
b Finding how much fencing you need to put a fence around a field.	
c Calculating what length of wood to use for a picture frame.	
d Working out how many tiles you need to tile a patio.	

2 Which measurements would you need to work out these areas and perimeters? Copy and complete the table.

Calculation	Measurement needed
Area of a square	
Area of a right-angled isosceles triangle	
Perimeter of a scalene triangle	
Perimeter of an equilateral triangle	
Area of a rectangle	

Practice questions

1 Find the perimeter of each of these shapes.

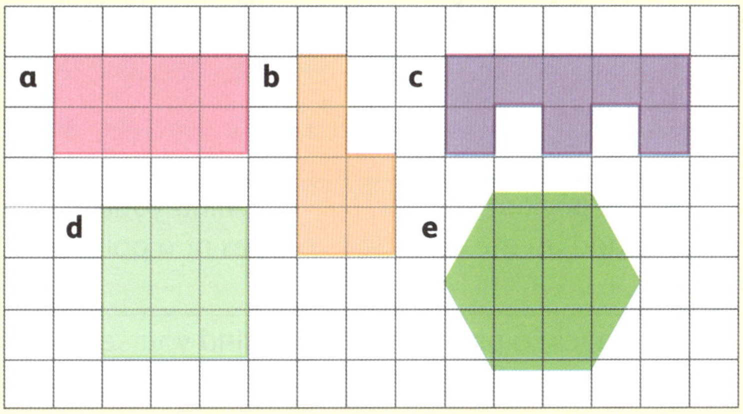

Shape **e** is a regular hexagon!

2 a The perimeter of a regular pentagon is 550 mm. What is the length of one of its sides?

b The perimeter of another regular shape is 28 cm. One of the sides is 7 cm. What shape is it?

3 Which of the following units is not used for measuring area?

cm^2 m^2 cm hectare

4 What is the difference between perimeter and area? Explain this in your own words.

5 A rectangular garden has a length of 4.5 metres and a width of 3 metres.

a What is its perimeter?

b What is its area?

6 Which of these shapes has the largest area? Calculate to find out.

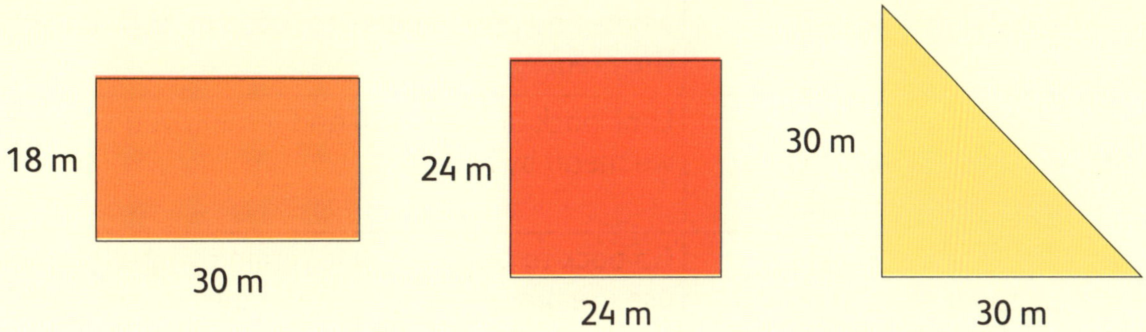

12 Self check

Check you can do this:	If you have difficulty, try this:
1 Calculate the perimeter of any shape by adding up the length of the sides.	Look back at page 176 and remind yourself how to add up the lengths of a sides of a polygon.
2 Use a formula to help calculate the perimeter of a rectangle.	Go back to page 178 and remind yourself of the different ways of calculating perimeter (addition or multiplication.)
3 Calculate the perimeter of regular polygons.	Look back at page 180. Remind yourself of the definition of regular polygons (shapes with all their sides equal in length) and use multiplication to calculate perimeters.
4 Explain the difference between perimeter and area.	Remind yourself of the definitions of perimeter and area, and the difference between them.
5 Find the area of shapes by counting squares.	Go back to page 182 and remind yourself what square units are. Look at shapes drawn on grids, and work out the areas based on the square units.
6 Count squares and use estimate to help find the area of irregular shapes.	Look at the shapes with curved outlines on page 182, and remind yourself how to add up the whole and part squares to reach an estimate of the area.
7 Use the square units of cm^2, m^2 and hectare to show measurements of area.	Look back at page 184 and remind yourself what we mean by hectares.
8 Use a formula to help calculate the area of a rectangle or square.	Go back to page 186 and remind yourself how we use the properties of these shapes to calculate their areas.
9 Use a formula to help calculate the area of a triangle.	Look back at page 188. Remind yourself how the areas of rectangles help us to work out the areas of triangles, and how this leads to the formula for calculating the area of a triangle.
10 Solve problems relating to perimeter and area.	Look back at the problems you solved on pages 189–192.

Unit 3: Geometry

Chapter 13 Properties of triangles and quadrilaterals

In this chapter you will explore this focus question:

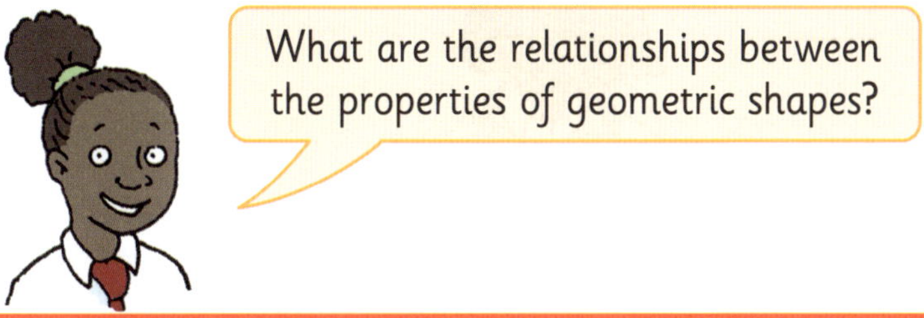

To answer this question, you will:

- draw, make and explore geometric shapes (polygons and non-polygons)
- solve problems involving shapes
- classify triangles as right, equilateral, isosceles or scalene
- identify opposite and adjacent sides of a quadrilateral.

Starting point

1 **a** Cut two squares in half to make four identical triangles.

b Join the four triangles together to make new shapes.

c Explore what different shapes you can make by joining the four triangles in new ways.

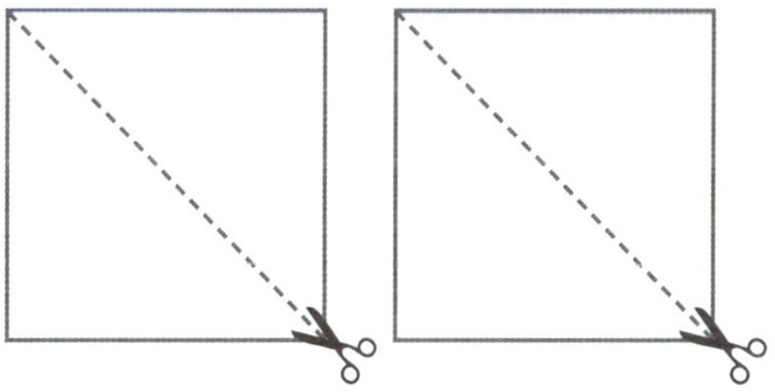

2 Explore the shapes you can make from these two pieces.

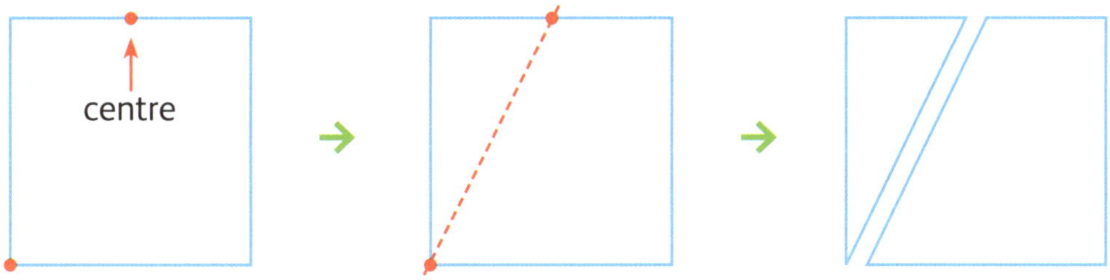

3 Akim made these two triangles on a 9-pin geoboard.

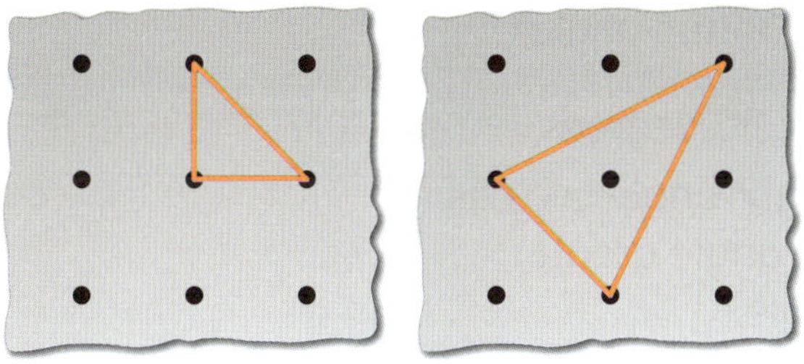

Compare the two triangles. What is the same and what is different?

Draw your own triangles on square grid paper or make them on a geoboard.

How many different ways can you make triangles on the board?

Talk about

Describe your triangles to a partner. Do you remember the different types of triangles? Try to make them.

4 Play a game. Name a type of triangle. Your partner then makes that triangle. Make as many different triangles as you can.

Triangles

Key maths idea

We **classify** things when we sort them into groups with similar **properties**. We can sort **triangles** in different ways. One way is to look the length of the sides:

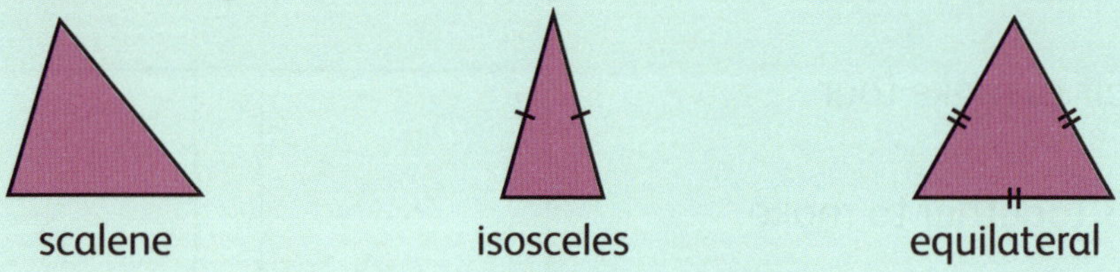

A **scalene** triangle has 3 sides of different length.
An **isosceles** triangle has 2 sides of equal length.
An **equilateral** triangle has 3 sides of equal length.
You show the equal sides using markings like the ones in the diagrams.
Another way to classify triangles is to consider the **angles**.

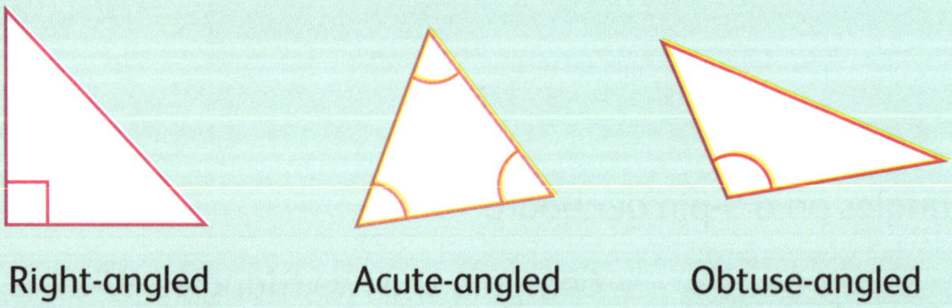

A **right-angled** triangle contains one angle of 90 degrees.
If all angles in a triangle are smaller than a right angle, we call it **acute-angled**. A triangle that contains one angle greater than a right angle is an **obtuse-angled** triangle.

Key words

classify
properties
triangle
scalene
isosceles
equilateral
right-angled
acute-angled
obtuse-angled

1 Write the properties of each triangle by looking at its side lengths and angles. Use a ruler and a protractor to measure accurately.

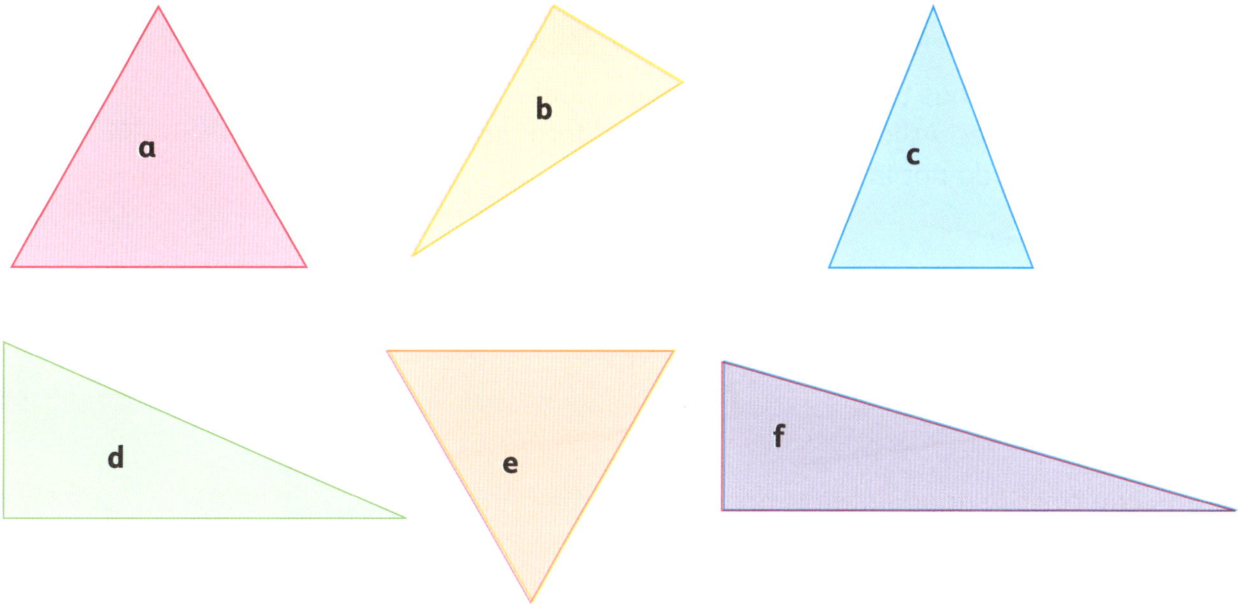

2 **a** Copy this empty Venn diagram into your notebook.

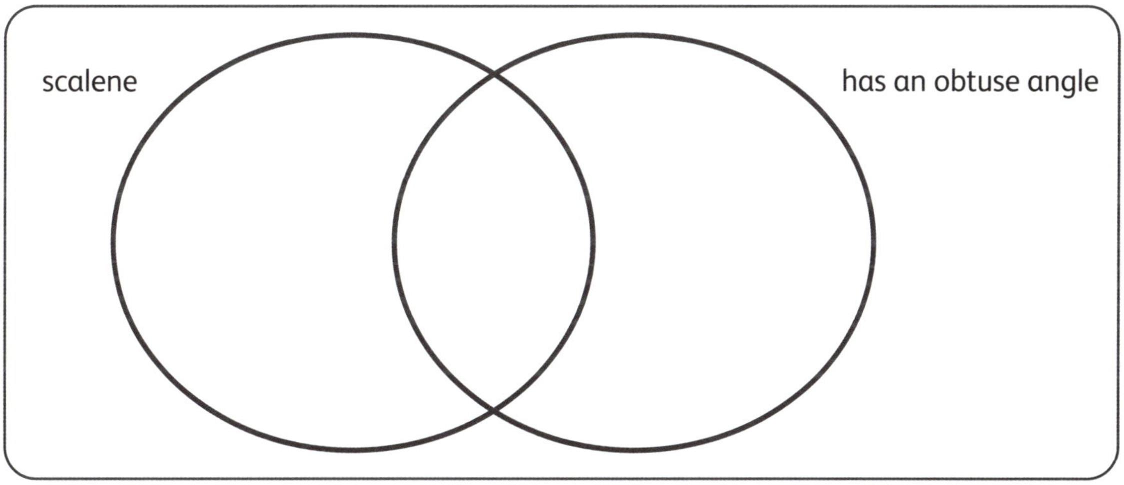

b Look at the diagrams of the triangles in the key maths idea. Which triangle can you draw in the left circle? Draw it in.

c Which triangle can you put in the right circle? Draw it in.

d Which triangle can fit in the overlap of the two circles? Draw it in.

e Which triangle(s) do not fit into the circles? Draw them in the rectangle part outside the circles.

3 Use a ruler to draw your own isosceles triangles.

Draw two lines of equal length, then join the two lines to form a triangle.

Accurately measure all of the angles of each isosceles triangle. What do you notice?

Cut out your isosceles triangles. Fold them in half to check that the two sides match in length. What do you notice about the angles when you fold?

Quadrilaterals

Key maths idea

A **quadrilateral** has two pairs of **opposite** sides and two pairs of **adjacent** sides. Adjacent sides meet and form an angle. Opposite sides do not intersect.

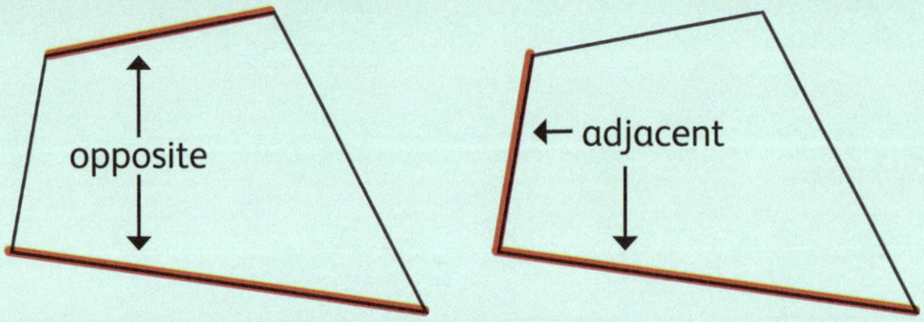

You can classify quadrilaterals according to their properties:

- the lengths of the sides
- number of pairs of parallel or perpendicular lines.

Parallelogram

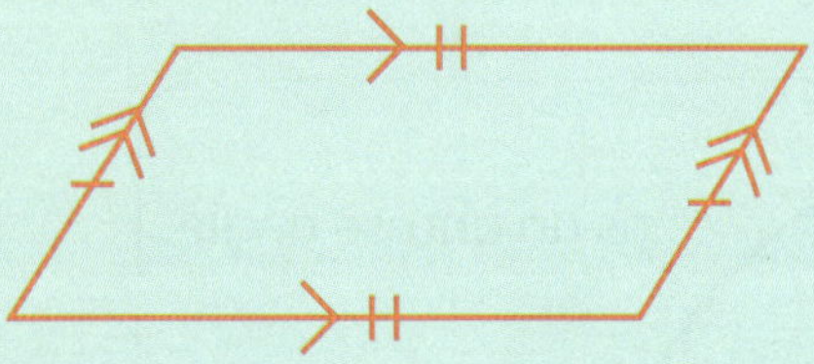

opposite sides equal and parallel

Rhombus

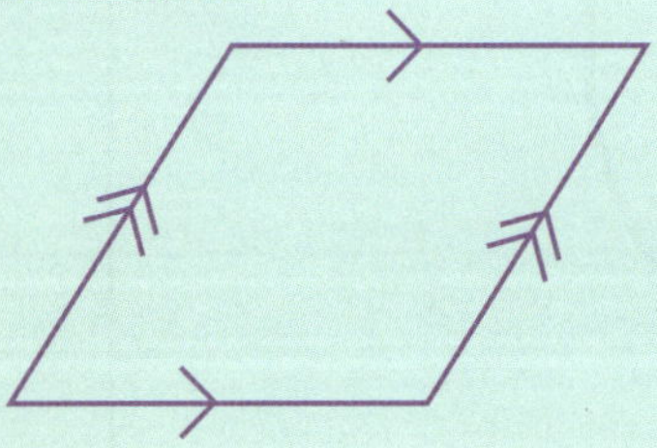

4 equal sides; opposite sides are parallel

Kite

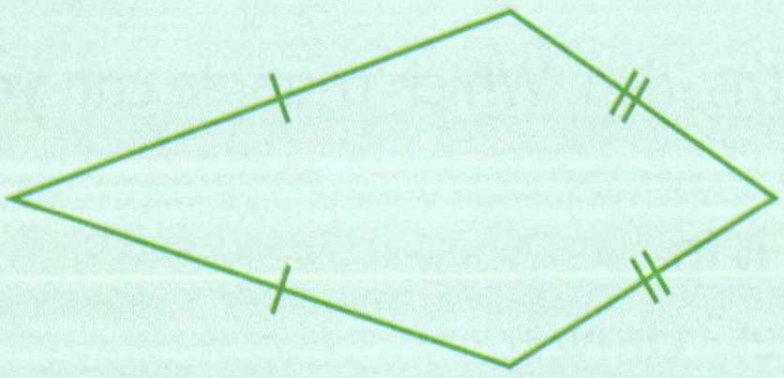

2 pairs of equal sides

Trapezium

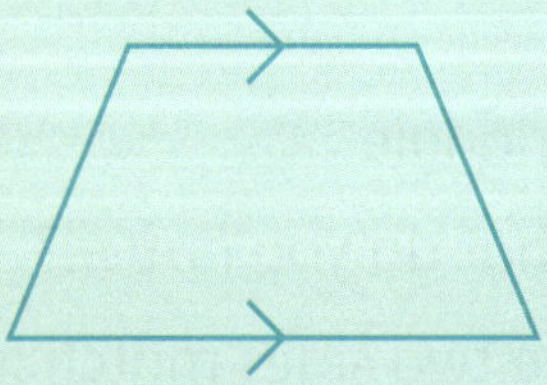

1 pair of parallel sides

Key words

quadrilateral
opposite
adjacent
parallelogram
rhombus
kite
trapezium

1 You can make one fold to turn a rectangle into a trapezium.

Investigate the quadrilaterals you can form with 1 fold, 2 folds or 4 folds if you begin with a rectangle.

2 a Accurately measure the sides and angles of these parallelograms. What do you notice about the angles?

b Describe your findings in part a.

c Create two parallelograms of your own to test your findings.

3 a This quadrilateral is called a kite. Measure the two angles marked f.

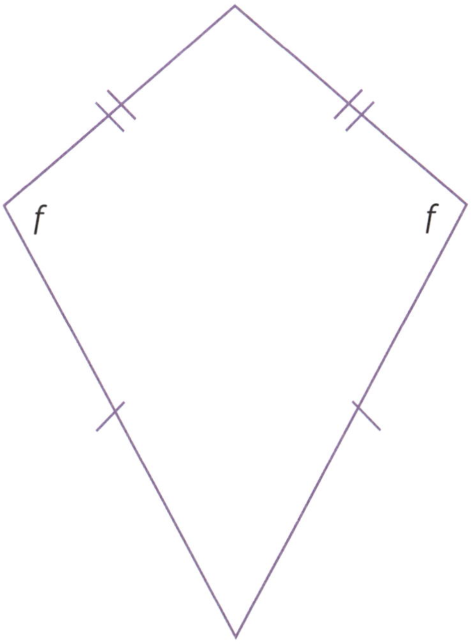

b What do you notice? Describe your findings in your own words.

c Test your findings by drawing two more kites and measuring its angles.

Maths detective: Angles in triangles

Now that you know more about triangles and quadrilaterals, use your maths detective skills to solve these cases.

Case 1

Use a geoboard. If you do not have one, you can also draw your answers or search for an interactive geoboard online.

1 How many different right-angled triangles can you create on a 16-pin geoboard?

2 How many different isosceles triangles can you create?

3 Is it possible to create an equilateral triangle on this board? Measure very accurately.

Case 2

1 Investigate the angles where the diagonals of this rectangle meet. What do you notice?

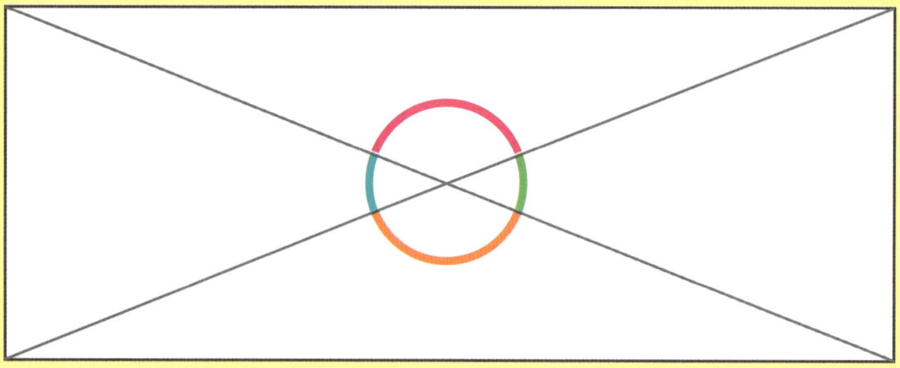

2 Now draw a square and add the diagonals. Explore the angles in the diagonals. Explain what you find.

3 Then do the same for a rhombus, a kite and a trapezium.

4 Write a short summary of your findings.

Case 3

1 Create a triangle with one obtuse angle.

2 Now try to create a triangle with two obtuse angles.

3 Then try to create a triangle with one obtuse, one acute and one right angle.

4 Explain what you found to a partner.

5 Copy this table. Sketch a triangle that fits in each cell. Which cells are impossible to fill?

	Equilateral	**Isosceles**	**Scalene**
Right-angled			
Acute-angled			
Obtuse-angled			

6 Work in pairs. One person gives a detailed description of a shape by describing its properties. The other must draw or use paper to fold the given shape. For example:

- a regular polygon with three sides
- a trapezium with two of its sides 4 cm in length
- an isosceles right-angled triangle.

Think of your own examples!

Real-world maths : Shapes around you

Search for triangles or quadrilaterals in your school environment. Take a photograph of them or sketch them. Then label them with the correct properties and classifications. Use the correct markings to show equal lengths.

What I have learned

Read the sentences. Copy them and fill in the missing words to summarise what you learned in this topic.

1 Explain the properties of each type of triangle as fully as you can.

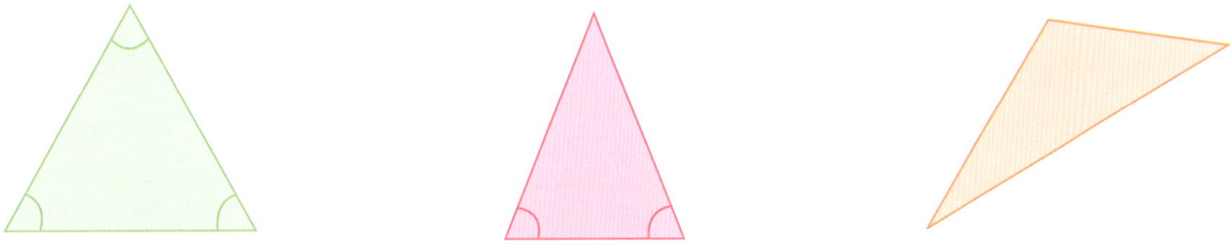

equilateral   isosceles   scalene

2 Use these words to describe each quadrilateral:

parallel angle equal opposite adjacent

Quadrilateral	Description
Parallelogram	
Trapezium	
Kite	
Rhombus	

Practice questions

1 Two squares overlap to form a triangle. What properties could a triangle have when two squares overlap?

2 What different quadrilaterals can you form on a 16-pin geoboard? Sketch and label the properties fully.

3 This is a tangram. The seven shapes have been joined to form a large square. What other quadrilaterals could you make by joining two of the shapes together? Cut out your own tangram pieces and rearrange them to make other shapes.

13 Self check

	Check you can do this:	If you have difficulty, try this:
1	Name and explore the properties of quadrilaterals.	Look back at page 198, and remind yourself of the different types of quadrilaterals and their properties.
2	Draw and make quadrilaterals.	Look back at the shapes you made from folding.
3	Classify triangles based on their properties.	Go back to page 196 and remind yourself of the different types of triangles.
4	Make and draw different types of triangles.	Use a ruler and protractor to draw the different types of angles, and join them to make triangles of different types.

Chapter 14 Properties of polygons

In this chapter you will explore this focus question:

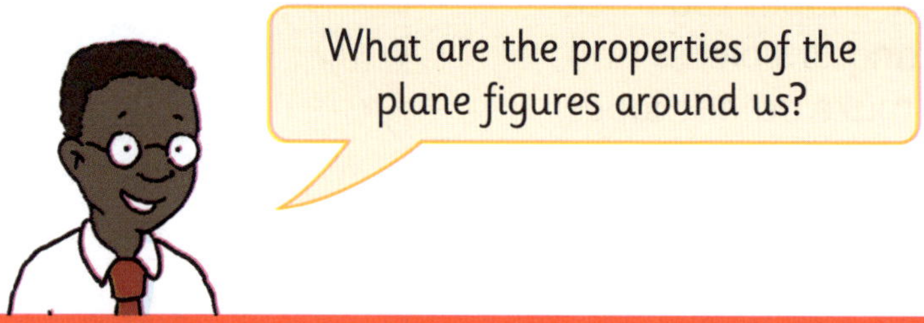

To answer this question, you will:

- identify and classify common shapes and objects
- identify and count the number of lines of symmetry in plane figures
- create shapes given the line of symmetry.

Starting point

Look at these tiles. Describe the shapes and patterns you can see in each tile. What is similar and what is different about the different tiles?

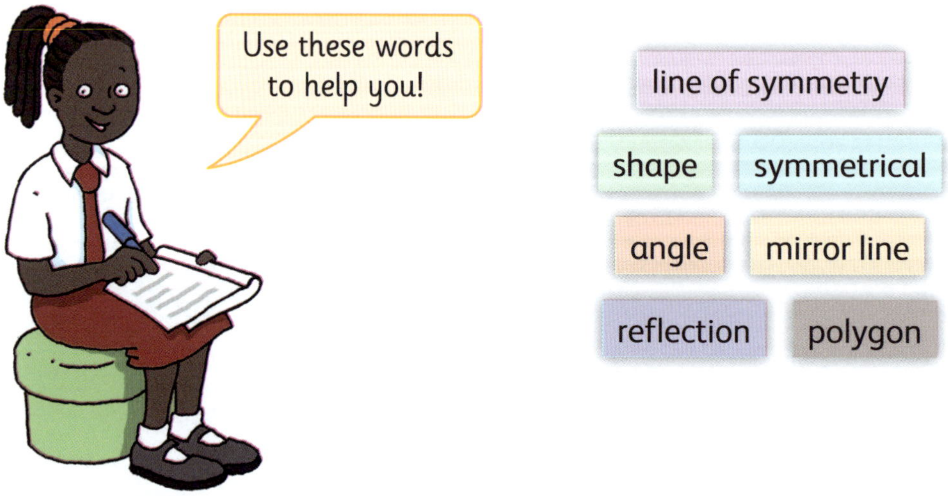

Regular and irregular polygons

Key maths idea

Polygons are closed plane figures with straight sides. A **regular** polygon has all its sides equal in length and inside angles equal in size. Here are some common regular polygons.

pentagon 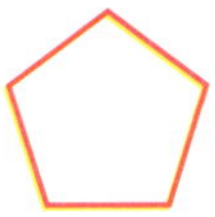 5 sides 5 angles	**hexagon** 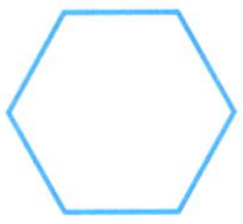 6 sides 6 angles	**heptagon** 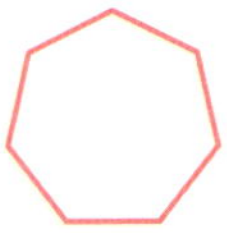 7 sides 7 angles
octagon 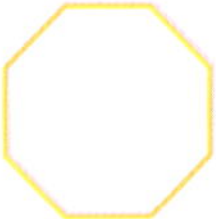 8 sides 8 angles	**nonagon**  9 sides 9 angles	**decagon**  10 sides 10 angles

This hexagon is **irregular**.

All angles are equal in size, but the sides are not equal in length.

Key words

polygon
regular
irregular

Remember: a regular triangle is called equilateral.

1 Measure the sides and angles of these triangles. Which triangles are regular?

a

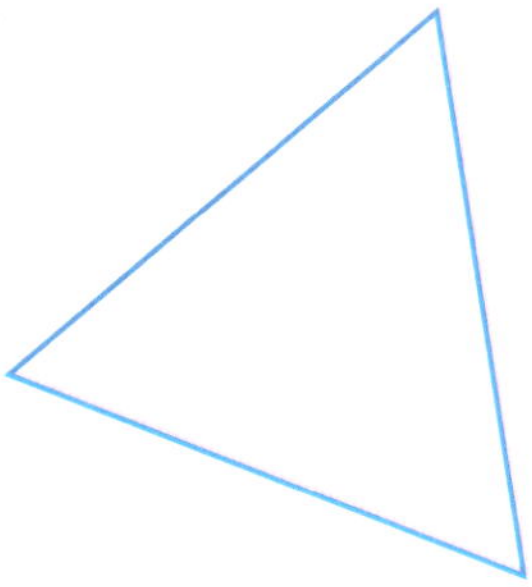

b

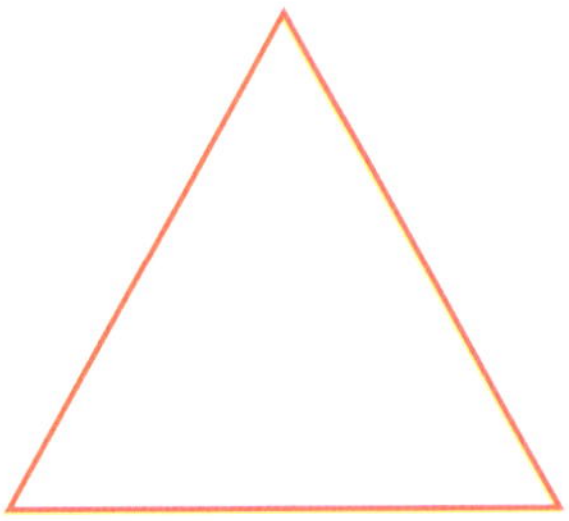

c

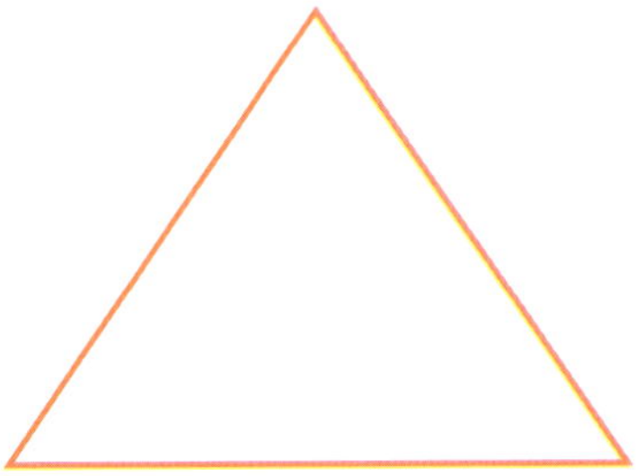

d

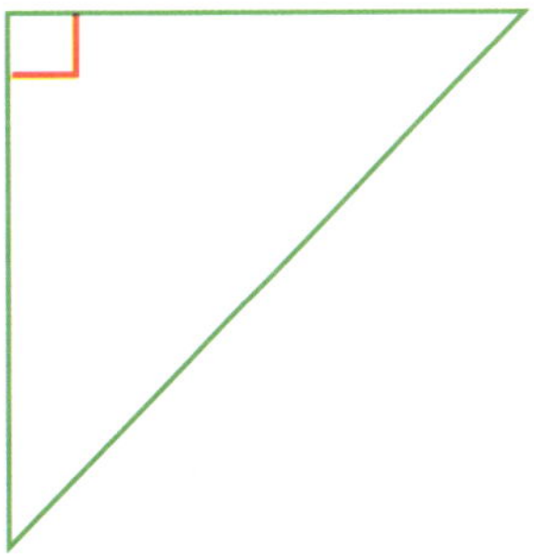

2 Look at these shapes.

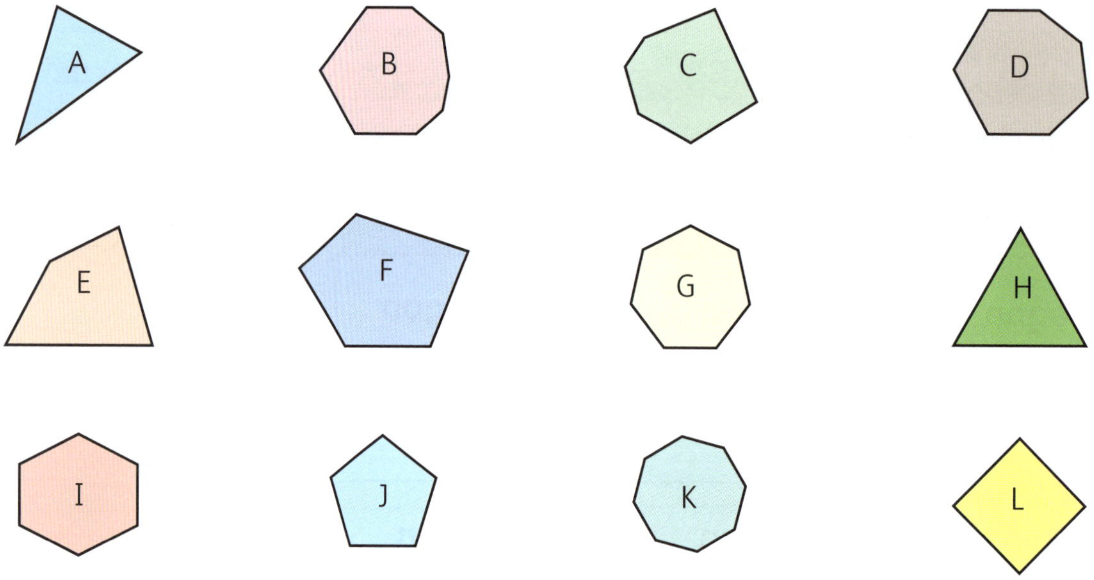

- **a** Write the letters of all the regular shapes.
- **b** Write the letters of all the shapes that have 4 sides.
- **c** Name the other shapes according to how many sides they have.

3 Name each irregular polygon according to its number of sides.

a

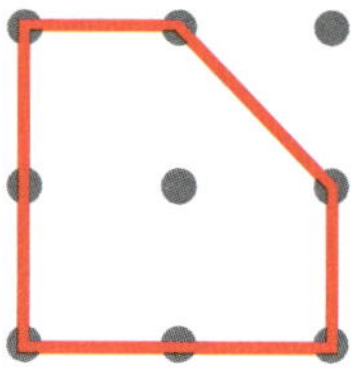

b

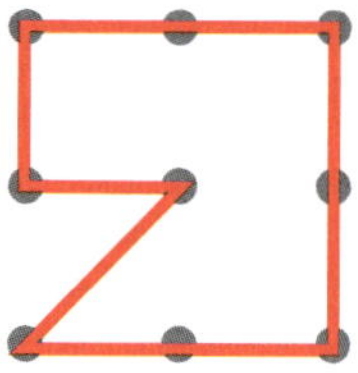

c

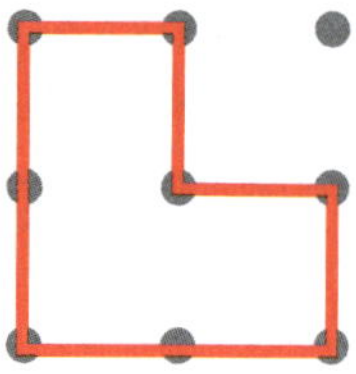

d

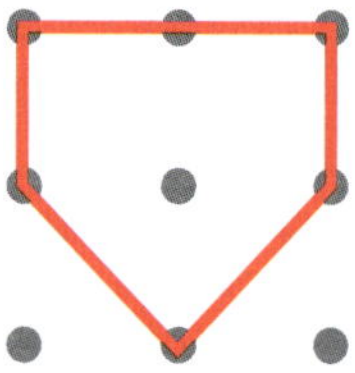

e

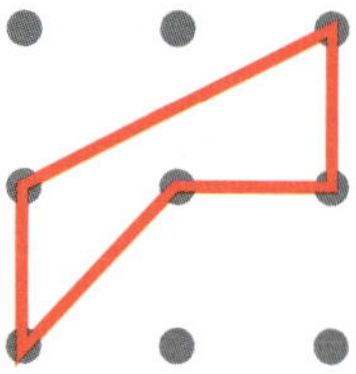

f

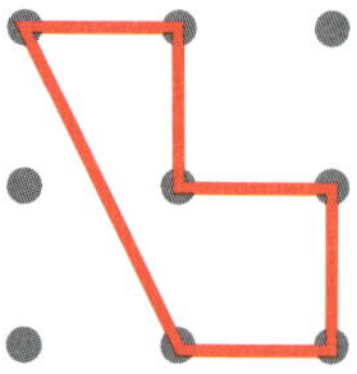

4 Draw at least four 16-pin boards in your notebook, or use an online interactive geoboard. Then make your own polygons. Which regular polygons can you form?

Symmetrical figures

Key maths idea

Some objects have a **line of symmetry** when we view them from a certain position.

Plane shapes can be **symmetrical**, like this picture of the front view of a car.

The broken line is called the **mirror line**. If you place a mirror on the line, the two halves will match. It is also called the **axis of symmetry**.

The side view of the car is not symmetrical. If you place a mirror on the mirror line, the two halves will not match.

Plane shapes can have vertical, horizontal or diagonal mirror lines.

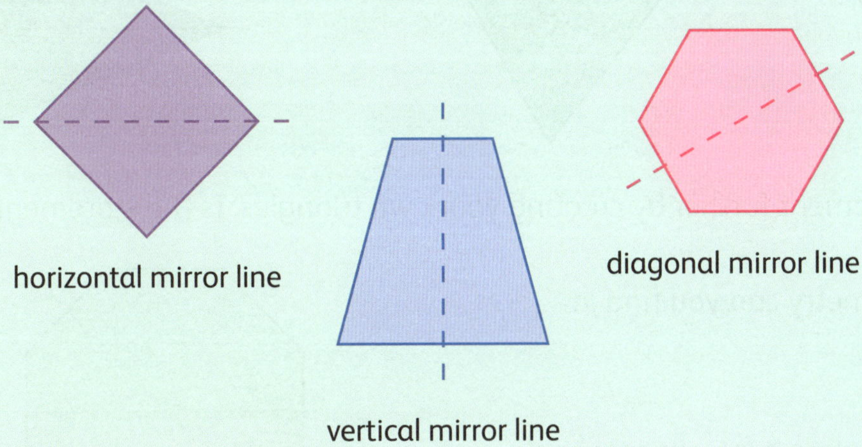

You can check for **symmetry** by cutting out a shape and folding it in different ways.

Key words

line of symmetry
symmetrical
mirror line
axis of symmetry
symmetry

Geometry

1 Use a mirror or copy and fold the shapes to check if the dotted lines are mirror lines.

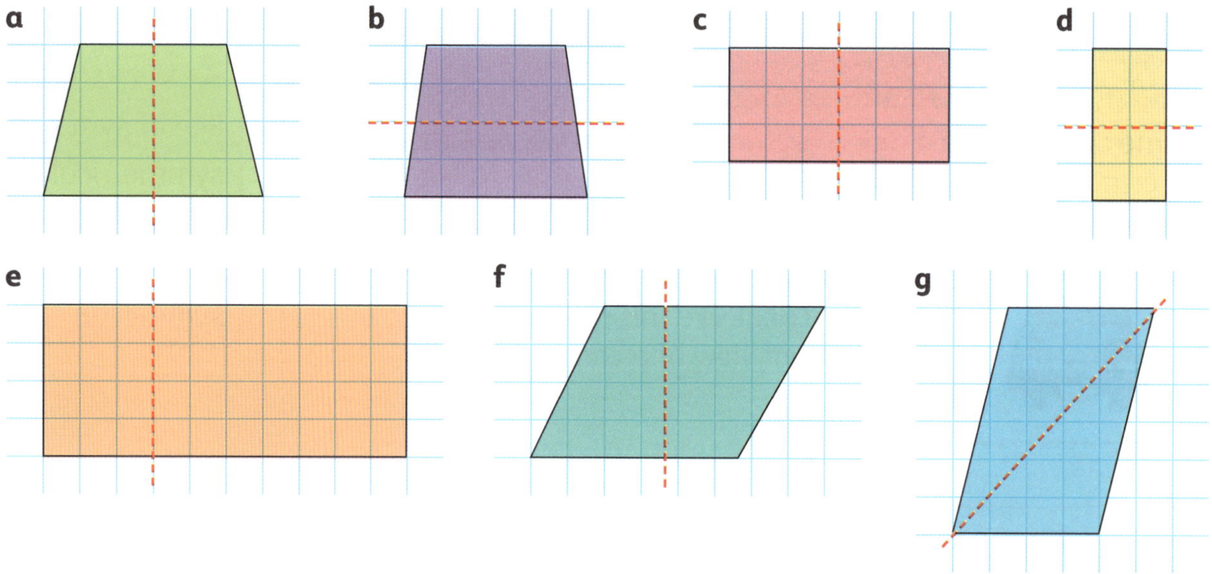

2 Look at the diagrams.

a Check this statement: 'All triangles have one line of symmetry.' Is it true or false for triangles A to E?

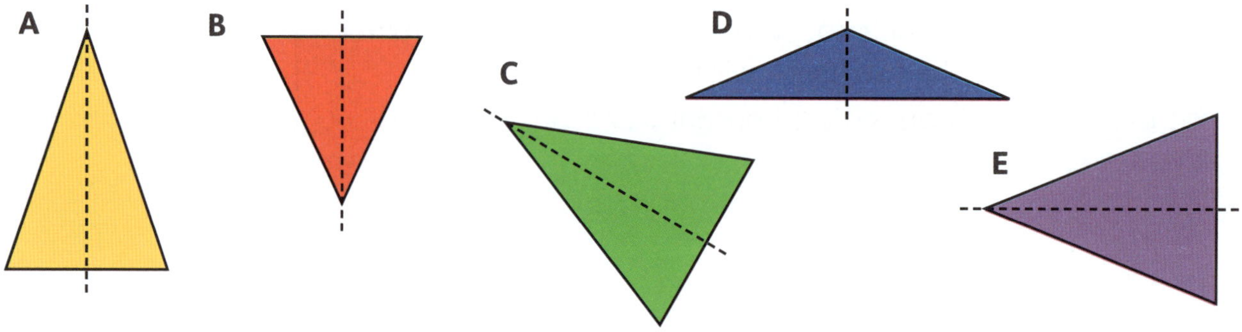

b Investigate the statement further by creating your own triangles. Is the statement always true?

3 How many lines of symmetry can you find in this regular polygon?

Two are marked.

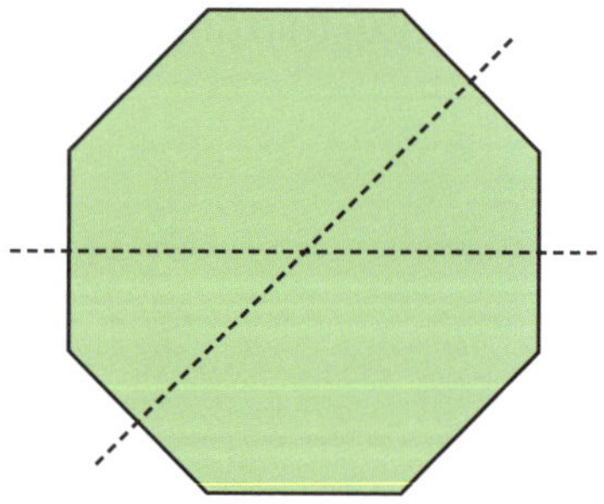

4 Copy the table and complete it in your notebook. Continue the table with rows for shapes with 7, 8, 9 and 10 sides.

Regular shape	Number of sides	Number of angles	Lines of symmetry
Triangle	3	3	3
Quadrilateral			
Pentagon			
Hexagon			
…			

5 Copy and complete these symmetrical figures. Predict the name of the completed shape. Guess how many sides or angles the completed figure will have.

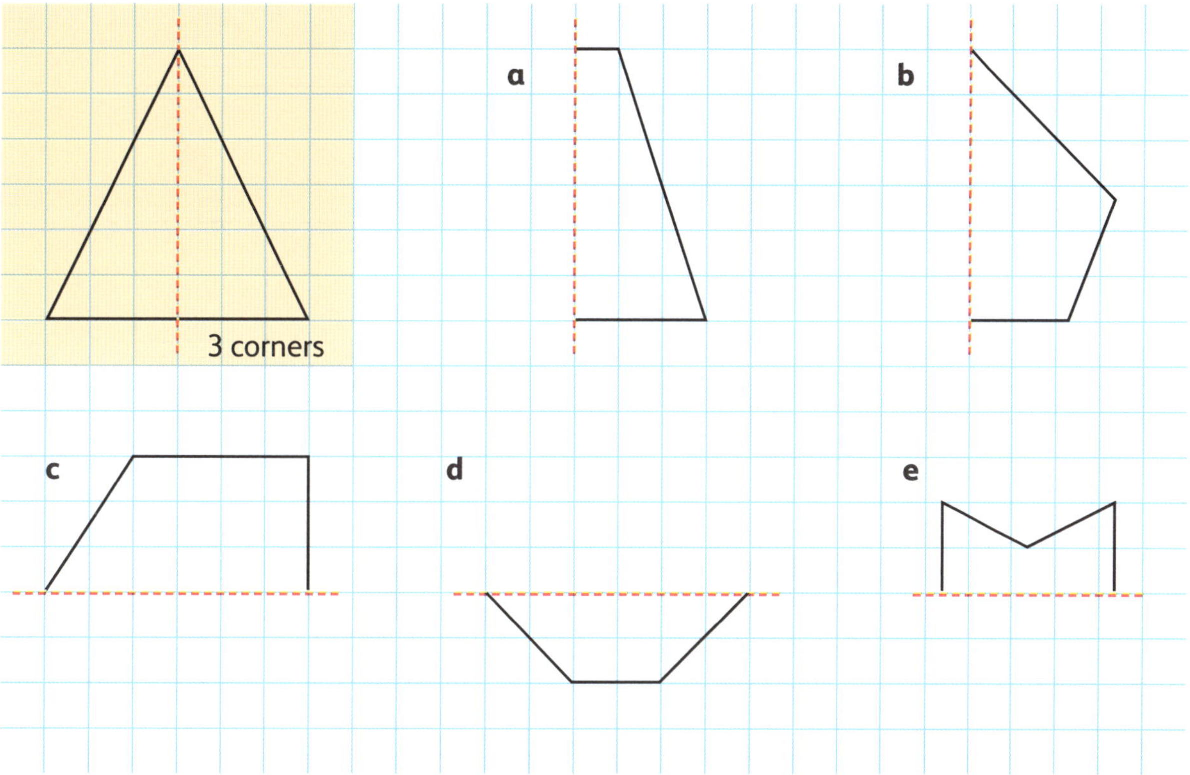

Maths detective: Lines and planes of symmetry

Use the clues to solve these cases of symmetry.

Case 1

The capital letter X has two lines of symmetry.

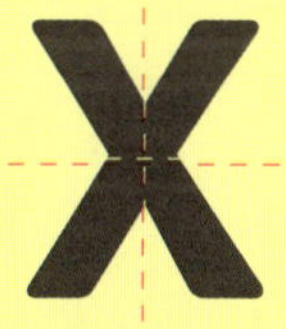

1 Investigate the capital letters below.
Do they have 0, 1, 2 or more lines of symmetry?

A **B** **C** **D** **E**

F **G** **H** **I** **J**

2 Write a short word that has one line of symmetry. Use any suitable letters of the alphabet.

Case 2

Copy and complete these designs that have two lines of symmetry.

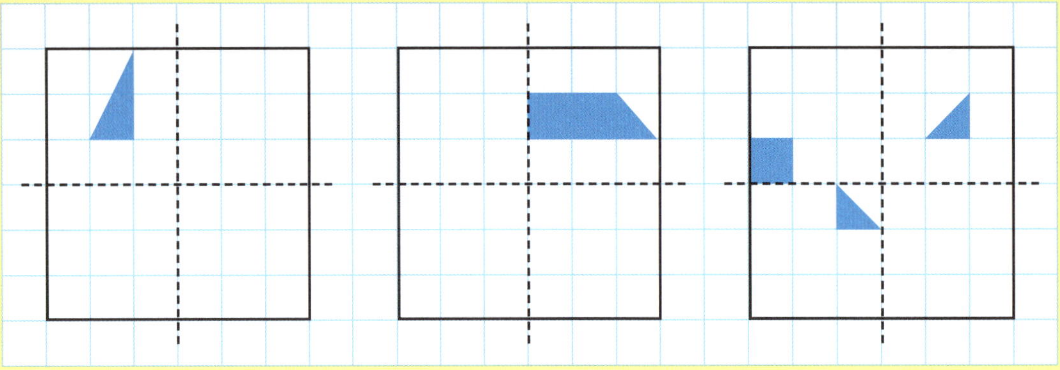

Case 3

1 You have learned about the lines of symmetry in flat shapes. In 3-D solids, we talk about **planes of symmetry**. A plane is an imagined flat surface. Imagine a slice across a 3-D shape. The slice is a flat surface that cuts the shape into identical parts. A solid has **plane symmetry** if you can slice it into identical parts along a plane.

Use these diagrams to help you answer the questions below.

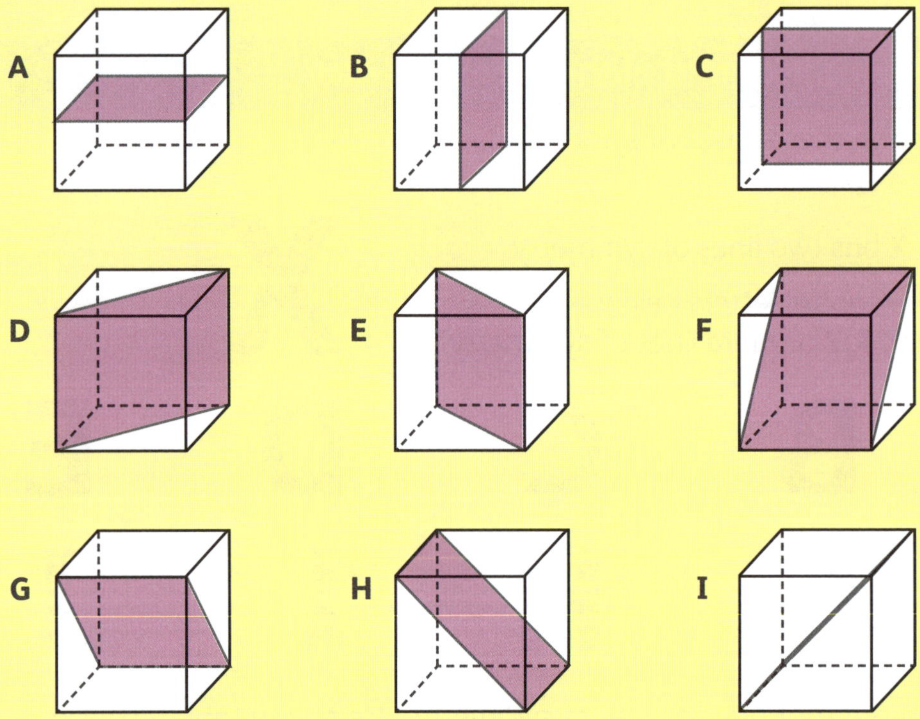

a Which diagram shows the horizontal plane of symmetry of a cube?

b How many vertical planes of symmetry does a cube have?

c How many planes of symmetry pass through each pair of opposite edges?

d Are there more than 9 planes of symmetry for a cube? How do you know?

2 Research the following solids on the internet:

a tetrahedron
b octahedron
c icosahedron
d dodecahedron.

- Discuss with a partner: Do these solids have the same number of planes of symmetry as the cube? How can you work it out?
- Create a chart to show the planes of symmetry of each shape.

Key words

plane of symmetry

plane symmetry

Real-world maths: Symmetry around you

In this photograph, the reflection in the water forms a horizontal mirror line.

1 a Collect at least three examples of symmetry from your school environment or find three photographs that show symmetry.

b Say if the mirror line is horizontal, vertical or diagonal.

2 Butterflies are regarded as really beautiful insects. Look at this picture.

a Use what you have learned about shape, pattern and line symmetry to describe what makes butterflies beautiful.

b Design your own images of butterflies. You may choose to draw or create prints using ink blots and folded patterns.

What I have learned

1 Name and describe the properties of each shape.

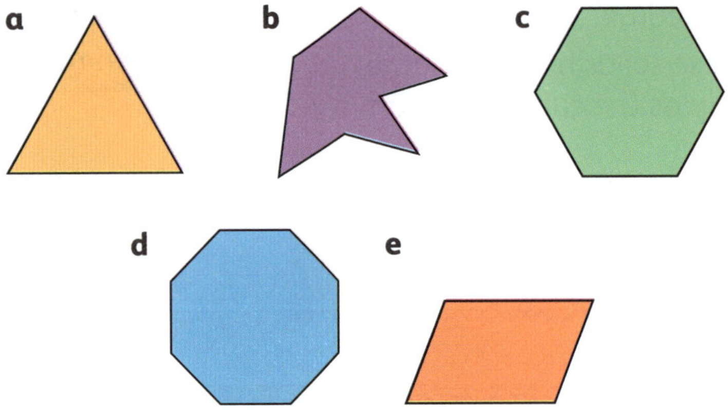

Think: How many angles? How many sides? Is it regular?

2 Copy this mind map about plane shapes. Sketch an example of each concept in the mind map.

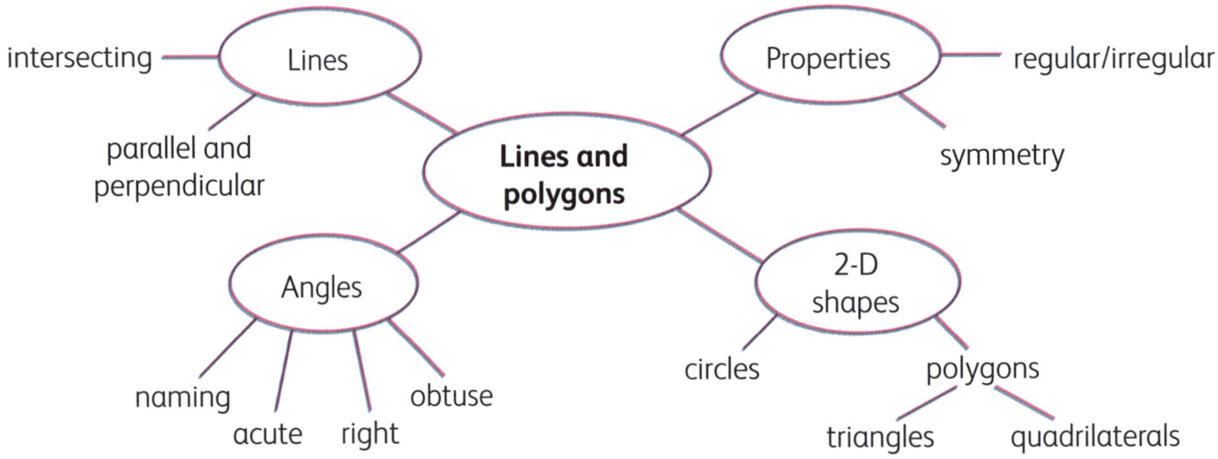

Practice questions

1 Are these shapes polygons? Write down a reason for your answer.

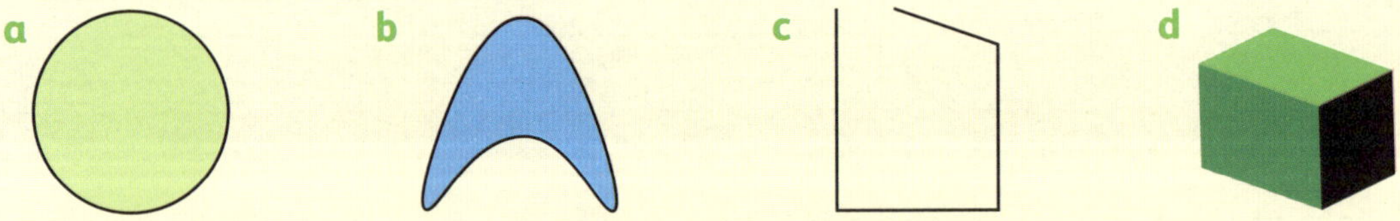

2 There are some irregular polygons hidden in this diagram.

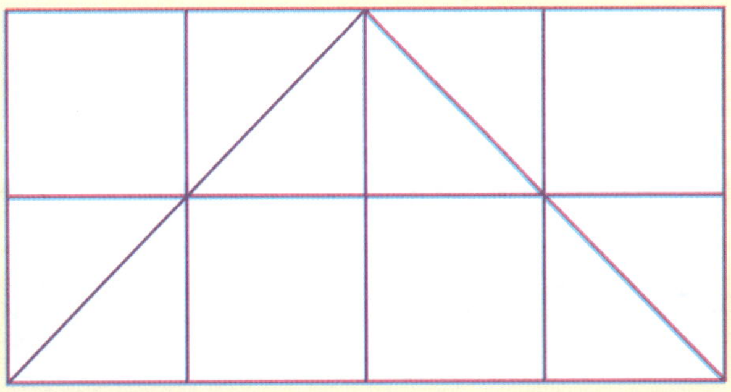

Find and sketch the polygons you can find.

Then write the name of each one.

Mark all the mirror lines on your polygons.

3 Patterns like these two can have mirror symmetry.

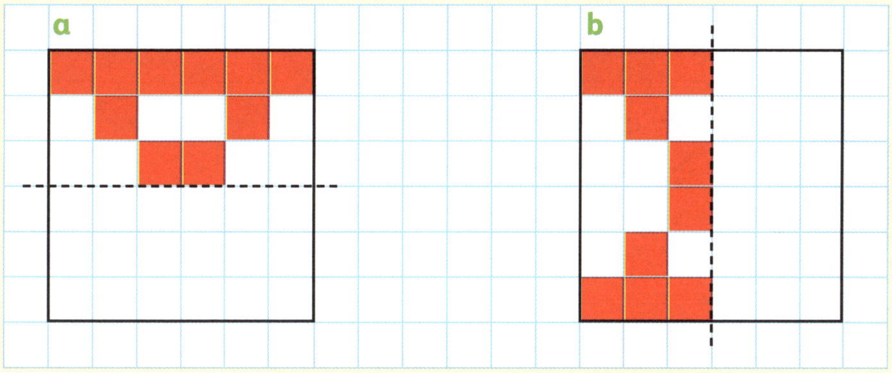

Copy and complete these two symmetrical designs.

14 Self check

	Check you can do this:	If you have difficulty, try this:
1	Recognise and name polygons.	Go back to page 205 and remind yourself of the names of the different types of polygons.
2	Define and identify regular and irregular polygons.	Remember that regular polygons have all their sides equal in length, and inside angles equal in size. Use this information to identify regular and irregular polygons.
3	Recognise lines of symmetry.	Revisit page 207 and remind yourself how to identify a line of symmetry. Use paper folding and mirror reflections to help you.
4	Complete symmetrical shapes or patterns.	Practise folding a piece of paper in half, and then cutting out a shape along the fold line. Place the folded line onto another piece of paper and try to draw what the shape will look like when you unfold it. Unfold the paper to check how well you did.

Chapter 15 Transformations

In this chapter you will explore this focus question:

To answer this question, you will:

- associate the 4 major cardinal points with quarter, half, three-quarter and full turns
- use grids to describe positions and movements
- describe and perform transformations on 2-D shapes (flips, turns and slides)
- identify designs and shapes drawn in different orientations.

Starting point

The arrow is pointing South.

If it rotates a half turn, where will it point?

Describe all the turns the arrow needs to make to point in the different directions.

Describe each turn as a half turn, quarter turn or three-quarter turn.

Translations

Key maths idea

In geometry, we describe the position of a point or shape by using points on a grid. We can move or change the position of the shape on the grid. These changes are called **transformations**.

The first type of transformation you will learn about is a **translation**. It is also called a **slide**, because when you translate a point or a shape, you slide it along a line to a new position.

Example 1

Point X has been translated 5 units right and 2 units up. We describe a translation with the **horizontal** movement first, and then the **vertical** movement.

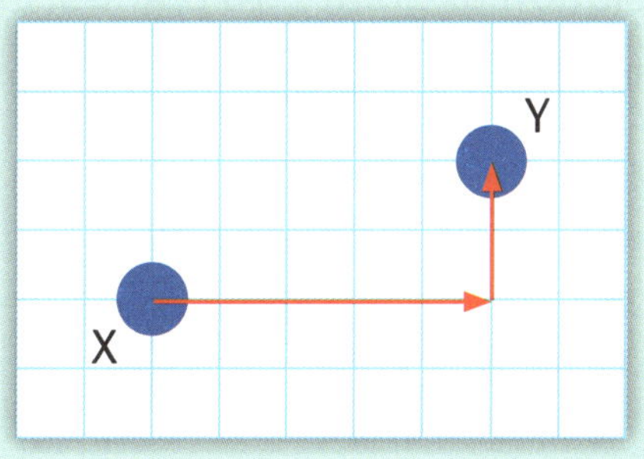

Example 2

Triangle A has been translated. Each vertex of triangle A has moved 4 units left, and 1 unit down.

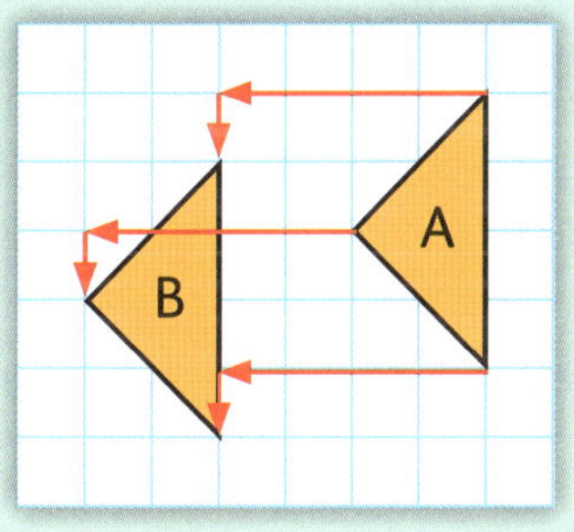

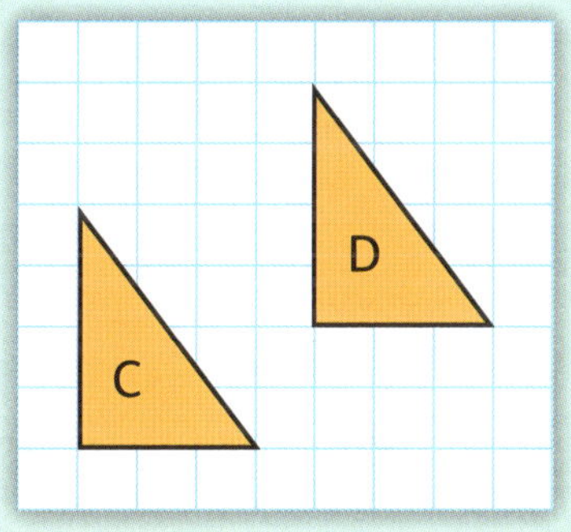

When a shape is translated, the shape remains the same size and in the same **orientation**. The shapes are **congruent** after a translation.

Shape C is a translation of shape D.

But shape C is not a translation of shape A.

Two shapes that are congruent are also **similar**. They have the same shape.

Mirror images are similar. Shapes that have been translated or rotated are similar too.

> ! Enlarged or reduced shapes are similar too. That means all circles are similar and all squares are similar, no matter what their size.

Key words

transformation
translation
slide
horizontal
vertical
orientation
congruent
similar

1 Point P has been translated to new positions.

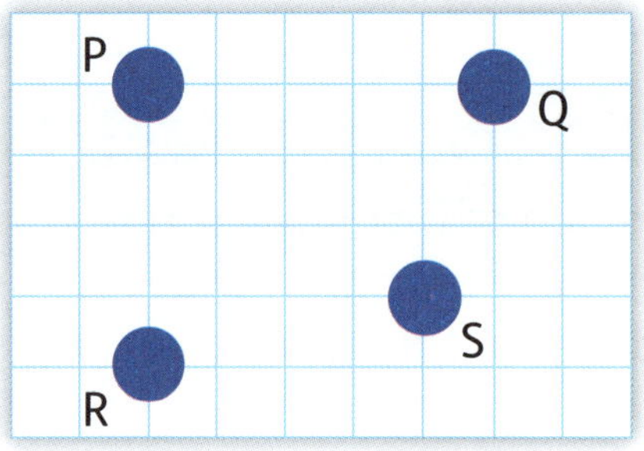

a Describe the translation of point P to point Q.

b Describe the translation of point P to point R.

c Describe the translation of point P to point S.

2 Describe the translation from the shape in its original position to each new position.

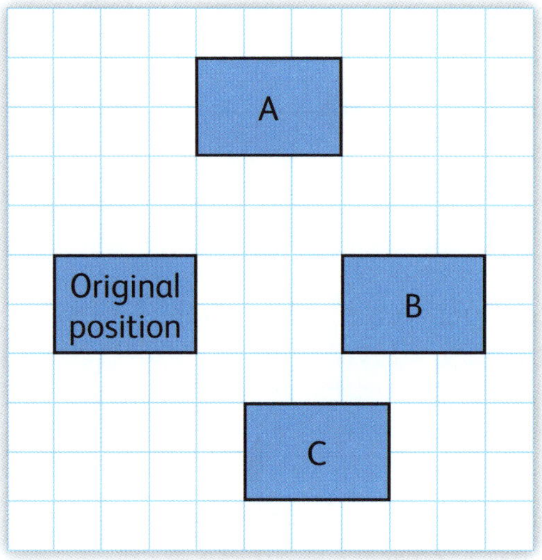

3 a Describe the translation of shape A to shape B.

b Now describe the translation from shape B to shape A.

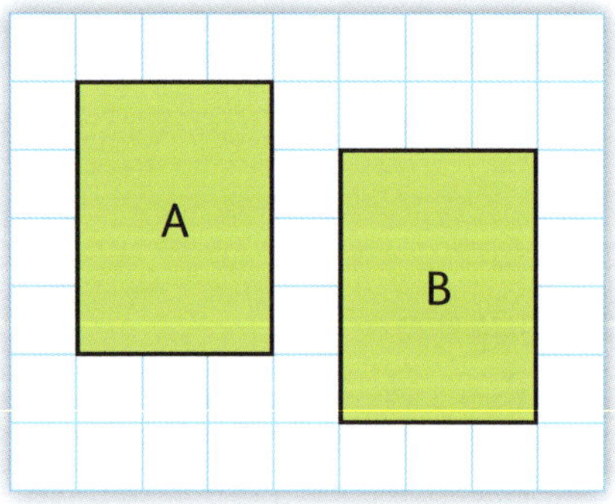

4 a Draw your own shape on a square grid. Translate it to a new position. Write the translation on your diagram.

b Translate the shape to another new position. Again, write the translation on your diagrams.

c Do this for another new position.

Rotation and reflection

Key maths idea

When a figure is rotated, the size and proportion of the shape do not change. Only its position and orientation change.

It's as if the shape turns or spins.

You can use tracing paper to see how a shape looks when it rotates.

This type of transformation is called a **rotation**.

A shape can also be reflected to a new position.

Triangle A stays the same size, but it is flipped from left to right.

This type of transformation is called a **flip** in a vertical line of reflection.

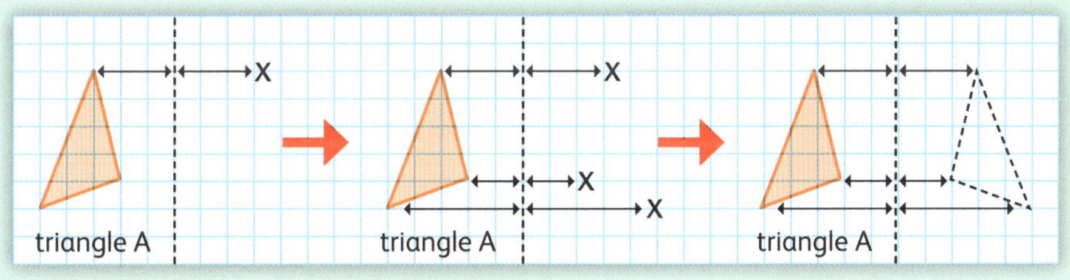

A shape can also be flipped from right to left. Triangle B stays the same size, but it is flipped from right to left.

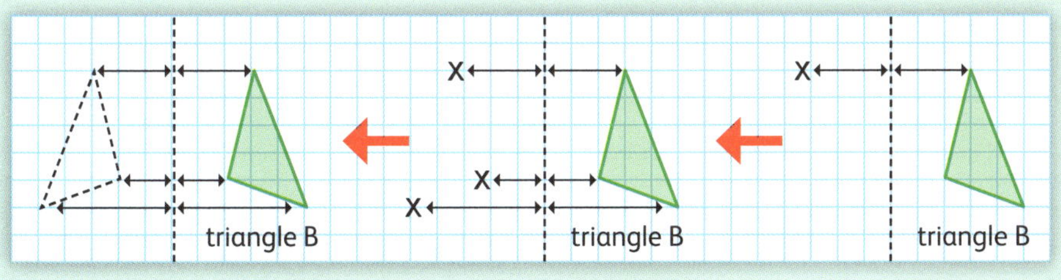

This shape is flipped in a horizontal line of **reflection**. Triangle C stays the same size, but is flipped downwards.

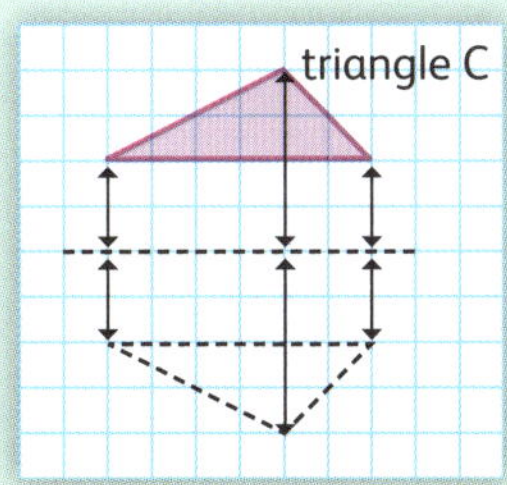

A rotated or reflected shape is congruent to the original.

Key words

rotation

flip

reflection

1 Copy each shape on grid paper and show it in its new rotated positions. Use tracing paper to help you.

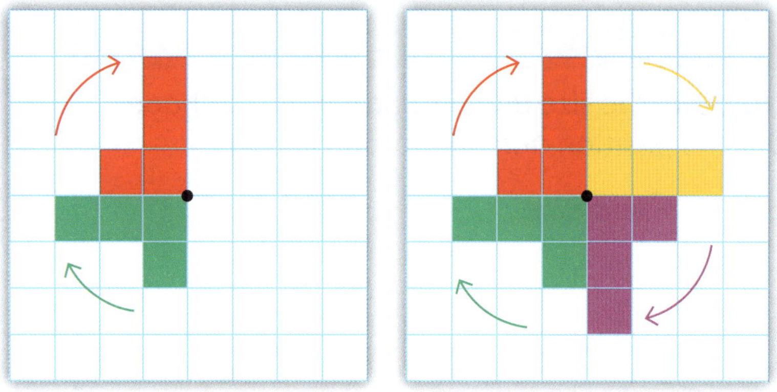

2 Copy each shape on grid paper. Then rotate each shape to new positions around the dot.

a

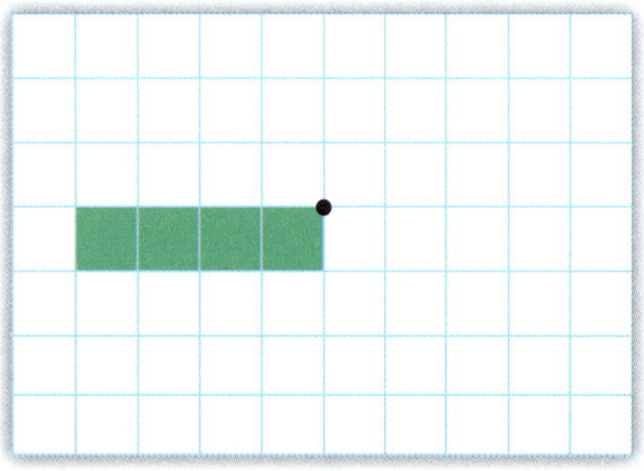

b

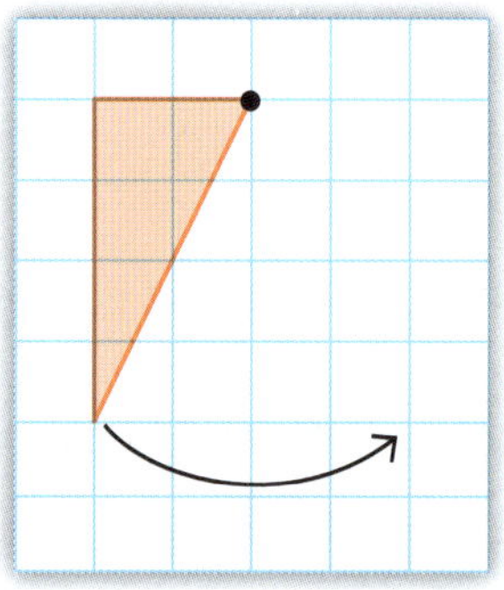

3 Copy the shapes. Then reflect each shape in the mirror line to complete it.

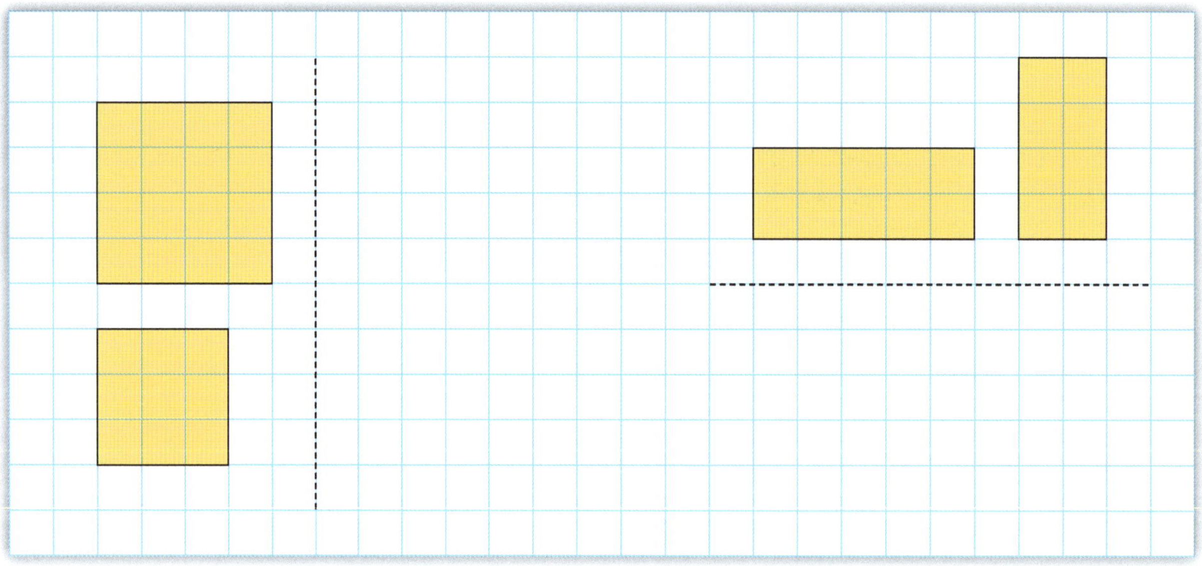

Maths detective: Repeated reflections

Use the clues to discover what happens when you reflect a shape again and again.

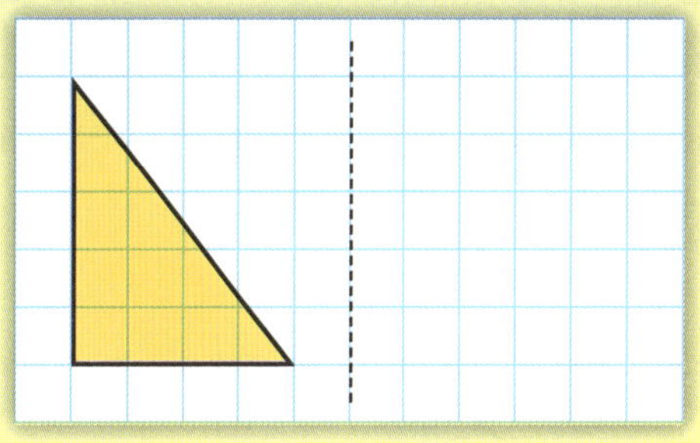

Case 1

Reflect the triangle in the mirror line. What stays the same and what changes?

Case 2

Reflect the triangle in the first mirror line. Then reflect the new triangle in the next mirror line. Keep going. What do you notice?

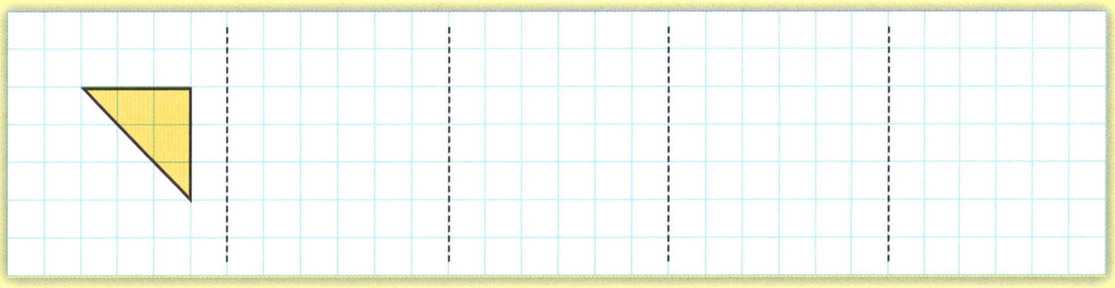

Case 3

Invent your own repeating reflection pattern.

Real-world maths: Symmetry in art and design

Symmetry is a very important concept in art. Look at this sketch of a face. Artists use symmetry to help them sketch accurately and to measure proportions.

Symmetry and rotation are also used in logos and advertising. Graphic designs that use principles of symmetry and rotation can produce fascinating images.

Challenge: Design a logo for your class. Use symmetry and rotation to make your design appealing.

What I have learned

Explain the difference between a translation, a reflection and a rotation.
What stays the same and what changes in each different type of movement?
Use the diagrams to help you explain.

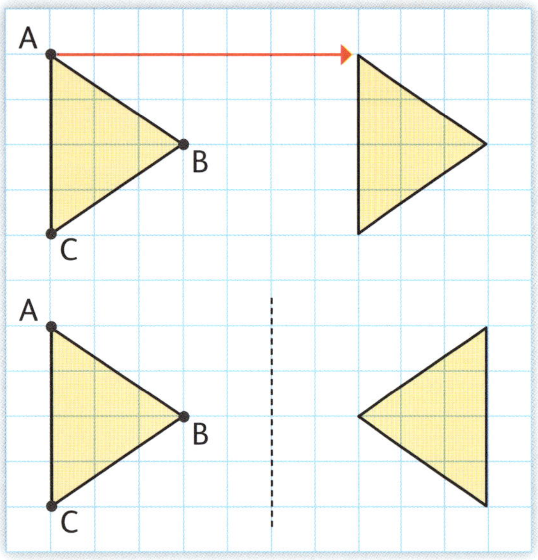

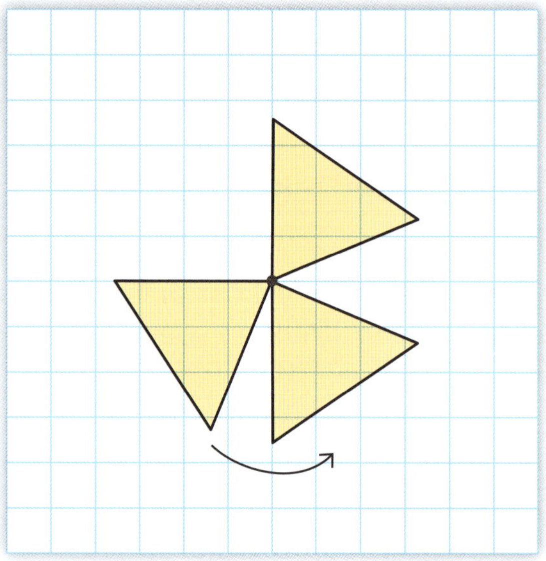

Practice questions

1 a Describe the translation of shape X to shape Y.
b Then describe the translation of shape Y to shape X.
c Now translate shape Y to shape Z.
d Predict the translation of shape Z to shape Y.

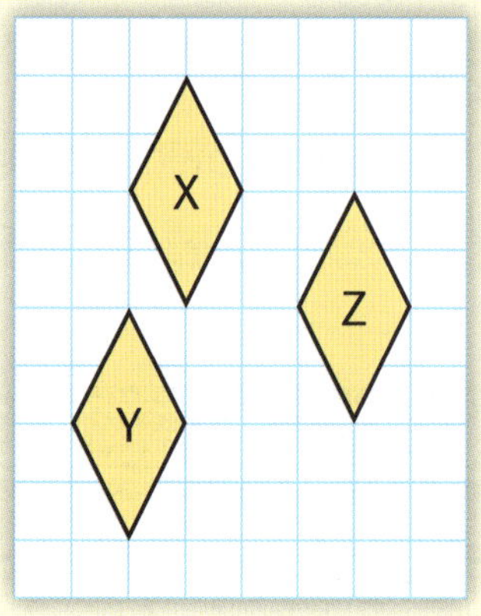

2 a Which shapes are congruent to the shape in the original position?
b Which shape could be a translation of the original?
c Which shape could be a reflection of the original?

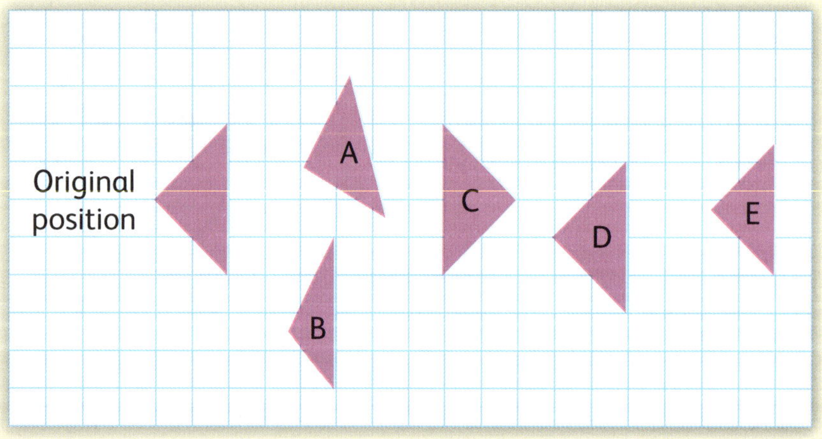

15 Self check

	Check you can do this:	If you have difficulty, try this:
1	Use compass directions to describe positions and turns.	Go back to page 214. Remind yourself of the four cardinal points and how to describe turns using direction (right or left, or clockwise or anticlockwise), and angles (a quarter turn, half turn, three-quarter turn and full turn).
2	Describe and complete a translation.	Look back at page 215 and remind yourself how we describe sliding a shape across a plane.
3	Reflect a shape in a line of reflection.	Go back to page 217 and remind yourself of the work on rotating a shape about a given point (turns).
4	Rotate a shape to a new position.	Go back to page 217 and remind yourself of the work on reflecting a shape in a mirror line (flips)
5	Identify similar shapes in different positions or orientations.	Remind yourself of the definition of similar shapes. Remember, similar shapes do not have to be the same size. However, they have the same number of sides, and their corresponding inside angles are equal.

Unit 4: Algebra

Chapter 16 Patterns and rules

In this chapter you will explore this focus question:

To answer this question, you will:

- investigate patterns and create tables
- describe the relationships between variables in a pattern rule
- use tables of values to make predictions and statements about patterns.

Starting point

Glenmore knows a maths trick. He gives his classmates these instructions:

Think of any number. → Double it. → Add 20. → Halve your total. → Now take away the starting number. → I can guess the number that's left!

- Why do you think Glenmore can guess the number you chose?
- Try choosing some different starting numbers.
- Collect your results.
- What number do you finish on each time?
- Try to explain what you noticed.

Explore and predict patterns

Key maths idea

This shape **pattern** is made from sticks. Each shape is larger than the previous one.

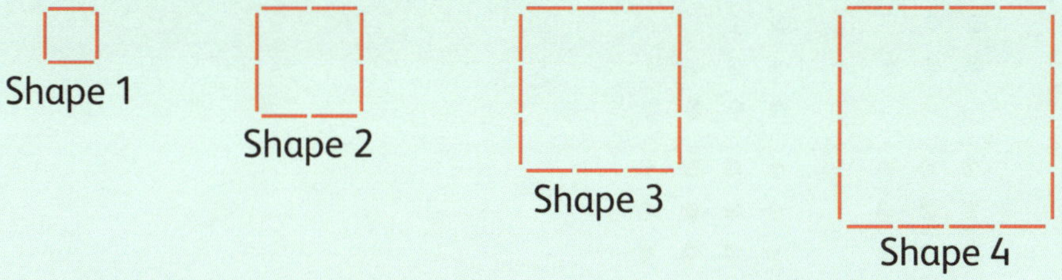

You can draw up a table of values to record the number of sticks in each shape.

The shape number and the number of sticks in each shape are called **variables**.

Variables are values that can change.

The shape number starts at 1. As the shape number increases, the number of sticks increases.

Shape number	1	2	3	4	5	6		
Number of sticks	4	8	12	16	20			

We can now look at the number pattern for the number of sticks and develop a **general rule** about it. That means we predict the next value for the number of sticks.

We can use this information to complete the table and make **predictions** about further shape numbers.

Shape number	1	2	3	4	5	6	10	100
Number of sticks	4	8	12	16	20	24	40	400

Key words

pattern
variable
general rule
prediction

1 Draw the next three shapes in each pattern.

a 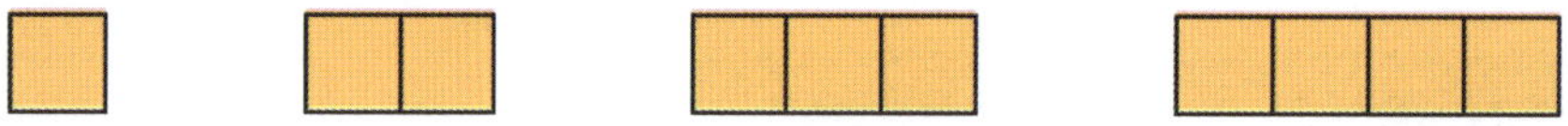

b 

c 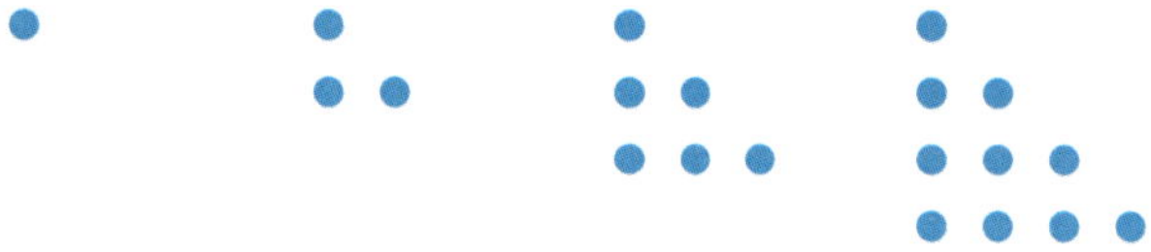

d 

2 Describe how each pattern in Exercise 1 grows. Express what you notice as general rules.

3 Complete a table of values for each pattern.

a 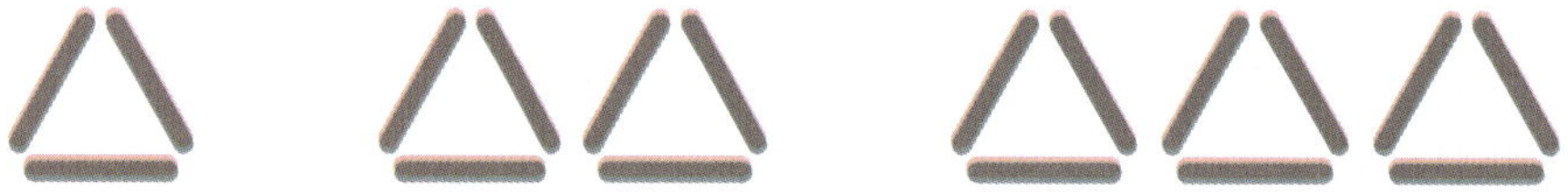

Shape number	1	2	3	4	5	20	99	100
Number of sticks	3	6	...					

b 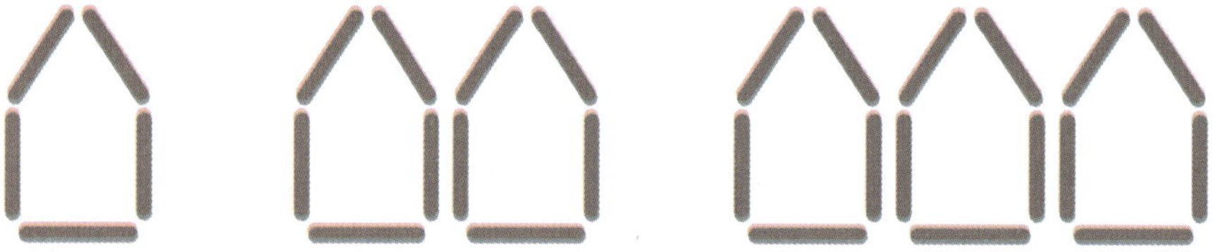

Shape number	1	2	3	4	5	20	99	100
Number of sticks								

4 Sometimes a pattern involves more than one operation. Generate your own number pattern using these rules:

a Start with 1. To get the next term, add 1, and then double. Write the first eight terms in the pattern.

b Start with 1. To get to the next term, halve, and then add 3. Write the first eight terms in the pattern.

c Start with 12. To get to the next term, multiply by 10, then subtract 90. Write the first eight terms in the pattern.

5 a Invent your own shape pattern. Use sticks or drawings to make the patterns and challenge your partner to continue your pattern.

b Create a table of values for your pattern.

c What is the general rule for your pattern?

Using symbols

Key maths idea

You often use the = sign in mathematical statements. You can use this to help you work out a value. For example:

$9 + \square = 20$

You can see that the $\square$ represents the difference between 20 and 9.

$20 - 9 = 11$, so $\square = 11$.

Sometimes we use other signs.

$>$ **greater than**

$<$ **less than**

$\neq$ **not equal to**

Example

n is a whole number that is divisible by 3 and $n < 10$

What are the possible values of n?

Numbers divisible by three that are less than 10: 3, 6, 9

So n could be 3, 6 or 9.

Key words

greater than

less than

not equal to

1 A bag contains a number of counters, n. The value n is more than 1 but not greater than 6. I add five more counters to the bag.

- **a** What are the possible values of n?
- **b** What are the possible totals after I add the 5 counters?
- **c** Write three different equations, using + and = to describe possible totals in the bag.

2 A bag contains a number of blue counters and a number of grey counters. There are more blue counters than grey counters, and the total number of counters is 30. Let the number of blue counters = b and the number of grey counters = g.

True or false? Give reasons for your answers.

- **a** $b - g = 30$
- **b** $b \neq g$
- **c** $b = 30$
- **d** $g < 30$
- **e** $b + g > 30$
- **f** $30 - g = b$
- **g** $2g + 2b = 60$

3 Jamie had 25 markers in a pencil case, but discovered that his little sister took some out. The remaining number of markers in the bag is greater than 10.

- **a** Write a statement describing how many markers Jamie's sister took. Use the symbols $<$ and $>$ and let the number she took = s.
- **b** Write another statement that contains s, and the symbols + and =.

Order of operations

Key maths idea

In mathematics, you may sometimes need to solve number sentences that have multiple operations or steps. Look at these examples:

a Think of a number. Double it, then add 5.

b Think of a number. Add 5, then double it.

- Why do these instructions give different answers?
- How can we write them as algebraic expressions?

Remember: an algebraic expression uses a letter to represent unknown numbers or variables.

In mathematics, we always solve anything in brackets first.

Next, we solve multiplication and division steps, as well as 'of' (for example, '$\frac{1}{2}$ of 5'). This is because 'of' is a form of multiplication.

Last, we solve the addition and subtraction steps. You will learn more about these rules in the next chapter.

1 Use the rules above to help you solve these number sentences.

a $\frac{1}{4}$ of 20 + $\frac{1}{2}$ of 20

b $4 \times 3 + 10 \times 3$

c $(2 + 17) \times \frac{1}{10}$ of 100

d $(45 - 10) \div 5$

e $\frac{1}{2}$ of 180 + (120 − 30)

2 Copy these number sentences and insert brackets to make each statement true.

a $\frac{1}{2}$ of 10 + 20 = 15

b $60 + 60 \div 2 + 1 = 40$

c $300 + \frac{1}{4}$ of 800 + 200 = 550

3 Write your own number sentences. Each number sentence should use an example of addition or subtraction in brackets, and an example of multiplication or division, to make the given answer.

a 200

b 27

c 1600

d 0.5

4 Explain each mistake.

a $\frac{1}{3}$ of 45 + 18 = 21

b $25 + 100 \div 25 + 1 = 6$

c $1800 \div 100 + 100 = 9$

5 The following expressions represent a set of steps to carry out on a number, y. The value of y may be 10, 7 or 12. Work out the possible values of each expression.

a $2y + 5$

b $(y - 2) \times 10$

c $(y + 15) \times 3$

Maths detective: A tower pattern

Use the clues to investigate this pattern.

One level high   Two levels high   Three levels high

Case 1

Predict the number of blocks to make a tower that is four levels high.

Make or draw the tower to test your prediction.

Case 2

Complete a table of values up to six levels.

Case 3

Predict the number of blocks needed to make a tower with 25 levels. Discuss your reasoning with a partner. How will you convince yourselves and each other that your reasoning works?

Case 4

Try to predict how tall the tower would be if you had 100 rectangular building blocks. Investigate your predicted answer. Justify your reasoning with calculations.

Case 5

Now push your investigation further. Copy and complete a table like this one.

Number of blocks	Predicted number of levels	Calculated number of levels	Number of blocks not used
16	3	3	1
20			
100			
150			
1 000			

Real-world maths: Tiling with hexagons

There are many patterns in science and engineering.

Tiling patterns are common in architecture and design.

Look at this hexagon tiling pattern.

It begins in the centre and grows outwards.

Investigate the number of tiles you need to increase the size of the tiling pattern.

What I have learned

1 Look at this pattern.

Describe, predict and explain how this pattern grows.

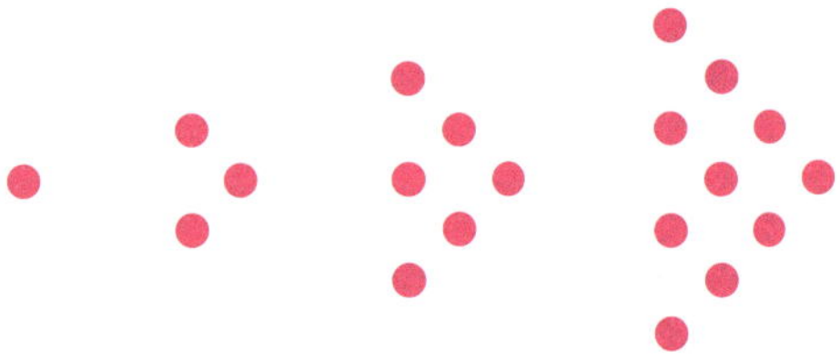

What do you notice? What skills have you learned that can help you?

2 Write your own steps to show how you know which operations to solve first in a number sentence with multiple operations. Write your own example as an illustration.

3 Write your own number sentences to illustrate the use of these symbols:

$>$, $<$, $=$ and $\neq$.

Practice questions

1 Make or draw the next shape in this pattern.

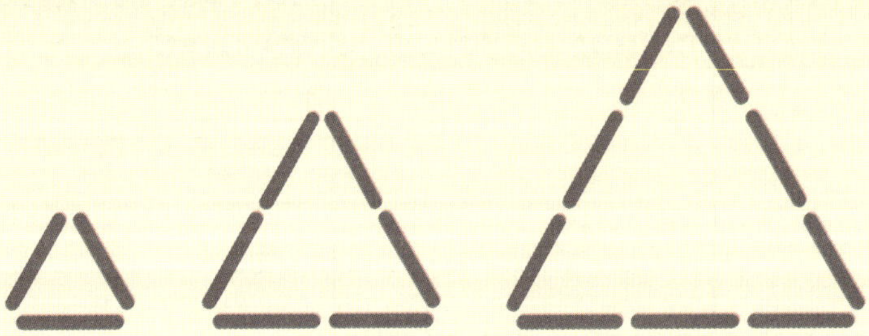

2 Complete the table of values for this shape pattern.

Shape number	1	2	3	4	5	6	7	8
Number of sticks								

3 Predict the number of sticks you would need to make shape 10. Now draw or make the shape to check your prediction.

16 Self check

Check you can do this:	**If you have difficulty, try this:**
1 Predict the next part of a pattern.	Go back to page 223 and look at the shape patterns. How did you decide what came next?
2 Complete a table of values to explore a pattern.	Go back to page 223 and look at the tables you created, where you filled in the shape number and the number of sticks (or shapes) in each new term.
3 Describe rules to explain how a pattern changes.	A rule usually tells you what to do. The rule in a pattern is usually an operation, for example, ‘add 3’ or ‘add 1, and then double’. Go back to the work you did on page 223, and think about what number operations you could use to come up with the next number in the table.
4 Use mathematical symbols.	Go back to page 225 and remind yourself of the symbols $<$, $>$, $=$ and $\neq$.

Chapter 17 Expressions and variables

In this chapter you will explore these focus questions:

To answer these questions, you will:

- investigate the order of operations when evaluating expressions
- use the symbols <, >, +, −, = and ≠ in number sentences
- substitute a number for a variable in a simple mathematical sentence
- use substitution in formulae to solve word problems.

Starting point

Discuss these calculations. Are there any mistakes? Why do the answers differ?

Order of operations in expressions

Key maths idea

An **expression** is a **mathematical statement**.

These are examples of numerical expressions:

$3 + 4$

$5 \times 1\frac{1}{2}$

$24 + 8 \div 4$

To **evaluate** an **algebraic expression** means to calculate its value.

$3 + 4$ gives the value 7

$5 \times 1\frac{1}{2}$ gives the value $7\frac{1}{2}$

To evaluate $24 + 8 \div 4$, you must remember the correct **order of operations**.

You can use the word **BODMAS** to help remember the order of operations:

Brackets	First work out the operations in brackets.
Of	Next work out operations with 'of'.
Division/**M**ultiplication	Now work out multiplication and division. It does not matter whether you divide or multiply first.
Addition/**S**ubtraction	Last, work out addition and subtraction. Again, the order does not matter.

Example

$(12 + 10\frac{1}{2}) + \frac{1}{3}$ of $6 - 3 \times 6$ — Work out the brackets first.

$= 22\frac{1}{2} + \frac{1}{3}$ of $6 - 3 \times 6$ — Next work out 'of'.

$= 22\frac{1}{2} + 2 - 3 \times 6$ — Now multiply.

$= 22\frac{1}{2} + 2 - 18$

$= 22\frac{1}{2} + 2 - 18$ — Finally, add and subtract.

$= 6\frac{1}{2}$

Key words

expression

mathematical statement

evaluate

algebraic expression

order of operations

BODMAS

Algebra

1 Evaluate each expression.

a $3 + 0.5 - 2$ **b** $30 \times 5 - 3$ **c** $30 - 3 \times 5$ **d** $10 \div 5 + 5$

e $10 + 5 \div 5$ **f** $2 \times 15 - 5 \times 4$ **g** $\frac{1}{2}$ of $3 - \frac{1}{2}$ **h** $5 - \frac{1}{2}$ of 3

Remember to use BODMAS!

2 Now evaluate each of these expressions.

a $(3 \times 5) + 5$ **b** $3 \times (5 + 5)$ **c** $100 - (50 \times 2)$

d $(100 - 50) \times 2$ **e** $(\frac{1}{2}$ of $40) + 10$ **f** $\frac{1}{2}$ of $(40 + 10)$

3 Write five different expressions that evaluate each value. Use all the operations in BODMAS, including brackets where needed.

a 10

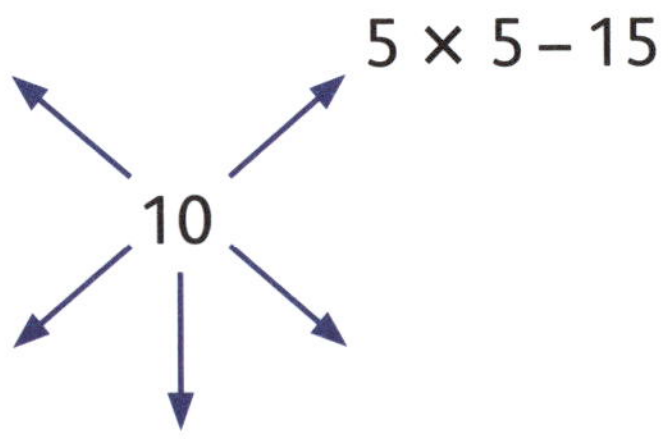

b 100 **c** 30 **d** 49 **e** $2\frac{1}{2}$

Variables and substitution

Key maths idea

You can write expressions that include unknown values or **variables**.

$4 + n$ is an expression that means:

'the sum of 4 and a variable, which we have called n'.

The letter n represents a variable in this expression. The variable n could have different values:

If n is 1, then the expression $4 + n$ has a value of $4 + 1$.

If n is 100, then the expression has a value of $4 + 100$.

If n is 0.25, then the expression has a value of $4 + 0.25$.

When we use a specific value for a variable, we call this **substitution**.

We can say: 'Substitute $n = 5$ in the expression $4 + n$.'

When $n = 5$, $4 + n$ has the value 9.

Key words

variable

substitution

1 Evaluate the value of each expression when $a = 5$.

a $a + 10$ **b** $10 + a$ **c** $10 - a$ **d** $20 + a - 10$

2 Evaluate the expression for each substitution $100 - b$.

a when $b = 20$ **b** when $b = 99$ **c** when $b = 12.5$ **d** when $b = \frac{3}{4}$

3 Write an algebraic expression for:

a *p* subtract 2 **b** 2 minus *p* **c** 10 plus *q*

d the sum of *r* and 2.5 **e** 20 more than *x* **f** *y* less than 20

Expressions involving multiplication and division

Key maths idea

In **algebra**, $4a$ means the product of 4 and a.

$4 \times a = 4a$

$a \times 4 = 4a$

When you multiply a variable by a number, the number always appears first in the expression. (Never write $a4$ to mean $4 \times a$).

$4 \div b = \frac{4}{b}$. Say: '4 divided by *b*' or '4 over *b*'.

$b \div 4 = \frac{b}{4}$. Say: '*b* divided by 4', or '*b* over 4'.

Expressions can include more than one operation.

$100 + 3a$ means '100 plus the product of 3 and *a*'

When $a = 5$, then $100 + 3 \times 5$ can be evaluated as:

$100 + (3 \times 5)$

$= 100 + 15$

$= 115$

Key word
algebra

1 Write algebraic expressions for these statements.

a the product of 2 and *j* **b** the product of *k* and 5

c the division of *m* by 2 **d** the result of sharing 100 into *n* equal parts

e 10 less than the product of 4 and *p*

2 Evaluate each expression when $a = 4$.

a $2a$ **b** $2a + 1$ **c** $2a - 1$ **d** $10 + 3a$

e $3a - 10$ **f** $\frac{a}{2}$ **g** $\frac{a}{2} + 3$ **h** $10 - \frac{2}{a}$

3 Evaluate each expression for different values of x, by substituting $x = 1$, $x = 2$, $x = 3$, $x = 4$ and $x = 5$.

a $\frac{60}{x} + 100$ **b** $100 - 20x$ **c** $\frac{x}{10} + 0.25$

Algebra

Expressions in puzzles and problems

Key maths idea

You can use variables to help you solve story problems.

I sell lemonade for \$2 a glass. If I sell 1 glass, I make \$2. If I sell 10 glasses I make \$20.

Amari sells a different number of glasses each day. The amount of money she makes also varies, depending on how many glasses she sells.

Amari chooses n as the variable to represent the number of glasses she sells.

If I sell n glasses, I make $2n$ dollars.

Now you can substitute different values for n to work out the total made for different number of glasses sold.

If $n = 100$, the total money earned is $2 \times \$100 = \$200$.

If $n = 175$, the total money earned is $2 \times \$175 = \$350$.

You can also use variables to represent values in puzzles.

Think of a number. Multiply by 2 then add 3.

Start: n

Multiply by 2: $2n$

Add 3: $2n + 3$

The variable n represents the starting number.

The expression $2n + 3$ represents the final number. We can call this an ***n*-sentence**.

If $n = 5$, the final value is $2 \times 5 + 3$, which evaluates to 13.

We can complete a table like this:

Operations	Arithmetical	*n*-sentences	Concrete form
Think of a number	5	n	■
Multiply by 2	10	$2n$	■ ■
Add 3	13	$2n + 3$	■ ■ ● ● ●

For the concrete form, a square symbol is used to represent one n, and a circle represents a number 1.

Key word

n-sentence

1 Write an algebraic expression to represent the total profit.

a I saved x dollars, then I received \$10 from my grandmother. The total I have now is ______ .

b I had y dollars, then I spent \$12.99. Now I have ______ .

c I get paid \$300 for washing a car. If I wash p cars, then I earn ______ .

d One block is 25 cm in length. What is the total length of q blocks?

e I had 25 litres of water in my water tank. I used r litres to water all my plants. Now the water that remains in the tank is ______ .

2 Write an algebraic expression for the total perimeter of each regular polygon.

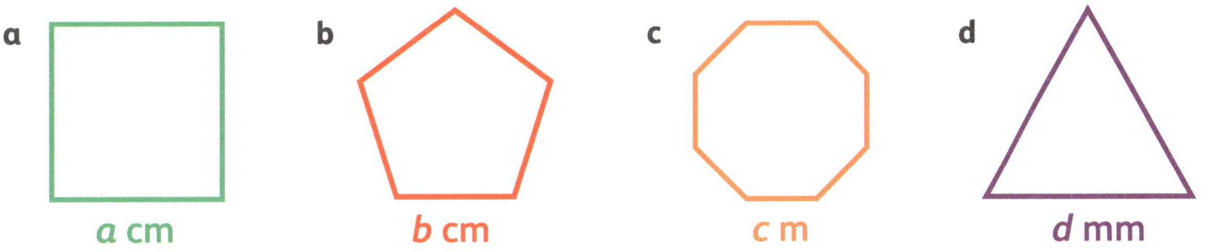

3 Choose a letter variable for the starting number in these puzzles. Write an expression for the final value.

Complete a table for each, including n-sentences.

a Think of a number. Multiply it by 5 and then subtract 3.

b Think of a number. Divide by 10. Then add 2.5.

c Think of a number. Multiply it by 2. Subtract it from 100 and then add 50.

Maths detective: Three Threes and Four Fours

Use different values as clues to find the solution to these challenges!

Case 1

1 Solve the Three Threes challenge!
Use three threes and then choose operations and brackets to make different totals.
For example,
$(3 + 3) \times 3$ gives a value of 18.

2 Evaluate these expressions:

a $3 + (3 \times 3)$ **b** $3 - (3 \div 3)$ **c** $3 \times (3 - 3)$

Case 2

Now do this Three Threes challenge!

Choose your own operations and brackets, and use three threes.

3 3 3

Find a way to use the three threes to make these values:

1 3 2 4 3 $\frac{1}{3}$

Case 3

Now do the Four Fours challenge!

4 4 4 4

For example: $4 + (4 - 4) \times 4$

Choose your own operations and brackets.

How many values from 1 to 20 can you make using four fours?

Real-world maths: The orange orchard

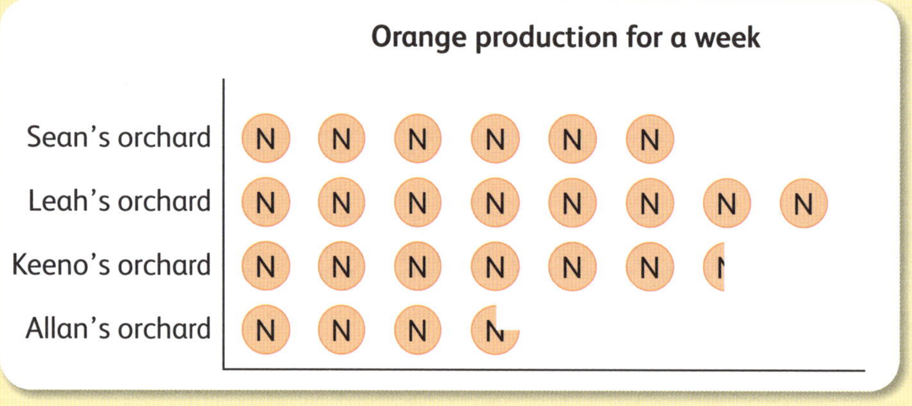

This pictograph represents orange production in different orchards.

a The key has not been included. Write an expression for the number of oranges grown in each orchard.

b 30 of the oranges from Sean's orchard were rotten and had to be thrown away. Write an expression for the number of oranges that remain.

c Keeno produced 60 fewer oranges than Leah. Use this information to calculate the number of oranges in each orchard.

d Allan miscounted. He found 50 more oranges. Write an expression for the total number of oranges in all orchards.

What I have learned

Use some or all of the words in the word clouds to explain the meaning of each expression in the boxes.

sum | product | quotient | less than | more than

$a + 10$ | $10 - a$ | $2a$ | $5 + \frac{a}{2}$

Practice questions

1 Which expressions have a value of 2?

a $2 \times 5 - 8$ b $2 + 5 \times 8$ c $24 \div 8 + 4$ d $24 \div (8 + 4)$

2 Each packet contains n sweets.
Which expression represents the total number of sweets?

a $2 + n$ b $5n$

c $2n + 5$ d $5n + 2$

3 Evaluate each expression when $x = 3$.

a $5x$ b $3 + 2x - 5$ c $100 - 20x$ d $\frac{45}{x} + 25$

Algebra

17 Self check

Check you can do this:	If you have difficulty, try this:
1 Use the correct order of operations to calculate a problem.	Go back to page 231, and remind yourself of the BODMAS rule.
2 Write expressions involving variables.	Revisit page 231 to remind yourself of the terms expression and algebraic expression. Remember that an algebraic expression uses numbers, letters and operation symbols.
3 Substitute values to evaluate expressions.	Go back to page 232.
4 Write expressions to represent problems and puzzles.	Look at page 234 and remind yourself how to use letters to represent an amount in a story problem or puzzle.

PEP task: Fundraising fun

This performance task has 4 parts. Your teacher will tell you when to do each part.

A Grade 5 class has been asked to raise funds for a school outing. The students were placed in groups and each group chooses its own project to raise funds.

You will need:
- a ruler
- a pencil
- a calculator.

Part 1

Group A decides to set up a stall at the local Farmer's Market and charge people to play a game of chance.

The first idea they have is to make a large spinner. They will give each person three spins of the arrow for $50. If the arrow lands on 1 three times in a row you will win a large bag of candies.

1 These are the spinners the students suggest.

- **a** Which spinner is most likely to give a result of 1?
- **b** Which spinner is least likely to give a result of 1?
- **c** Which of these spinners do you think they should use? Why?
- **d** The students decide that all of these spinners would make it too easy for people to win. Help them design a spinner that has about a 5% chance of landing on 1. Draw your spinner.

2 At the fundraiser 250 people played the game.

- **a** How much money did the students raise from entry fees?
- **b** If 4% of the people won prizes, how many prizes did they give away?
- **c** The students bought 12 bags of candy at a cost of $180.50 per bag. They also bought a large sheet of thick cardboard that cost $490 to make the spinner. How much money did they raise after deducting the costs?

3 The group decides to hold a raffle at school to raise some extra money. The prize is the left-over candy. They sell 172 raffle tickets.

a Fynn buys 4 tickets. What is the probability that he will win? Give your answer as a fraction in simplest terms.

b How many raffle tickets would you need to buy to have at least $\frac{1}{12}$ chance of winning?

Part 2

Group B decides to build wooden toys for children. They find this design for a pull-along pick-up truck online.

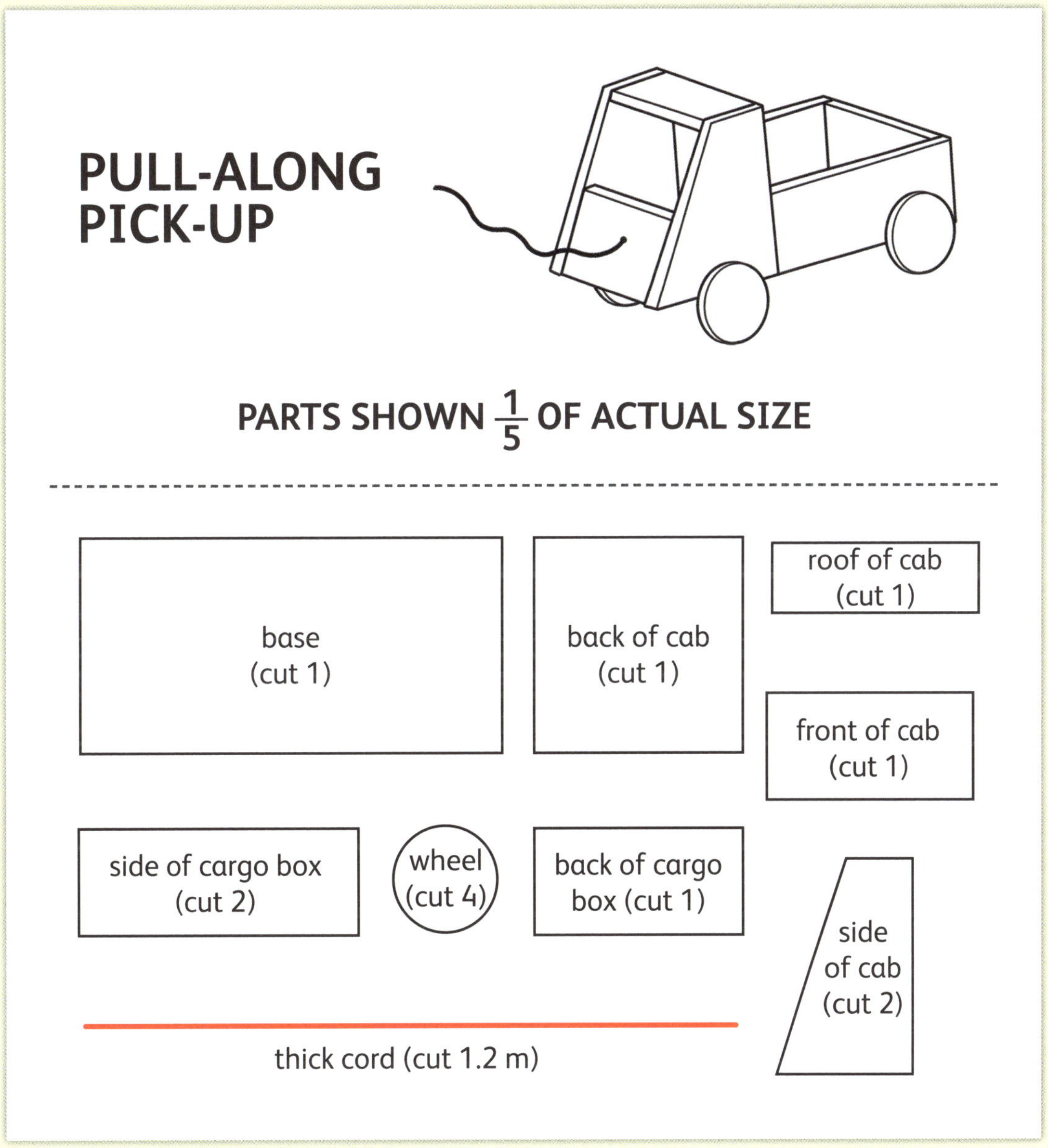

1 Work out the approximate area of wood needed to make 1 truck. Show your calculations and give your final answer in square metres.

2 The students decide to make the trucks out of dark brown hardboard. One sheet of board is 1.3 m wide and 2.4 m long and costs $1 883.30.

  a Is one sheet of board enough to build 20 trucks? Show clearly how you work this out.

  b The students decide to make 18 trucks to sell. How much thick cord will they need?

  c The cord comes in 5 metre lengths and costs $295 per packet. How many packets will they need to buy? How much will this cost?

3 All the other items the students need are donated to them.

  a Work out the cost of making one truck using the information above. Round the costs to the nearest dollar.

  b The students aim to raise $25 000. At what price should they sell the trucks? Show how you work this out.

Part 3

Group C decides to make a laminated poster showing local attractions to sell to tourists.

They find the following information from their local copy shop.

Cost to print a 28 × 44 cm coloured sheet:

1–5 copies	$100 each
6–20 copies	$60 each
21–100 copies	$15 each
+1 000 copies	$13.50 each
Laminating	$220 per sheet

1 What is the total cost of making 75 posters?

2 The group spends $15 275 on making their posters. $14 300 of this is for laminating. How many posters do they make?

3 Marlon says they should make 1 000 posters as it will be much cheaper to print them. Explain to him why this is not a good idea.

4 If the group sells all of the posters for $525 each, how much profit will they make?

Part 4

1 Group D has a set of three novelty dice. The six faces of each die are shown here.

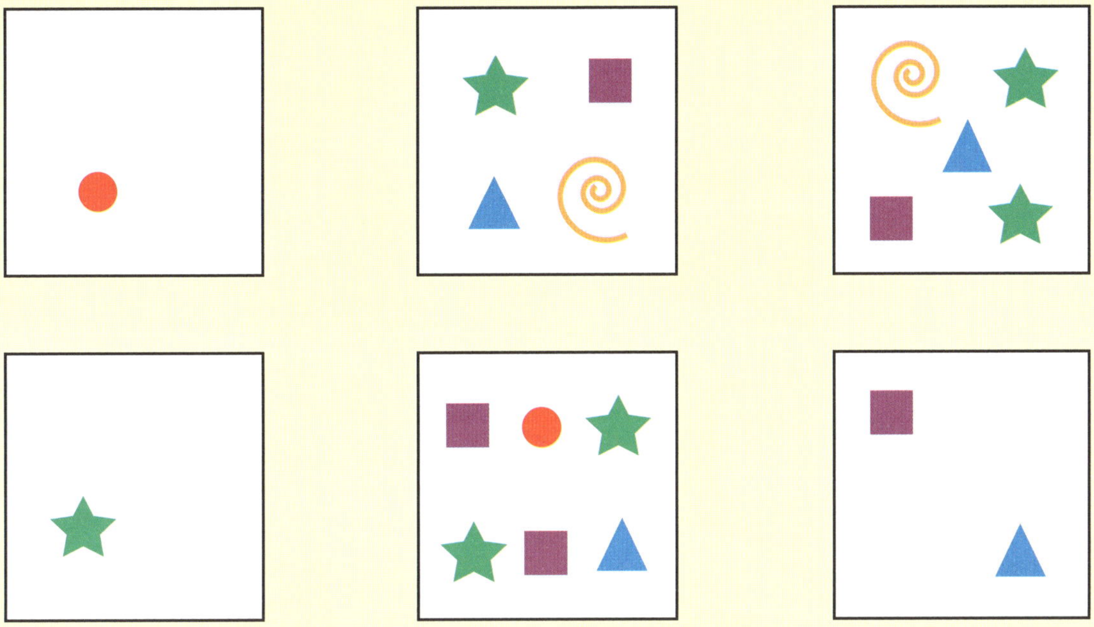

How could they use these dice to play a game with a small chance of winning? Write instructions for the game and explain how it works.

2 The students in group D end up with four different ideas. They ask some other students to rate these to help them decide which one to use.

These are the scores they get.

Idea A	−2	3	2	−1
Idea B	−4	2	−3	1
Idea C	3	−2	−2	2
Idea D	−3	−3	4	1

Which idea should they choose? Why?

TERM 3

UNIT 1: NUMBER
p243

Multiply and divide larger numbers

Solving real-life problems

UNIT 2: GEOMETRY
p279

Circles

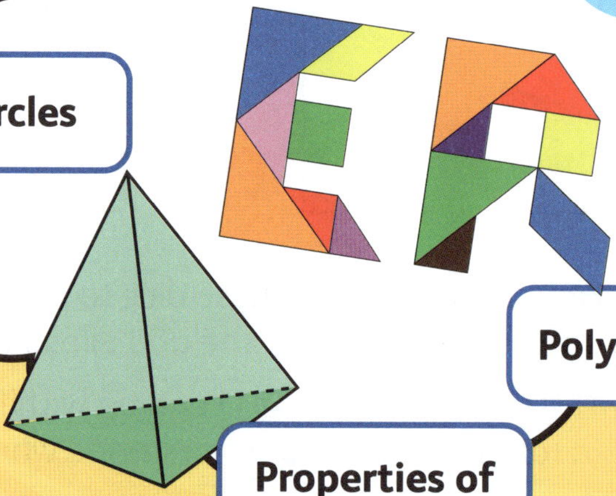

Polygons

Properties of pyramids

UNIT 3: ALGEBRA
p304

Variables and equations

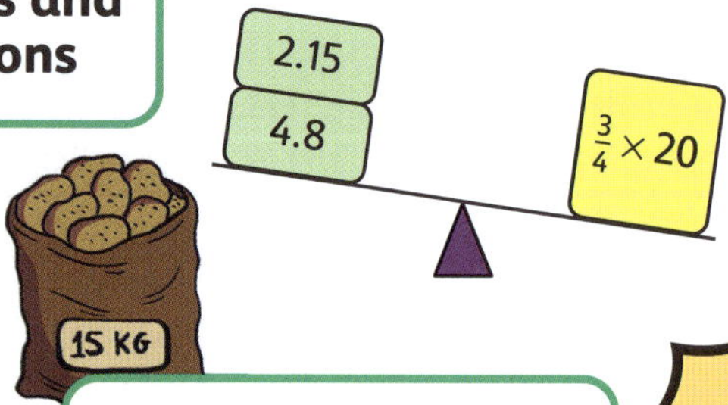

Solving problems with equations

UNIT 4: STATISTICS AND PROBABILITY
p320

Averages

Probability

Unit 1: Number

Chapter 18 Multiply and divide larger numbers

In this chapter you will explore this focus question:

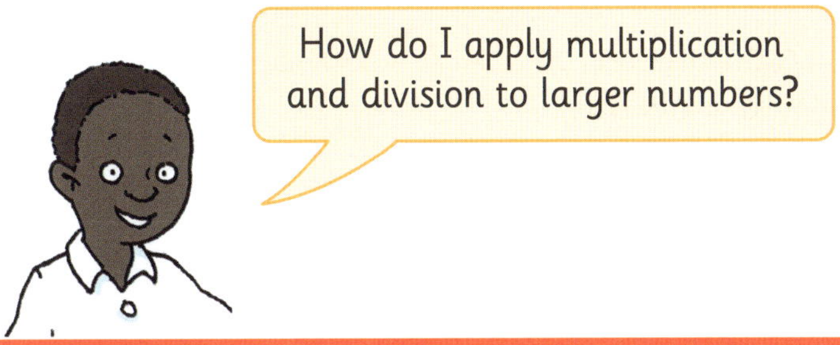

To answer this question, you will:

- make calculations using multiplication and division
- solve real-life problems using multiplication and division
- estimate answers and judge whether your calculated answers are reasonable
- apply inverse relationships between multiplication and division
- explore multiplication facts and divisibility rules for multiples of 5, 8 or 9.

Starting point

1 Hattie is a gardener. She earns the same amount each day. At the end of a week, she is paid \$11 396. She works for more than 1 day but not for as long as a week.

- How much could she have earned each day?
- Which solutions will give answers as a whole number of dollars? Explain how you can make decisions before calculating.

Hattie saves the \$11 396 for a trip to the USA. The current rate of exchange is 1 US dollar = 154 Jamaican dollars.

How can you work out how many US dollars Hattie will get for this money?

Work with a partner. Use a calculator to try out your idea.

Revising what you know about multiplication and division

Key maths idea

Remember:

- **Multiplication** works like repeated addition.
  5×7 means '5 sets of 7'. The **product** of 5 and 7 is equal to the sum of $7 + 7 + 7 + 7 + 7$.
- Multiplication is commutative, so $5 \times 7 = 7 \times 5$.
  You can use arrays to show this.

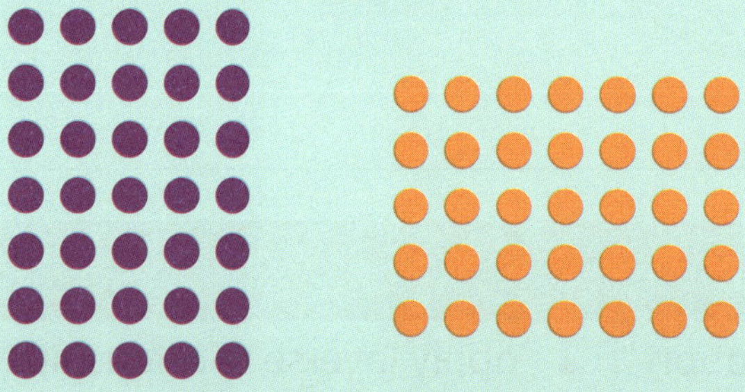

- Multiplication and **division** are inverse operations.
  $5 \times 7 = 35$, so $35 \div 7 = 5$ and $35 \div 5 = 7$.
  These division and multiplication facts make a fact family.
- In the sentence $35 \div 7 = 5$, 35 is the **dividend**, 7 is the **divisor** and 5 is the **quotient**.
- You can express a division sentence as an improper fraction, for example $\frac{35}{7}$. When you simplify it, you will reach the quotient.
- Zero divided by any number is equal to zero. You can't divide by zero.
- When the divisor is not a factor of the dividend (or the dividend is not a multiple of the divisor), you are left with a remainder.
  So, $38 \div 7 = 5$ r 3.

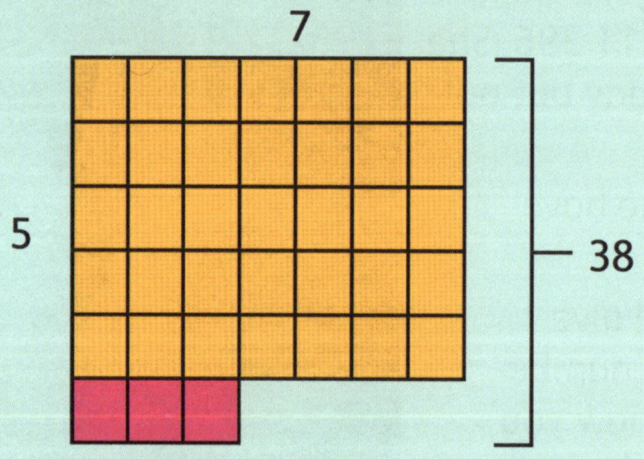

- When a division leaves a remainder, you can express the answer as a mixed number.

Remember: $38 \div 7 = \frac{35}{7}$.
You can simplify this to $5\frac{3}{7}$.

Key words

multiplication
product
division
dividend
divisor
quotient

1 Work in pairs. Make a poster about multiplication and division.

  a Include definitions of the terms quotient, dividend, divisor, product, commutative, inverse, remainder and fact family.

  b Include an array that shows a fact family of multiplication and division facts.

2 Make up a 5-digit number. With a partner, find the quotient when you divide your number by 5, by 12 and by 25. Is it divisible by any of those numbers? If not, express the answer with the remainder and then as a mixed number.

3 Kenisha says that a quotient cannot have a zero as one of its digits. Use this example to explain why she is incorrect:
30 720 ÷ 15.

4 Melisha uses these six cards each time to make up division calculations.

Her first calculation is 4 265 ÷ 10.

  a Explain how Melisha knows that there will be a remainder.

  b How should she give the answer as a mixed number?

  c Melisha makes up another division calculation. There is no remainder when the divisor is 2. What could the division be?

  d Explore some other divisions that Melisha can make with a 1-digit or 2-digit divisor.

What is the smallest answer you can find?

What is the largest answer?

Can you find a number that Melisha can use for a calculation that is divisible by 4?

Multiplication and division facts

Key maths idea

What facts do these images represent?

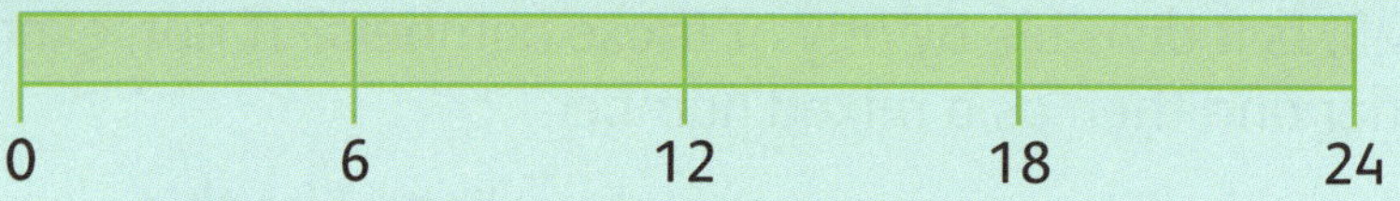

×	1	2	3	4	5	6	7	8	9	10
1	1	2	3	4	5	6	7	8	9	10
2	2	4	6	8	10	12	14	16	18	20
3	3	6	9	12	15	18	21	24	27	30
4	4	8	12	16	20	24	28	32	36	40
5	5	10	15	20	25	30	35	40	45	50
6	6	12	18	24	30	36	42	48	54	60
7	7	14	21	28	35	42	49	56	63	70
8	8	16	24	32	40	48	56	64	72	80
9	9	18	27	36	45	54	63	72	81	90
10	10	20	30	40	50	60	70	80	90	100

We can use arrays to represent facts on a multiplication square.
The array shows that 6 × 4 = 24 and 4 × 6 = 24.
We say that the product of 4 and 6 is 24.

The same array can be used to find the related division facts.
24 is divided equally into six rows. We can count four squares in each row.
So 24 ÷ 6 = 4.
24 is also divided equally into four columns. We can count six squares in each column.
So 24 ÷ 4 = 6.

Let's think about how we can use place value to make other sets of related facts.
6 × 4 = 24
6 tens × 4 = 24 tens so 60 × 4 = 240
6 tens × 4 tens = 24 hundreds because 10 tens make one hundred
So 60 × 40 = 2 400.

What other related multiplication and division facts can you find?
For example, 24 tens ÷ 6 = 4 tens, so 240 ÷ 6 = 40.
Do you notice any **patterns**?

Key word
pattern

Notice how the language of place value (tens) is used with the multiplication facts to help make connections.

1 Copy and complete these facts. You can use the multiplication square to check your answers.

	a	b	c	d
	7 × 4 = □	8 × 9 = □	6 × □ = 48	5 × □ = 0
	□ × 7 = 28	9 × □ = 72	□ × 6 = 48	0 ÷ 5 = □
	28 ÷ 4 = □	72 ÷ 9 = □	48 ÷ 6 = □	5 × □ = 5
	28 ÷ 7 = □	□ ÷ 8 = 9	48 ÷ □ = 6	5 ÷ 5 = □

2 Arrange the digits 1 to 9 in the circles so that pairs multiply together to make the products in the triangular sections.

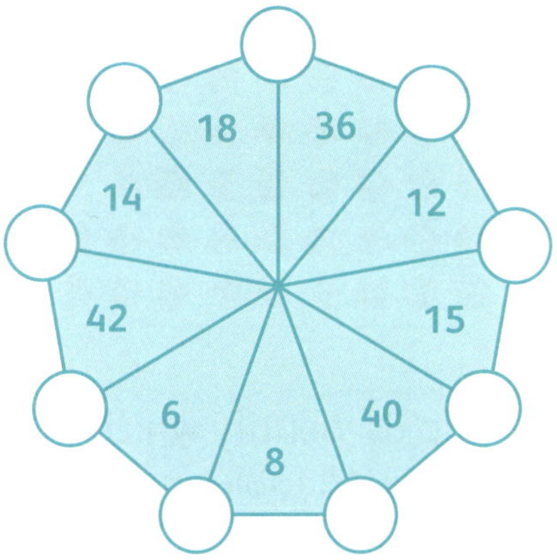

3 Choose a multiplication fact from the numbers in Exercise 2.

a Use what you know about multiplication, division and place value to make at least six related facts.

b Now choose another fact. What related facts can you make this time? Can you find any 4-digit products or use a 4-digit dividend? How about 5-digits?

4 Imagine adding another two columns and another two rows to the multiplication square on page 246.

a Here is the start of the two rows. Copy and complete them.

11	11	22										
12	12	24	36									

b Now use the rows to write at least ten different facts for the multiplication tables of 11 and 12.

c Then write the related division facts.

Divisibility tests for 5, 8 and 9

Key maths idea

Look at these arrays. Which ones represent numbers that are **divisible** by 5, 8 or 9?

We can use **divisibility** tests to help us identify which numbers beyond the products in our multiplication tables are divisible by 5, 8 or 9.

Look at these numbers.

2 736	5 895	3 600	3 681	12 588

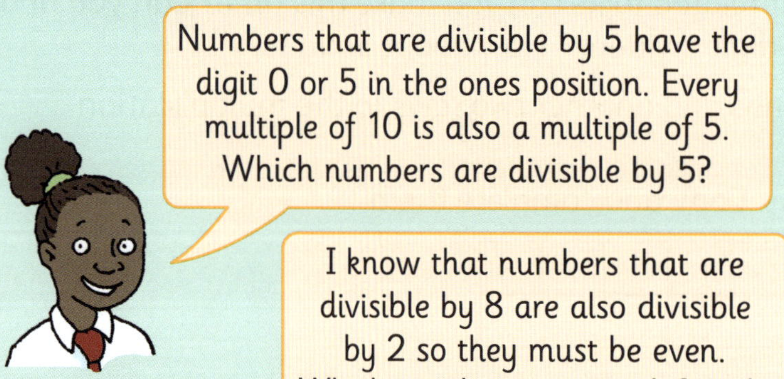

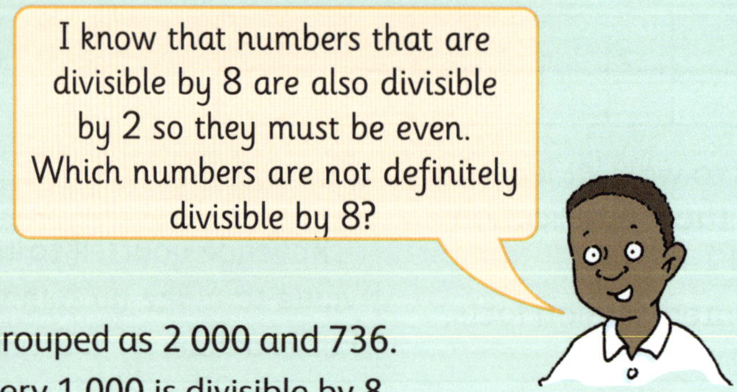

2 736 can be regrouped as 2 000 and 736.

We know that every 1 000 is divisible by 8 because $125 \times 8 = 1\ 000$, so we only need to check the last three digits of the number.

Use a method of your choice to calculate $736 \div 8$. Does it divide exactly by 8?

Now check the other numbers.

Key words

divisible

divisibility

We can use digit sums to check if a number is divisible by 9.

To find a digit sum, we add all the digits and continue until a 1-digit number is reached.

Let's try the first two numbers:

2 736	5 895
$2 + 7 + 3 + 6 = 18$ $1 + 8 = 9$	$5 + 8 + 9 + 5 = 27$ $2 + 7 = 9$

What do you notice?

Numbers that are divisible by 9 have a digit sum of 9.

Now check the other numbers.

1 **a** Use divisibility tests to decide which numbers are divisible by 5, 8 or 9. Copy and complete the table.

9 352 23 445 3 930 15 336

43 659 5 692 14 312 8 100

> Some numbers may go in more than one part of your table.

Divisible by 5	**Divisible by 8**	**Divisible by 9**

b Some numbers are left over. Explain how you know that they are not divisible by 5, 8 or 9.

2 Look at this number line.

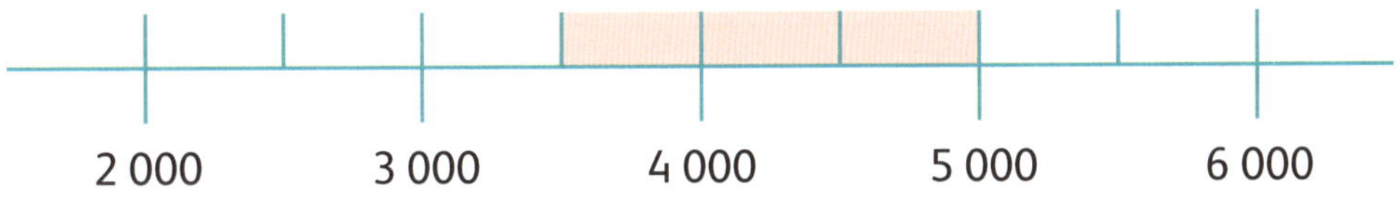

Write five numbers each time you can position in the shaded part that are:

a divisible by 5 **b** divisible by 8 **c** divisible by 9

3 A school organises a fun run to raise funds for charity. It raises $95 364 in total. Can the money be divided equally between nine charities so they each get a whole number of dollars? Explain how you know.

4 Lloyd adds two 4-digit numbers together.
The total is 4 389.
One of the numbers is divisible by 8, the other is divisible by 5 but not by 8.
What could the two numbers be?
Find at least three different solutions.

Working with division

Key words
estimate
inverse

Key maths idea

A group of children are going on a school trip.
They sit in rows of four on the buses.
How many rows do 132 children need?

Let's think about the problem.
We are trying to find out how many groups of 4 are in 132 so we are describing the division as grouping.
The dividend 132 represents the number of children. What does the divisor 4 represent?

Explain to a partner why 132 ÷ 4 will not have a remainder.

We know that 100 ÷ 4 = 25 so we can use this as an **estimate**.
The division can be solved in different ways. Discuss how each method works.

Mental method with regrouping	Repeated subtraction	Short division
$132 \div 4 =$ $100 \div 4 = 25$ $32 \div 4 = 8 +$ 33	-4×25 (0 to 100), -4×8 (100 to 132) 0, 100, 132 $25 + 8 = 33$	$\begin{array}{r} 3\ 3 \\ 4\overline{)1^{1}3^{1}2} \end{array}$

So $132 \div 4 = 33$. This is close to our estimate.
What does the quotient 33 represent?
Let's check the answer using the **inverse** multiplication:

$$33 \times 4 = 30 \times 4 + 3 \times 4$$
$$= 120 + 12$$
$$= 132$$

The 132 children are divided equally among the three school buses.

How many children are on each bus?

This time the context of the problem describes division as equal sharing.

The dividend 132 still represents the number of children. What does the divisor 3 represent this time?

A bar model can be used to represent this new problem.

132		
☐ children	☐ children	☐ children

We know that $120 \div 3 = 40$ because $40 \times 3 = 120$, so 40 is a useful estimate.

Let's try the mental regrouping method.

$132 \div 3 =$

$120 \div 3 = 40$

$12 \div 3 = 4 +$

44

Now try the short division method and then use multiplication to check your answer.

Can you now work out how many rows of 4 children there will be on each of the three buses?

1 The whole has been divided equally each time. What are the values of the shaded bars? What method of division will you use? Remember to make an estimate first.

a
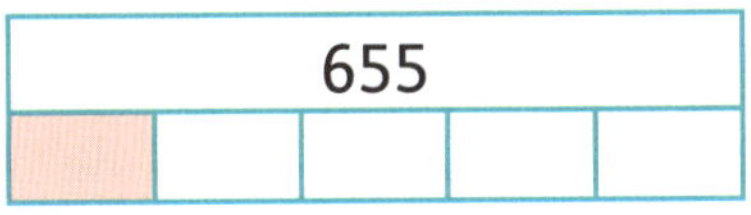

b

c

d
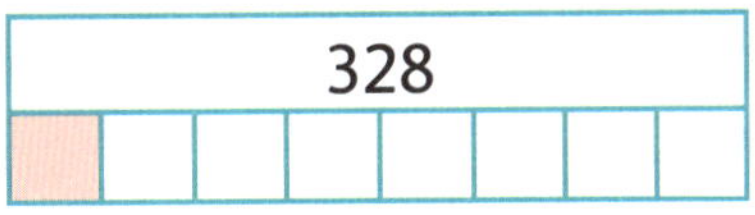

2 Tamila estimates that the quotient for two of these divisions will be approximately 20.

$65 \div 5$ $117 \div 9$ $176 \div 8$ $256 \div 8$ $133 \div 7$

a Which of these are the two divisions that Tamila esimated?

b Make estimates for the other divisions.

c Calculate the answers for all five calculations. What method will you use? Check your answers against your estimates and then by using multiplication.

3 Solve these problems. Tell a partner whether the context describes division as equal grouping or equal sharing.

- **a** Mrs Brown packed 152 apples into bags with 8 apples in each bag. How many bags did she need?
- **b** A warehouse stocks 5 kg bags of pet food. The total mass of the bags is 765 kg. How many bags is this?
- **c** A sponsored run raises $868 for charity. The money is divided equally among seven charities. How much do they each get?

4 Look at these divisions.

$864 \div 6$ $\quad$ $864 \div 3$ $\quad$ $864 \div 9$ $\quad$ $864 \div 8$

- **a** Use reasoning to decide which division will give the largest quotient.
- **b** Use reasoning to decide which division will give the smallest quotient.

Explain your thinking to a friend. Check your ideas with a calculator to see if you are correct.

Division with remainders

Key maths idea

Sometimes a dividend is not always divisible by the divisor and a remainder is left.

$25 \div 4 = 6$ remainder 1

The context of the problem will help to decide how to represent the answer to a division.

Consider these examples for the division $25 \div 4$:

- 25 people arranged in groups of 4 is 6 groups with one person left over.
- 25 kg of sand divided equally between 4 bags. Each bag is $6\frac{1}{4}$ kg.

When we want to show the answer as a mixed number, we can use what we know about writing a division as a fraction to help us.

$$25 \div 4 = (24 \div 4) + (1 \div 4)$$
$$= 6 + \frac{1}{4}$$
$$= 6\frac{1}{4}$$

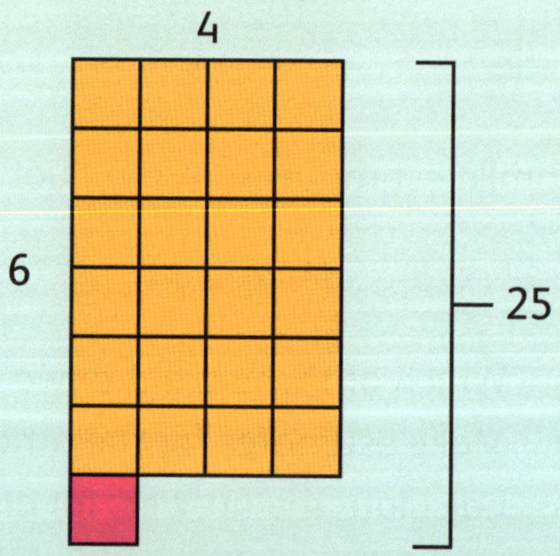

Remember: we are converting the remainder into a fraction of the divisor!

1 Complete these calculations. Convert any remainders to fractions of the divisor and give your answer as a mixed number. Use multiplication to check each calculation.

a $65 \div 3 = \square$ **b** $66 \div 4 = \square$ **c** $66 \div 5 = \square$

d $68 \div 7 = \square$ **e** $69 \div 8 = \square$

2 Look at each calculation and decide if there will be a remainder.

a $164 \div 4 = \square$ **b** $787 \div 5 = \square$ **c** $495 \div 9 = \square$

d $545 \div 9 = \square$ **e** $461 \div 5 = \square$

Now complete the calculations to see if you were correct.

3 Solve these word problems. Will you need to round the answer or convert any remainders to fractions?

a Balls come in boxes of 75. Akiel puts them in bags of four. How many bags of four balls are there?

b Jevaughn cuts a 48 m length of rope into five equal pieces. How long is each piece?

c Devon divides 380 ml of water equally between eight containers. How many millilitres of water are in each container?

d There are six eggs in a box. How many boxes does the farmer need for 94 eggs?

> Discuss other examples in real life where the quotient would be written as a mixed number or as a decimal. Also think about examples where we would show a remainder.

4 379 is divided into equal groups.

a What remainders can you find?

b Are any remainders the same? Why?

5 Melisha has a box full of animal stickers that she wants to sort into an album. She has more than 50 stickers, but less than 100.

She thinks about putting 7 stickers on a page, but there is 1 sticker left over.

When she thinks about putting 8 stickers on a page, there are 6 stickers left over.

Finally she decides to puts 6 stickers on a page. There are no stickers left over.

How many animal stickers did Melisha have? How many pages did she fill with stickers?

Dividing larger numbers by a 1-digit number

Key maths idea

We can use the method of short division to divide a larger dividend by a 1-digit divisor.

Think about the division 6 372 ÷ 3:

We know that the quotient will be approximately 2 000 because 6 000 ÷ 3 = 2 000 because 2 000 × 3 = 6 000.

The division has been represented with counters.

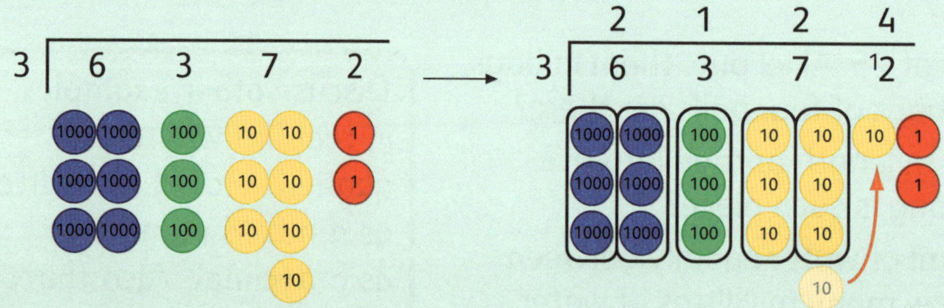

Look at the way the counters have been grouped in threes to find out 'How many threes are there in …?' each time.

The quotient is 2 124. This is close to our estimate.

We can use the multiplication 2 124 × 3 to check. We can also use a calculator.

Think about a much larger dividend:

16 346 ÷ 8

Will there be a remainder? Think about the test of divisibility to check.

We know that 16 000 ÷ 8 = 2 000 because 2 000 × 8 = 16 000, so the quotient will be approximately 2 000.

Talk to your partner about the steps for carrying out this division.

$$\begin{array}{r} 2\,0\,4\,3\text{r}2 \\ 8\overline{)\,1\,{}^{1}6\,3\,{}^{3}4\,{}^{2}6} \end{array}$$

So 16 346 ÷ 8 = 2043 r 2 or $2043\frac{2}{8}$ or $2043\frac{1}{4}$

Now check the answer as 2 043 × 8 + 2.

1 Solve these. Remember to make an estimate first.

a 1 374 ÷ 6 **b** 3 752 ÷ 8 **c** 5 481 ÷ 9 **d** 7 410 ÷ 5

2 **a** Use reasoning to help sort these calculations into the two sets. Do not calculate yet.

2 025 ÷ 5	7 110 ÷ 9	3 436 ÷ 8	8 395 ÷ 5	8 843 ÷ 9
5 216 ÷ 8	7 716 ÷ 4	9 999 ÷ 2	1 120 ÷ 8	2 727 ÷ 9

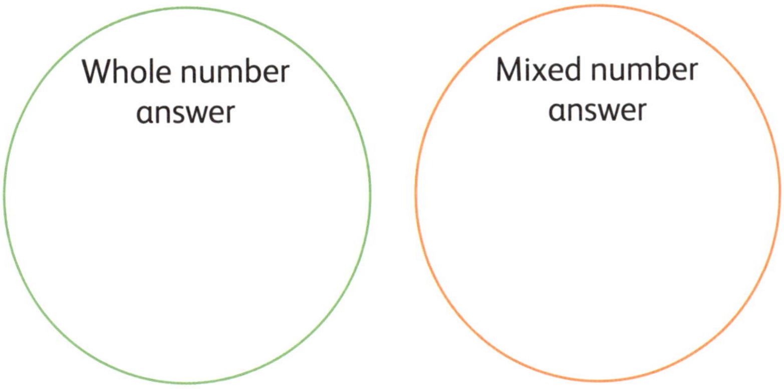

b Now calculate to see if you were correct.
You may need to give some answers as mixed numbers.

c Work with a partner to make up two new divisions for each set.

3 **a** Calculate the mass of each one of the shapes on the scale. Remember to estimate first.

b Then use multiplication to check each answer.

4 Solve these problems.

a It costs a fisherman \$1 435 per week to moor his boat in the harbour. What is the cost per day?

b The dividend is 5 694 and the divisor is 3. What is the quotient?

c The quotient is 193 and the divisor is 8. What is the dividend?

d A length of string is 2 138 cm. It is cut into five equal parts. What is the length of each part?

Working with 2-digit divisors

Key maths idea

Read these two problems.
What is the same and what is different about them?

3 165 pencils are packed into boxes of 15. How many boxes of 15 pencils are there?	$3 165 is split equally among 15 charities. How much money does each charity receive?

We can make an estimate for the division 3165 ÷ 15 by rounding 3 165 to 3 000.
3 000 ÷ 15 = 200, so the answer to each problem is approximately 200.

Let's use long division to calculate the answer.

Step 1: How many times does 15 go into 31?
15 × 2 = 30
31 – 30 = 1 (Remember these are hundreds)

Step 2: How many times does 15 go into 16?
15 × 1 = 15
16 – 15 = 1 (Remember these are tens)

Step 3: How many times does 15 go into 15?
15 × 1 = 15
15 – 15 = 0 (Remember these are ones)

Step 4: How many times does 15 go into 3 165 altogether?
211 times so 3 165 ÷ 15 = 211.

```
      2 1 1
15 | 3 1 6 5
    −3 0
      1 6
     −1 5
        1 5
       −1 5
          0
```

Let's check the **computed** answer against the estimate. Is the answer reasonable?

Now use a calculator to check the answer as a multiplication.

What does the answer 211 represent for each of the original problems?

> What would you do differently for the division 7 780 ÷ 10? Explain why.

Key word
compute

1 Work out these divisions. Show all your steps.
Remember to make an estimate.

a 9 660 ÷ 24 **b** 5 600 ÷ 25

c 6 831 ÷ 11 **d** 7 170 ÷ 15

e 1 750 ÷ 14 **f** 2 016 ÷ 21

g 3 430 ÷ 98 **h** 5 135 ÷ 79

> Remember to check your computed answers against your estimates to see if they are reasonable.

2 Solve these problems. Check your solutions using the inverse.

a The mass of 37 bags of cement is 3 885 kg. What is the mass of one bag?

b The phone account shows that James spent 4 410 minutes on his phone in the month of January. What was the average number of minutes James spent on his phone each day in January?

c 1 100 people are invited to a function. There are 18 seats at each table. How many tables will be needed so everyone can be seated?

3 The length of the day is different on each planet. Use the table to help you answer these questions.

Planet	Length of day (in hours)
Mercury	1 409
Venus	5 832
Earth	24
Mars	25
Jupiter	10
Saturn	10
Uranus	18
Neptune	19

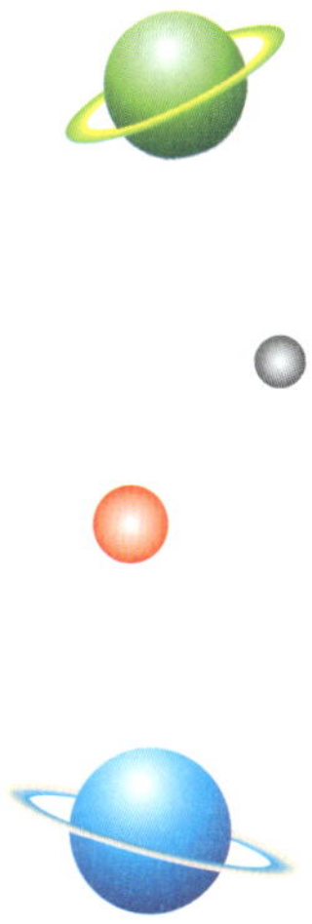

a The length of a day on Venus is about how many times the length of a day on Uranus?

b The length of a day on Venus is about how many times the length of a day on Earth?

> Here is a challenge for you! Use rounding to help work out about how many Earth days are equivalent to one day on each of the other planets.

4 Use multiplication and your own numbers to help make up three real-life problems for the division:

□□□□ ÷ □□

Share your problems with others so they can solve them.

Compare each other's methods of working. What improvements can be made?

Working with 3-digit divisors

Key maths idea

Let's look at an example of a division by a 3-digit divisor:

25 750 ÷ 125

We know that 26 000 ÷ 100 = 260, so the quotient will be approximately 260.

We can now use long division to calculate the answer.

Step 1: How many times does 125 go into 257?

125 × 2 = 250

257 – 250 = 7 (Remember these are hundreds)

Step 2: How many times does 125 go into 75?

The divisor 125 is larger than 75 so we go on to the next step.

Step 3: How many times does 125 go into 750?

125 × 6 = 750

750 – 750 = 0

Step 4: How many times does 125 go into 25 750 altogether?

206 times so 25 750 ÷ 125 = 206.

```
        2 0 6
125 | 2 5 7 5 0
     -2 5 0
        7 5
      -   0
        7 5 0
       -7 5 0
            0
```

The quotient 206 is close to the estimate.
Now use a calculator to check the answer as a multiplication.

1 Use long division to solve these calculations. Remember to make estimates first.

- **a** 9 734 ÷ 125
- **b** 3 186 ÷ 118
- **c** 8 717 ÷ 150
- **d** 2 987 ÷ 112
- **e** 6 395 ÷ 110
- **f** 7 050 ÷ 150

Remember to check your computed answers against your estimates to see if they are reasonable.

2 Solve these problems. Remember to make estimates and to check your solutions using multiplication.

- **a** Mark earned $9 750 as his monthly salary. If he earned $150 per hour, how long did it take him to earn this amount?
- **b** A school was built to hold 1 440 students. Each grade has 120 students. How many grades are there in the school?
- **c** A length of cord measures 1 550 cm. How many 1-metre lengths of cord can be cut from it?

d Mother puts away $450 each week for savings. If she has $5 400 in her account, how long did it take her to save this amount?

e Peter is saving $125 each week for a new TV that costs $4 250. How many weeks will it take until he has saved enough for the TV?

3 Use multiplication and your own numbers to help make up three real-life problems for the division:

□□□□□ ÷ □□□

Share your problems with a partner and let them solve them. Compare each other's methods of working. What improvements can be made?

Maths detective: More about numbers

Use your maths detective skills to discover more about numbers.

Case 1

Lloyd divides the dividend 565 by one of the numbers in the box.

4 | 5 | 8 | 2 | 10

He writes the answer as the mixed number □ $\frac{1}{2}$

Which dividend(s) could Lloyd have used? Convince others of your decision.

Case 2

Lloyd can pack pencils into boxes of 12, 14, 15, 16, 18 or 20.

Which size box is best for 1 265 pencils so that the fewest possible pencils will be left over?

Case 3

Tamila has completed two divisions but some digits have been smudged.

What should they be? Copy the calculations and write in the missing numbers.

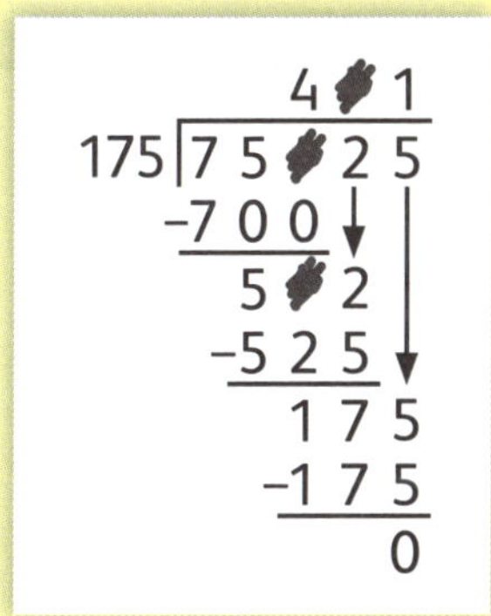

Real-world maths: Welcome to the Commonwealth Games!

1 The organisers of the Commonwealth Games had many problems to solve. Show how you would solve each problem.

a One stadium was able to seat 26 500 people. The stadium had four stands, each with the same number of seats. How many people could sit in each stand?

b If 7 560 tickets were sold in the first 24 hours, on average, how many tickets were sold each hour? How can the organisers use this information to predict how many tickets would be sold in the first week?

c Athletes are taken to the track by buses. Each bus has a fuel tank with a capacity of 275 litres. If 16 775 litres of fuel are used to fill the tanks of the buses, how many buses are there?

d Three meals a day were prepared for the athletes making a daily total of 5 850 meals. How many athletes were there?

Practice questions

1 The number in the centre of each wheel is the answer to the division calculations around it. Write different division calculations in the empty boxes.

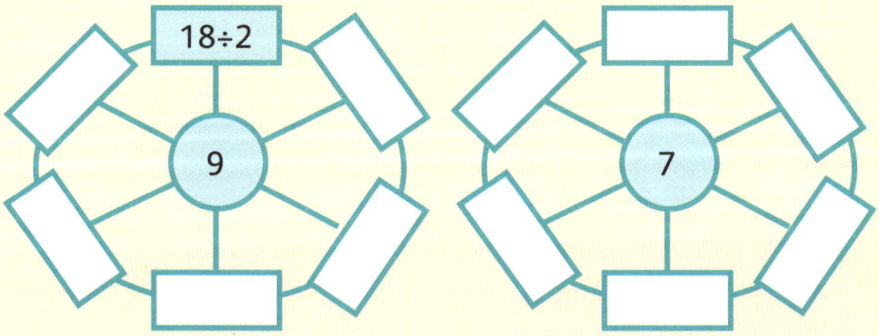

2 Here are some numbers.

800 456 1 845 2 976 720

List the numbers that are divisible by:

a 5 b 8 c 9

3 Use long division to solve these division problems. Remember to make estimates.

a 3 405 ÷ 5 b 7 239 ÷ 9 c 3 405 ÷ 15

d 6 796 ÷ 31 e 4 980 ÷ 25 f 20 095 ÷ 36

> Some answers will leave a remainder.
> Write these answers as mixed numbers.

4 Mrs Williams is running a virus check on her computer. Her display shows this information:

> Progress
> Status: Scan progressing ...
> Product: Fraud.Sysguard
> Estimated time left: 18 540 minutes

How many hours will the virus check take?

5 Complete these divisions. Make an estimate first. Use multiplication to check each one.

a 13 266 ÷ 201 b 27 608 ÷ 136

c 26 532 ÷ 402 d 55 216 ÷ 136

What I have learned

Copy and complete these sentences.

1 Multiplication and division are _____ operations.

2 You can use tests of _____ to decide whether a number is exactly divisible by another without doing the division

3 Numbers that are divisible by 9 have a _____ sum of 9.

4 To check if a number is divisible by 8, you can check to see if the last _____ digits are divisible by 8.

5 The answer to a division is called the _____ . The number being divided is called the _____ and the number you are dividing by is called the _____ .

6 When one number does not divide exactly into another, you are left with a _____ . The answer can also be written as a _____ number.

18 Self check

	Check you can do this:	If you have difficulty, try this:
1	Use the relationship between division and multiplication to help find related facts.	Go back to page 244 and remind yourself about inverse operations and number facts. Also look back at page 246.
2	Use tests of divisibility for 5, 8 and 9.	Go back to pages 248–249, and remind yourself of the patterns to test for divisibility.
3	Solve division problems involving equal sharing or equal grouping.	Look back at the work you did on pages 250–251.
4	Use rounding to help make estimates.	Go back to page 254 and look at examples where you used rounding to estimate your answer.
5	Use multiplication to check divisions.	Revise multiplication and division facts and fact families, and how to use the inverse to check each calculation.
6	Use mental and written methods to divide by a 1-digit number.	Go back to page 254 and remind yourself of different strategies you used to divide.
7	Give answers as a mixed number when a remainder is left.	Go back to page 252 and remind yourself how to express a division sentence as an improper fraction, and the answer as a whole or mixed number.
8	Use written methods of division to divide larger numbers by 2-digit or 3-digit divisors.	Look back at the work you did on pages 256–258 to remind yourself of the strategies you can use.

Chapter 19 Solving real-life problems

In this chapter you will explore this focus question:

What are the roles of financial institutions?

To answer this question, you will:

- identify missing facts and hidden questions
- solve one and two-step problems using different operations
- know and use terms associated with financial institutions, savings and loans
- estimate, calculate and check the solutions to real-life problems
- identify types of financial institutions and their roles/functions
- tell the importance of being honest.

Starting point

Candace scores 3 000 points on a computer game. Maliki scores 1 000 points more.
The children play the game for 2 hours. Amy scores double the number of points as Maliki.
How many points does Amy score?

Talk to your partner about:

- What information you know?
- What you need to find out?
- What information is useful and what is not?
- Is there a hidden question that you need to answer before you can find a solution?
- What calculations do you need to carry out to solve the problem? Which operations will you use?
- What methods will you use?

Using calculators

Key maths idea

Calculators can help you to perform bigger calculations quickly or to check your answers. You can also use them to explore mathematical **patterns** and rules. However, you still need to know what operations to use and in which order.

Look at this problem.

Dana has 384 marbles in her collection. Marbles are sold in bags of 12.

Each bag costs $25.

How much did Dana spend on marbles in total?

These three children have different strategies to work out the answer. Which strategies will give the right answer? Which would you choose?

Here is another type of problem where it is useful to use a calculator:

The product of three whole numbers is 540.

Two of the numbers are less than 10. The third number is between 10 and 25.

What could the three numbers be? Is there more than one solution?

Key words

calculator

pattern

Make sure you know how to use the basic function keys on a simple calculator. Do a two-step calculation, for example, $40 + 6 \times 10$. Do any of your calculators give different results? Remember that only an algebraic calculator will calculate according to the BODMAS rules. So always estimate your answers first.

You could easily try out some different examples with the calculator or use reasoning to help you first. Ask yourself questions such as:

- What do I know about even products?
- Can tests of divisibility help to identify one of the numbers?
- What numbers can I multiply that result in a multiple of 10?

Work together with your teacher to find all possible solutions.

1 Practise using the calculator to compute the following. Make estimates first.

a 1 949 × 12

b 1 345 ÷ 5

c 145 687 – 36 987

d 603 376 + 215 342

e 17 × 37.5

f 2 655 ÷ 9

g 85.368 – 18.67

h 56 782 + 37 106 + 4 841

2 Mrs Baker sells mangoes at the market. She buys the mangoes from a farmer at $500 per dozen. She sells them at $50 each. If she sells 60 mangoes, how much profit does she make?

3 Use a calculator to help you explore these problems.

a The sum of three consecutive odd numbers is 141. What are the three numbers?

b The sum of four consecutive numbers is 398. What are the four numbers?

Numbers that are consecutive come one after another.

4 Work with a partner. You each need a calculator. Choose ten numbers between 500 and 5 000. These are your target numbers. Write them in a table like this:

Target numbers	**Multiplication calculation:** ☐ × ☐

Now follow these steps for the first target number:

- Enter any number into the calculator.
- Press the multiplication (×) key.
- Within 5 seconds enter another number that will give a product close to the target number.
- Press the equal (=) key.

Who is closer to the target number? How might estimates help next time?

Repeat for the other target numbers.

The three-step problem-solving approach

Key maths idea

When you solve a problem, you follow a three-step **problem-solving approach**.

- The first step is to understand the problem. What is the situation? What information do you have? What are you trying to find out?
- The second step is planning. This might involve selecting an operation to use, writing a number sentence, drawing a diagram or sketch or even making a model.
- The third step is solving the problem. You use the method you chose to work out the answer. Check the answer. If it doesn't make sense, try again.

Example

368 children take part in a carnival. Each child needs a ribbon for the carnival.

Ribbons come in packs of eight. How many packs are needed?

UNDERSTAND: What do you need to find? What do you know?

You know there are 368 children in total and they each need 1 ribbon. We know that ribbons come in packs of 8. We need to find out how many packs of 8 ribbons are needed.

PLAN: How can you solve the problem?

You can write a number sentence to solve the problem. Write a division sentence to divide 368 into groups of 8.

Key word

problem-solving approach

SOLVE: Find the solution or solutions.

368 ÷ 8 =

Make an estimate first.

400 ÷ 8 = 50

Now calculate. For example:

368 ÷ 8 =

320 ÷ 8 = 40

48 ÷ 8 = 6 +

46

46 packs are needed.

This is close to the estimate.

You can check using multiplication 46 × 8 or with a calculator.

1 Use the three-step problem-solving approach to help solve these problems. Remember to make estimates and check your answers.

- a Devon rides his bicycle for 9 km each day for a week. How many kilometres does he cycle in total?
- b Susan saves an equal amount each month. She saves $36 000 in one year. How much did she save each month?
- c 315 tickets are sold for the school play. Each ticket is $100. How much money was taken in ticket sales?

2 Look at these problems. There is a missing piece of information in each one that is needed to solve the problem. What is it?

- a Kaci saves $5 from her lunch money each day. How much money does she save altogether?
- b Each student in Grade 5 took an apple to sell at the school fair. They sold the apples for $10 each. How much money did they collect altogether?
- c Bill earns $20 000 per week. One week, he receives a bonus for overtime. How much money does he take home, including his bonus?

Now make up the missing information and solve the problems in Exercise 2.

3 Solve these problems.

- a Altogether 70 boys and 30 girls ate lunch at the school canteen on Wednesday. The canteen received $15 000 for the lunches. Work out the cost of one lunch.
- b A vendor sells 40 bags of sugar cane at $20 per bag. How much money did he earn from selling the sugar cane? Check your answer with a calculator.

What do you notice about the information in the problems in Exercise 3?

Solving two-step problems

Key maths idea

For some problems, you need to work out some additional information before you can work out the answer.

A two-step problem involves using two of the four arithmetic operations to solve it. We will need to identify a 'hidden' question to help find a solution.

Look at this problem. We can use the three-step problem-solving approach to help us.

Example

Romano saves $50 from his allowance each week. At the end of 4 weeks he puts half of his **savings** in his bank account. How much money does he put in the bank?

UNDERSTAND: What do you need to find? What do you know?

We know that Romano saves $50 a week for 4 weeks and he then puts half in the bank. We need to find out how much he puts in the bank.

PLAN: How can you solve the problem?

Identify the hidden question as 'How much does Romano save after 4 weeks?'. We can write a number sentence for each of the two steps. A multiplication to find out how much Romano saves after 4 weeks and then a division to divide this amount by two.

SOLVE: Find the solution or solutions.

Step 1:

Find out the total amount saved in 4 weeks:

1 week [$50]

4 weeks [$50] [$50] [$50] [$50]

So 4 weeks = [$50] × 4 = $200

Key word

savings

Step 2:

Work out $\frac{1}{2}$ of the savings.

$200	
$100	$100

$\frac{1}{2}$ of $200 = $100

Or $200 ÷ 2 = $100

Finally, make sure to look at the original question again. Did you answer it?

Check the answer on a calculator.

Use the three-step problem-solving approach to help you with these problems. Talk to your partner about the 'hidden' question each time.

1 **a** Which two bar models – i, ii, iii or iv – represent the two steps in this problem?

A class has 28 boys and 19 girls. One Tuesday, 11 students were absent. How many students were present?

i

girls	
boys	class

ii

class	
boys	girls

iii

class	
present	absent

iv

present	
class	absent

b Now write the number sentences and solve the problem.

2 The table shows the amounts of money a group of friends collected for charity.

Gary	$750
Sandra	$600
Bev	$550
Roy	$900

The money was donated equally between two charities.

How much money did each charity receive? Sketch two bar models to represent each step in the problem.

3 Here is the first step in a two-step problem: Joy had $350.
She gave Lorna $\frac{1}{5}$ of her money.

a Which bar model – i, ii or iii – represents this first step? Explain how you know.

i

Lorna's money				
Joy's money				

ii

Joy's money			
Lorna's money			

iii

Joy's money				
Lorna's money				

b The next day, Joy was given $350. How much money did Joy have in total? Sketch a bar model to match this second step.

4 Solve this problem. A maths test lasted for 45 minutes. There were 9 questions to complete. A second test had 14 questions and allowed the same time per question as the first test. How long did the second test last?

Remember to use a calculator to check your answers.

Solving problems about patterns

Key maths idea

We can also use the three-step problem-solving approach to help us solve problems about patterns.

Read this example together.

Example

Malcolm writes a number pattern.
It begins with 0, 2, 6, 12, 22.
What will the next four numbers in this pattern be?

UNDERSTAND:
What do you need to find?
What do you know?

PLAN:
How can you solve the problem?

SOLVE:
Find the solution or solutions.

Look at the numbers in the pattern.

2 = 0 + *2* (previous number is 0, add 2 to make the next number 2)

6 = 2 + *4* (previous number is 2, add 4 to make the next number 6)

12 = 6 + *6* (previous number is 6, add 6 to make the next number 12)

22 = 12 + *10* (previous number is 12, add 10 to make the next number 22)

The number added increases by 2 each time.
The next four numbers are:

22 + *12* = 34, 34 + *14* = 48, 48 + *16* = 64, 64 + *18* = 72

So the pattern continues as 0, 2, 6, 12, 22, <u>34</u>, <u>48</u>, <u>64</u>, <u>72</u>

1 Here is part of a pattern.

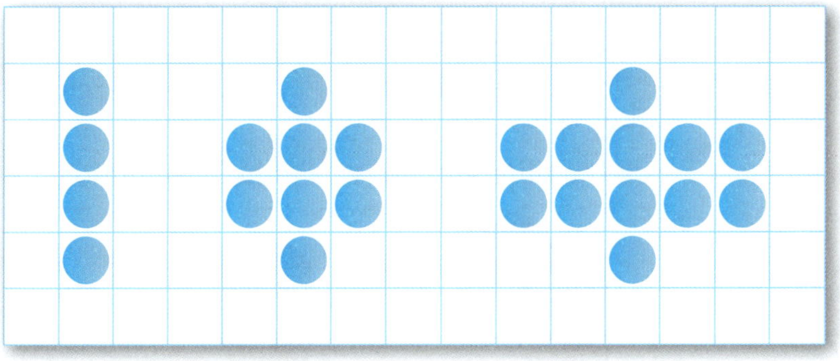

We can also call the shapes in a pattern like this Term 1, Term 2 and Term 3, or Train 1, Train 2 and Train 3.

a How many blue dots will be needed for next three shapes in the pattern?

b Work out how many dots the tenth shape will have.

2 Here are some number patterns. How does the pattern continue each time? Write the next four numbers.

a 6, 12, 24, 48

b 800, 400, 200, 100

c 12.575, 125.75, 1 257.5, 12 575

d $5\frac{3}{4}$, $5\frac{1}{4}$, $4\frac{3}{4}$, $4\frac{1}{4}$

3 Here is another pattern.

a What are the values of the empty squares in this pattern?

b What will be the value of the 4th triangle in the pattern?

c Lloyd says: '12 is in the sequence, so 120 must also be in the sequence.' Do you agree? Explain your thinking.

Try making up some patterns of your own for others to continue.

Other types of problems

Key maths idea

Here are some different types of problems. They require you to think about ways to arrive at a solution and will also require you to think logically.

Read each problem together and decide which of the children's suggested strategies on page 272 could help you reach a solution.

Romano arranges a truck, a train and a car in a line. *In how many different ways could he do this?*	*There are 20 fruits in a box. 5 are apples, 6 are oranges and 4 are bananas.* *What fraction of the box is made up of other fruits?*
Lloyd writes down a number. He adds 5 and then multiplies by 10. The answer is 450. *What number did Lloyd write down?*	*Kaci collects 5 leaves on Monday. Each day after that she collects twice as many leaves as the day before. How many has she collected in total by the end of the week on Sunday?*

Now work through the problems with your teacher.

What might you do differently for the following problem? Discuss your ideas with a partner.

Amy and Devon have 20 stickers between them.

Amy has 2 more than Devon.

How many stickers do they each have?

Work together with your teacher to try out some ideas and solve the problem.

Remember to use the three-step problem-solving approach to help you.

Solve these problems. Think about the approaches you will use.

1 100 people visit the library on Saturday. 50 are adults, 20 are teenagers and the rest are younger children. What fraction of the 100 people is made up of younger children?

2 Candace pays \$200 in total for a t-shirt and a scarf. The t-shirt costs \$80 more than the scarf. How much does each item cost?

3 Bev is double the age of Ann. Ann is three years older than Troy. Troy is 18 years old. How old is Bev?

4 There are some cows and chickens on a farm. A total of 44 feet can be counted. How many of each type of animal could there be?

5 Dana has 4 animal stickers to put on the page in her album.

In how many different ways can she arrange the 4 stickers on the page?

Financial institutions

Key maths idea

Banks are the most commonly used examples of **financial institutions**. Financial means to do with **money**. Financial institutions help us to save and invest money, and to keep track of how much money we have and what we are using it for. Insurance companies and investment companies are also financial institutions.

When you have a **bank account**, you can **deposit** money into the account and **withdraw** money from it. There are different types of accounts. A **current account** allows the account holder to make payments, deposits and **withdrawals** easily for everyday transactions. A **savings account** is designed to let you keep extra money in a safe place. The bank may pay you **interest** whilst the money is **invested**.

Here are some other words you need to know:

Loan: A **loan** is an amount of money that you borrow from a bank or a lending institution. The original amount of money you borrow is called a **principal** loan.

The financial institution charges interest on loans. This is how banks and credit unions earn money – by lending it out and charging interest.

Interest: This is the money that is paid on a loan from a financial institution.

Interest can also refer to the money that you earn when a bank is looking after your money.

We calculate interest as a **percentage**. For example, if the bank pays you 4 % interest per annum on your savings, you will get $4 for every $100 in your account each year. 'Per annum' means 'per year'.

Key words

financial institution
money
bank account
deposit
withdraw
current account
savings account
interest
invest
loan
principal
percentage

Talk about

Mrs Kingston applies for a job in a building society. She notices that the job description includes that the person should demonstrate '**honesty** and **trustworthiness**'. Why do you think that these qualities are important for someone working in a financial institution?

Key words

honesty

trustworthiness

1 Here is a list of some types of financial institutions. Use the internet to research them to find out about their roles and the services they provide.

Financial institution	Roles and services
Commercial bank	
Building society	
Credit union	
Friendly society	
Partner scheme	
Insurance company	
Cambio	

2 Mr Kingston has $5 000 to invest in a new savings account. He will not withdraw any money in the first year. He finds out about the interest rates and offers on some different accounts.

Account A: 4% per year

Account B: 5% per year

Account C: 6% per year

Account D: Earn $200 when you open an account and then 3% interest per year on total amount

a Calculate how much interest Mr Kingston will earn in the first year in the different savings accounts.

b Explain to Mr Kingston which savings account he should open and how much money he will have in total at the end of the year.

We also describe interest per year as 'per annum'. This is often abbreviated as p.a. so Account A, for example, gives interest of 4% p.a. (or per annum).

3 In the Talk about on page 274 you discussed why honesty and trust are important when dealing with money. Read these scenarios. What should each person do?

a Amy lends Devon $500. He promises to pay her back after a week. A week later, Devon has money to pay her back, but he notices Amy has forgotten.

Should he:

- keep the money – after all, it is not a lot and she has forgotten
- remind her and pay her back
- wait until she remembers?

b Candace buys 15 kg of mangoes. She notices that the cashier rang up 5 kg, not 15 kg. Should she:

- pay for 5 kg and keep the extra money
- pay for 5 kg and give the money to someone else as a gift
- tell the cashier his mistake and pay for 15 kg?

Talk about

What does it mean to be honest? Why is honesty important? What problems does it cause when people are dishonest?

4 A **cambio** is offering the following exchange rates to buy 1 unit of each of these **currencies**, for example, it will cost $154.20 to buy US $1.

US dollar (USD)	154.20
Barbados dollar (BBD)	77.11
UK pound sterling (GBP)	216.62
Euro (EUR)	186.61
Australia dollar (AUD)	119.77

How many Jamaican dollars does it cost to buy:

a 10 units of each currency, for example, US $10

b 100 units of each currency, for example, US $100

c 1 000 units of each currency, for example, US $1 000

d 500 units of each currency, for example, US $500?

e Now choose some amounts of your own, for example, 200 units.

Key words

commercial bank
building society
credit union
friendly society
partner scheme
insurance company
cambio
currency

Maths detective: Money matters

Use your maths detective skills to solve these cases involving money.

Case 1

You want to buy a game. It costs $580. You decide to save $20 each week for the first four weeks, $30 each week for the next four weeks and $40 each week for the four weeks after that. How much do you need to save each week for another four weeks so you now have enough for the game?

Case 2

You want to exchange some money at the cambio for a trip to Florida. Cambio A is advertising an exchange rate of US $1 = $152.35 and Cambio B is advertising US $1 = $153.50.

- Which cambio should you use? Why?
- How much less will you need to pay to buy US $200 at this cambio?

Case 3

You start with some money in your savings account. You then deposit another $2 000. 4% interest is paid on your account. After a year you have $3 120 in your savings account. How much money did you have in your account to start with?

Real-world maths: Imports and exports

In Jamaica, money is used to purchase goods from other countries (imports).

Money is also earned from goods sold to other countries (exports).

Money earned from exports is called foreign exchange.

Search on the internet to help answer these questions.

1 In pairs, find out the foreign exchange Jamaica gets from exporting goods to:

	Country/region	Foreign exchange
a	USA	$1 = US $ ________
b	United Kingdom	$1 = ________
c	Canada	$1 = ________
d	Eastern Caribbean countries	$1 = ________
e	European Union countries	$1 = ________

2 In small groups, find out at least three goods imported from and three goods exported to each of the countries or regions listed in the table.

3 Present your findings in creative ways, for example, by using PowerPoint, charts or models.

What I have learned

True or false? If the statement is false, give a reason.

1 Banks lend money without any conditions.
2 A deposit is putting money in a jar.
3 Lenders and borrowers take a risk of losing their money.
4 Savings accounts and credit accounts are types of financial accounts.
5 Interest is the price borrowers pay for using someone else's money.

Practice questions

1 Solve these problems:

a Bottled drinks cost ten dollars each at a tuck shop. In one day, the tuck shop sold one hundred and fifty bottled drinks. How much money did the tuck shop make from bottled drinks that day?

b Shamila leaves home at 8:00 a.m. She starts work one hour later and finishes as 5:00 p.m. She spent two hours answering emails. How many hours did she spend at work?

c Cara bought a television valued at $60 000 on hire purchase. She paid the deposit of $6 000, then made monthly instalments of $6 000 for one year. How much did she pay altogether for the television?

2 Here is the start of a pattern. What are the next five numbers? 5, 10, 20, 40

3 Solve these problems:

a Chanice leaves school and walks home. It takes her 35 minutes. She has dinner $1\frac{1}{4}$ hours after she gets home. She has dinner at 17:30. What time did she leave school?

b In a class of 30 children, $\frac{3}{5}$ of them are 10 years old. 5 of the children are 11 year old girls. How many of the children are 11 year old boys?

4 Match each term in the first column of the table to the correct definition.

Term	Definition
1 Deposit	a An account where you leave your saved money.
2 Credit card	b To take money out of an account, for example, from an ATM.
3 Savings account	c A card that uses the money you already have in your bank account.
4 Interest	d A card that allows you to use money borrowed from a financial institution.
5 Loan	e An account you use for day-to-day transactions.
6 Withdrawal	f To put money into an account.
7 Current account	g Money that is paid on a loan from a financial institution.
8 Debit card	h An amount of money that you borrow from a bank or a lending institution.

19 Self check

	Check you can do this:	If you have difficulty, try this:
1	Confidently use a calculator.	With a partner, remind each other which keys to press on a calculator to enter each of the different operations. Make sure you know where the operator keys, equal sign and 'clear' button are. Also look back at page 264 and remind yourself how to use a calculator to help you to solve problems.
2	Explain and use each step of the three-step problem-solving approach.	Look back at page 266, and remind yourself of the three steps – understand, plan, solve. Don't forget that part of solving is checking your answer and trying a new strategy if the answer does not look right.
3	Identify the 'hidden' question in a two-step problem.	Look back at pages 268–269. You will still need to use the three-step problem-solving approach. Your understanding and planning steps will help you to work out the 'hidden' question.
4	Write number sentences to help solve word problems.	Look back at any of the problems you have solved in this chapter and identify the number sentence that helped you to solve them. Remember, a number sentence uses numbers and symbols.
5	Make mental estimates, and use a range of mental and written methods to help calculate.	Look back at any of the problems you solved in this chapter. Without working out the answer, give an approximate idea of what the answer will be. Explain how you got that idea. Identify examples where you worked mentally (in your head) and where you used written methods (calculating using a pencil and paper).
6	Use appropriate strategies to solve word problems, for example, working backwards, making lists, etc.	Look back at pages 271–272. What strategies did you use to solve these problems?
7	Identify types of financial institutions and explain their roles.	Read over the information about banks and other financial institutions on pages 273–274 to remind yourself what these institutions do.
8	Explain terms such as: loan, interest, savings, etc.	Revise the terms pages 273–274.

Unit 2: Geometry

Chapter 20 Circles

In this chapter you will explore this focus question:

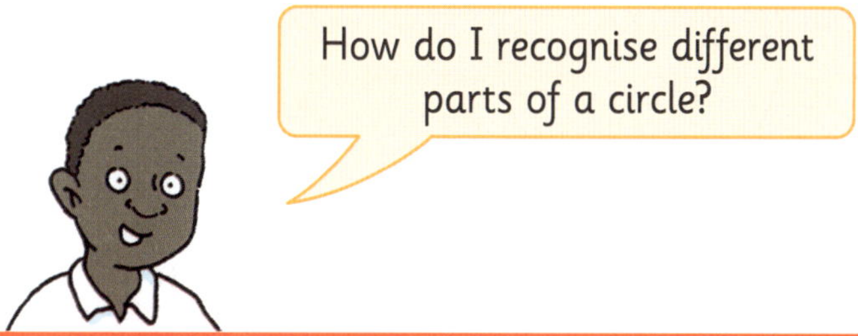

To answer this question, you will:

- make and explore geometric shapes
- solve problems using the properties of different shapes
- identify parts of a circle
- distinguish between circles and other curves.

Starting point

1 Follow these instructions for sketching different solid shapes.

Now think about and try sketching different models of 3-D shapes. Explore how you can use angles and shading to make them appear 3-D.

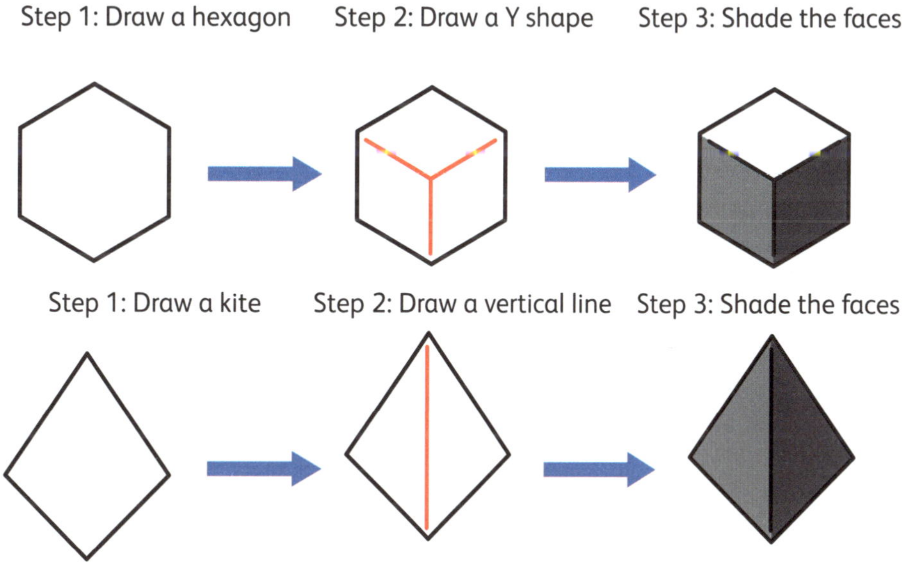

2 Place a cone or marker on the playground. Everyone in the class should stand 4 m from the cone.

What pattern or shape will be formed as more people stand 4 m from the cone?

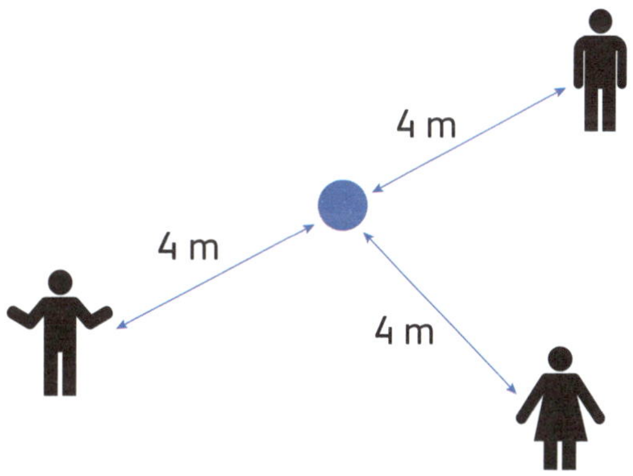

Drawing circles

Key maths idea

The curved outline of a **circle** is called its **circumference**. The instrument you use to draw a circle is called a **compass**. It looks like a pair of dividers in which one of the pointers is replaced by a holder for a pencil.

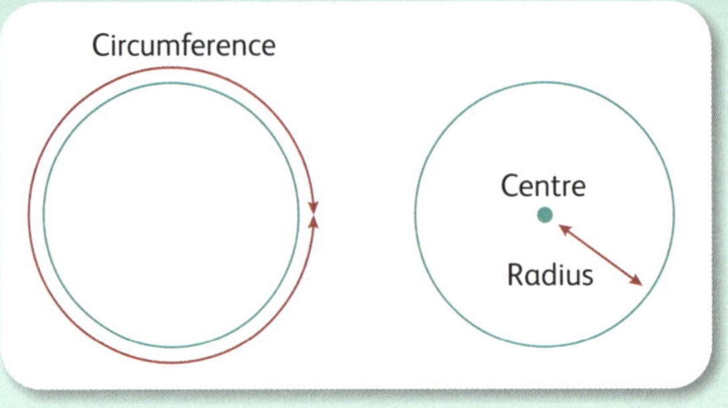

The distance from the **centre** to the circumference is called the **radius**.

To draw a circle with a given radius, follow these three steps:

Step 1: Choose a centre for your circle and mark it with an x or a clear dot.

Step 2: Measure the length of the radius you want to draw.

Step 3: Draw a smooth outline to form the circle around the centre.

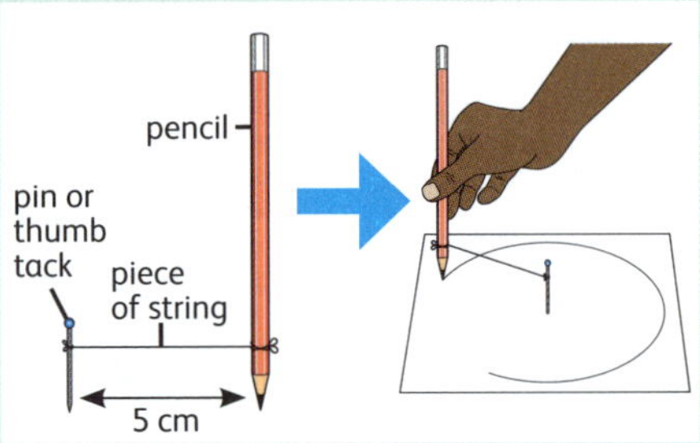

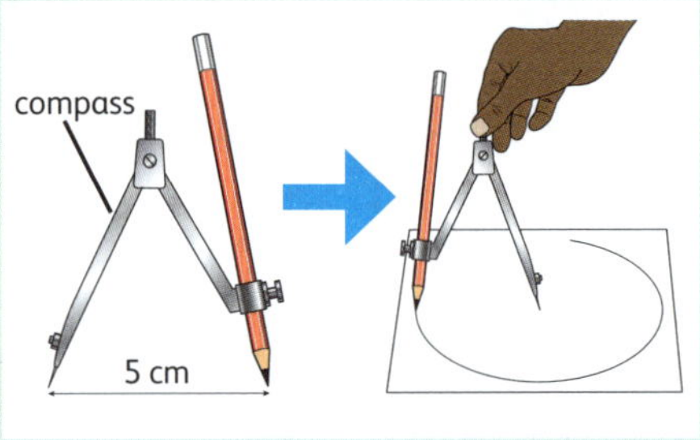

Here are two methods for drawing a circle with a radius of 5 cm. Have a class discussion about two different ways of thinking about a circle:

- as the curved outline
- as the area of space inside the outline.

Key words

circle
circumference
compass
centre
radius

1 Draw your own circle diagram. Label the centre, radius and circumference.

2 Draw circles with the following radius measurements.

a

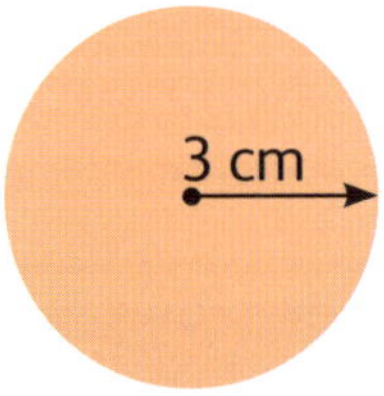

b

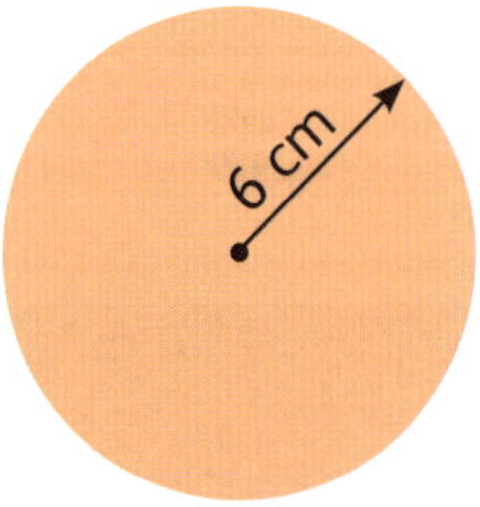

c

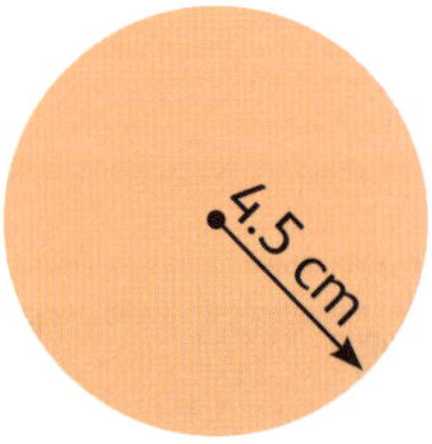

d

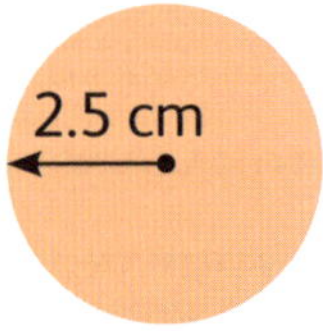

3 Draw concentric circles, like this one, by following the steps.

Step 1: Choose a centre point.

Step 2: Draw a circle of radius 3 cm.

Step 3: Now use the same centre point for circles with a radius of 4 cm, 5 cm, 6 cm and 7 cm.

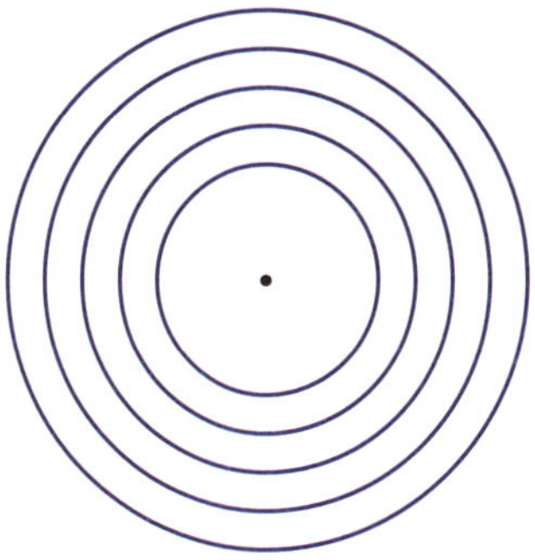

Lines through circles

Key maths idea

A **chord** is any straight line that crosses a circle, intersecting the circumference at two points.

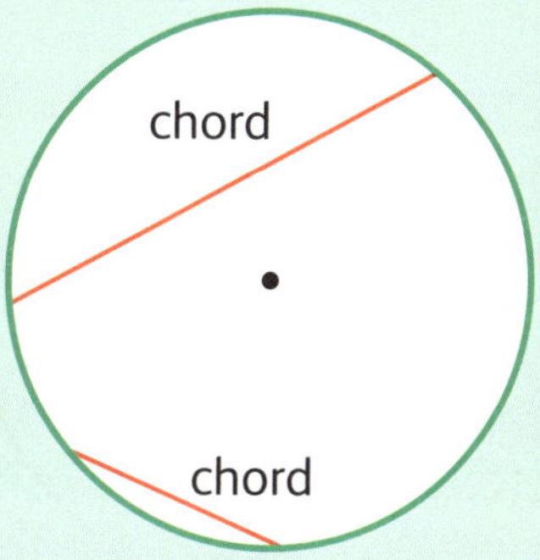

A **diameter** is any chord that passes through the centre point.

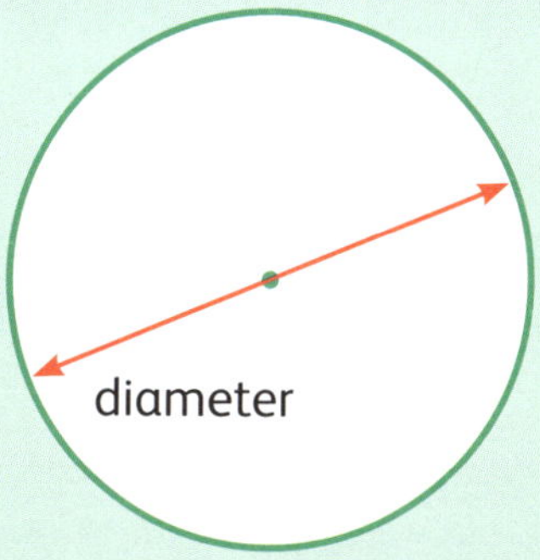

A diameter is the longest possible chord for each circle.

Key words

chord

diameter

1 Draw and label a circle diagram showing three different examples of chords, and one example of the diameter.

2 Measure the radius of each circle below. Then measure the diameter of each circle. What do you notice?

a

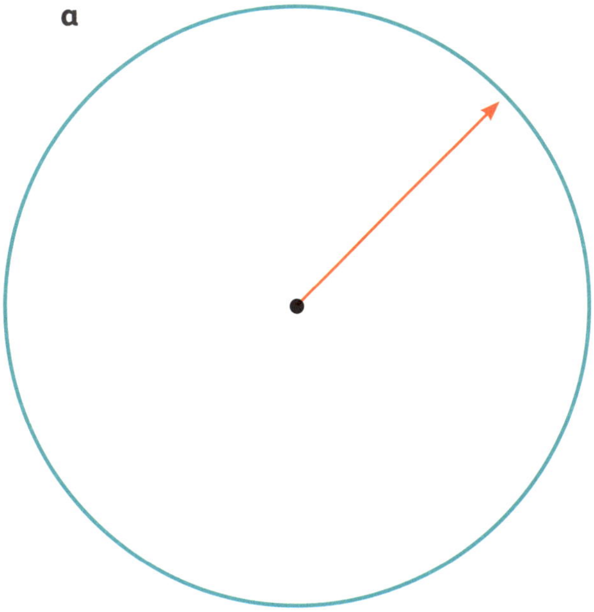

b

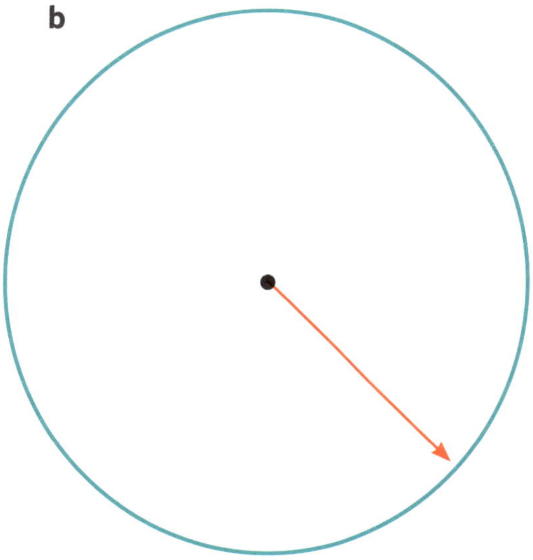

c

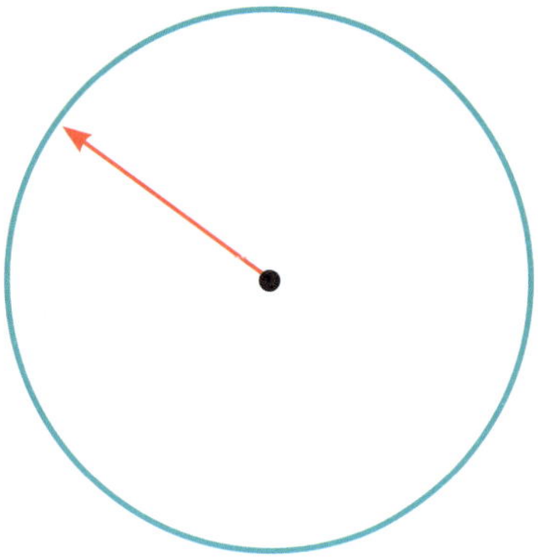

d

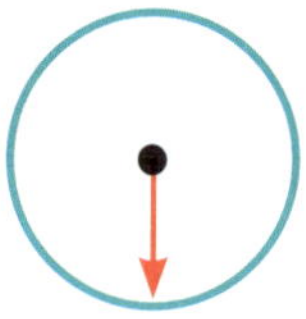

3 The centre point in these circles is not shown. Measure the longest possible chord for each circle. You may need to experiment with adjustments to your ruler as you measure. Write the diameter and radius for each circle.

a

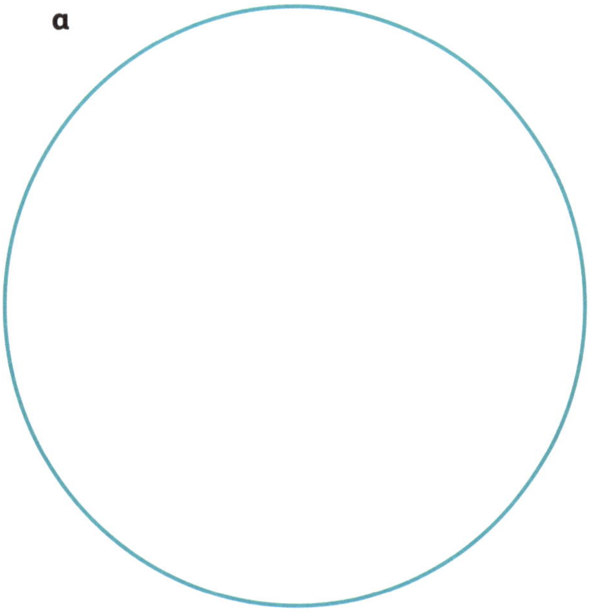

b

c

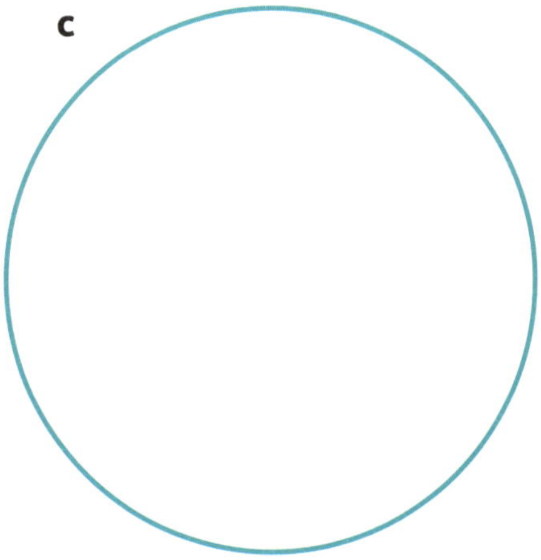

d

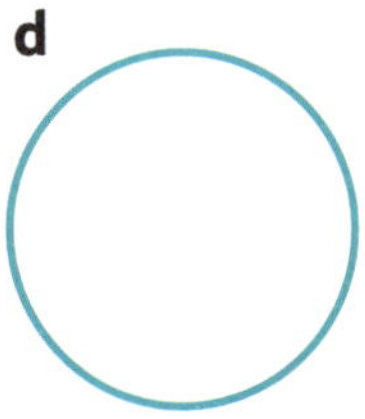

Segments, sectors and arcs

Key maths idea

A chord divides a circle into two **segments**.

A segment is a part of the area of a circle. The area enclosed between two radius lines is called a **sector**.

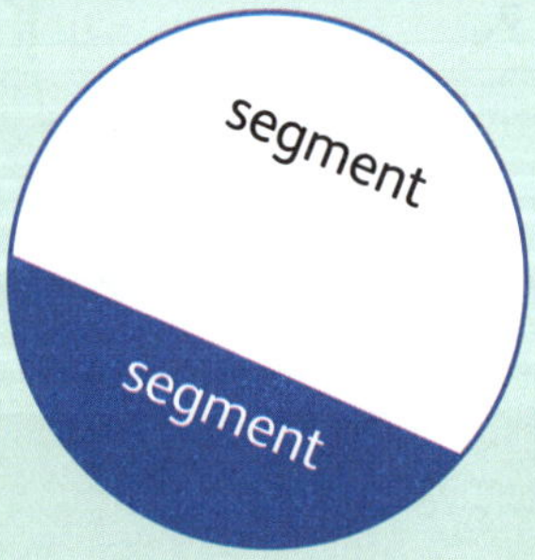

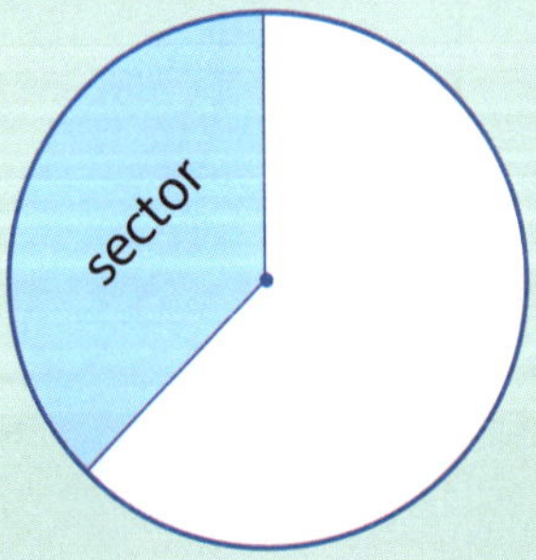

Key words

segment
sector
arc
curve

An **arc** is a **curve** formed as part of a circumference.

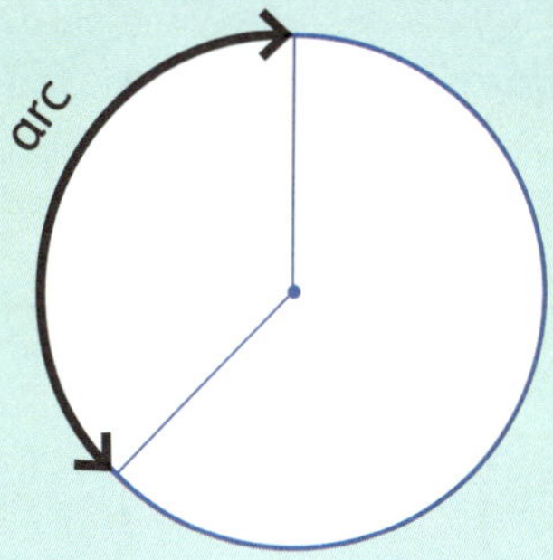

1 Draw your own diagrams to show the difference between a segment and a sector. Label the diagrams accurately.

2 Sectors are often used to show fractions.

Describe the fractions shown in this diagram which uses sectors.

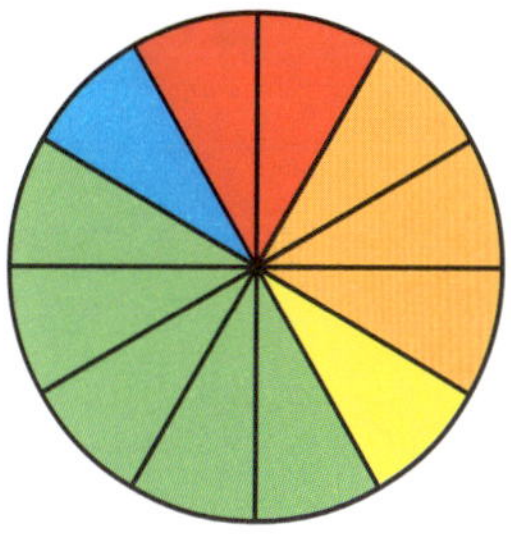

3 A semi-circle is a special kind of sector. A quadrant is a specific sector of a circle.

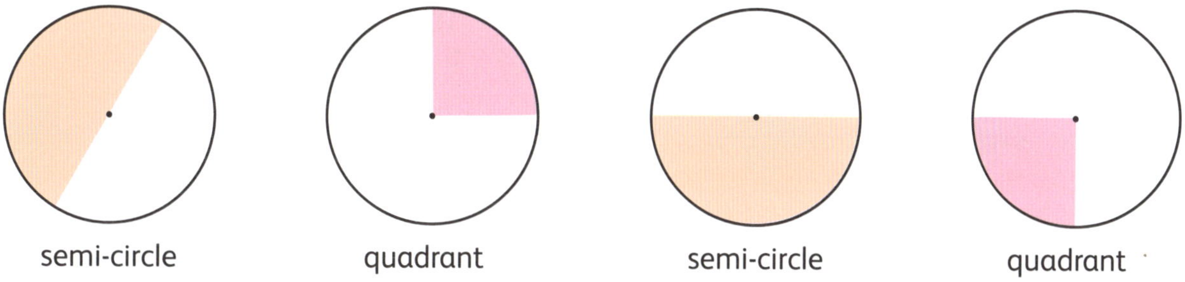

Draw, cut and fold circles to form semi-circles and quadrants. Join them together to form patterns.

4 Use compasses to draw arcs of different sizes. Try to form different capital letters using combinations of arcs and straight lines.

Maths detective: Circle puzzles

Use the clues to draw each diagram.

Case 1

Draw this diagram consisting of four equal circles.

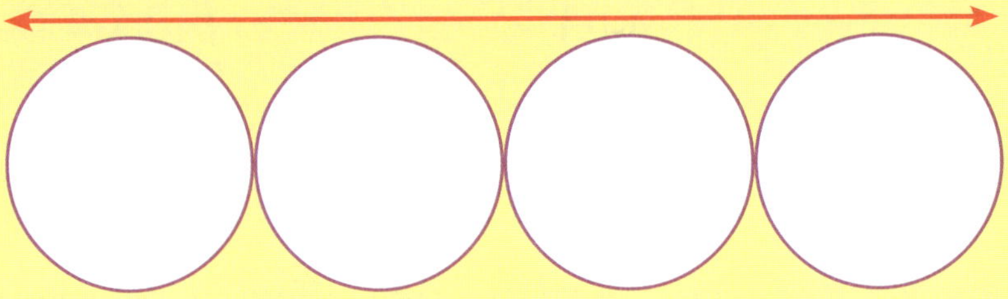

Case 2

Now draw this diagram of three equal circles.

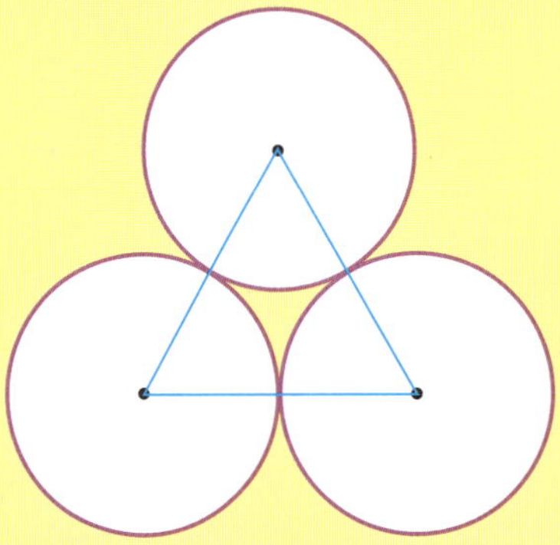

The perimeter of the triangle is 45 cm.

Case 3

Circles are used in art and architecture. Investigate how to make these circle patterns using drawings or cut-out and folded circles.

a

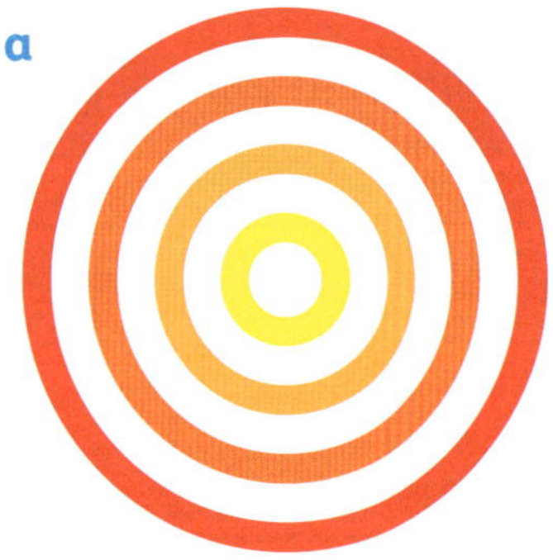

b

c

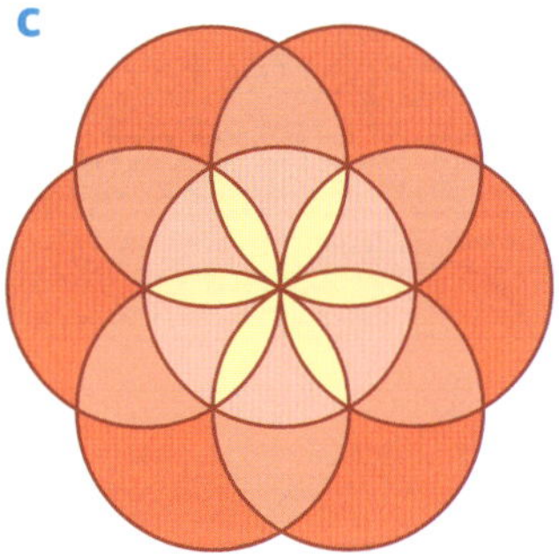

d

e

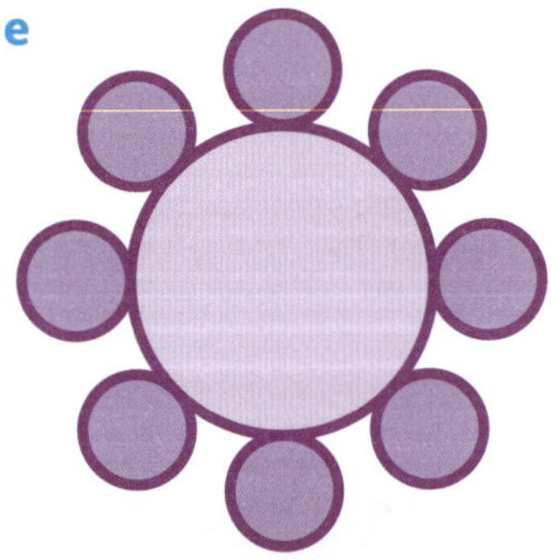

Case 4

Finding the area a circle covers is a problem in mathematics, and in engineering and design.

Estimate the area of each shape, and then try to find a method to approximate the area as precisely as you can.

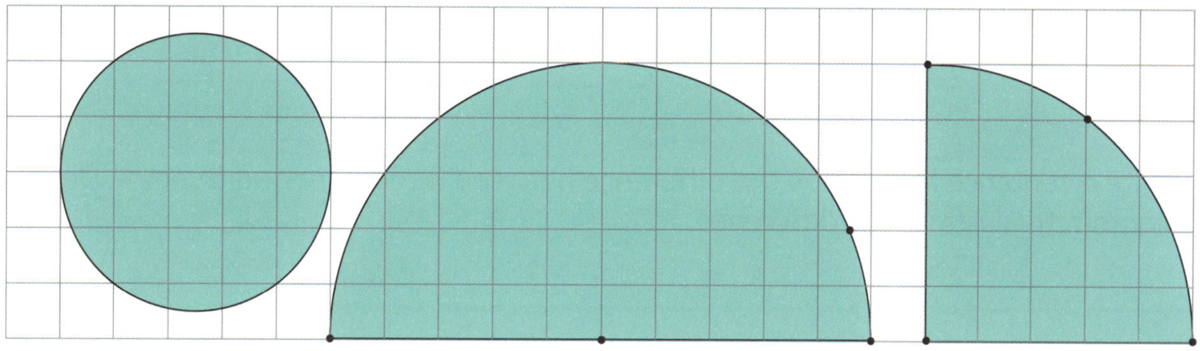

Geometry

Real-world maths: Colours of the spectrum

A circle is an excellent way to show the colours of the **spectrum**. Artists and designers use the **colour wheel** to choose colours.

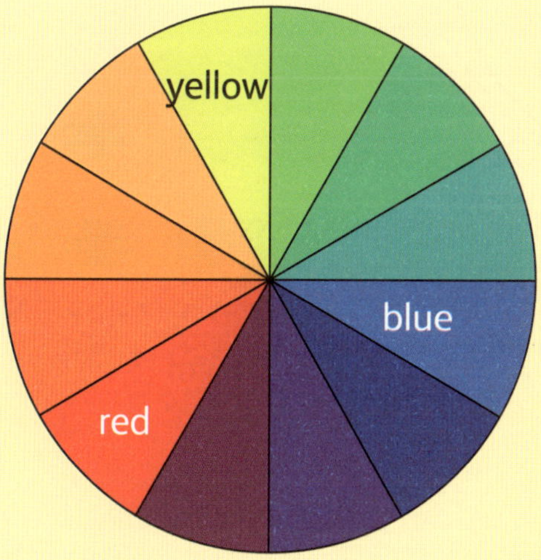

Key words
spectrum
colour wheel
primary colour
intermediate colour
complementary colours

Investigate Isaac Newton's research on colour wheels and the colour spectrum.

Use your circle skills to draw a colour wheel.
Start with the **primary colours** and mix them to form the wheel.
For example, mix red and blue to form purple.
Explain what **intermediate colours** are and how they are created.
Find out more about **complementary colours**.

What I have learned

1 Copy and label each diagram using the following vocabulary:

radius	centre	circumference	chord	diameter

a **b** **c** **d**

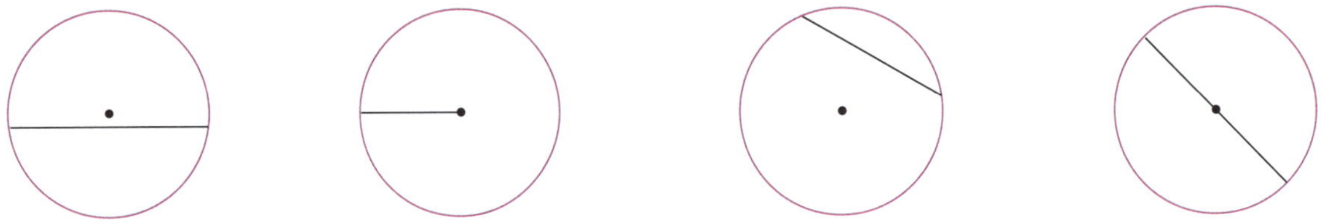

2 Write a sentence to describe the difference between a segment and a sector. Draw a diagram to illustrate the difference.

Practice questions

1 Label each part of the circle correctly.

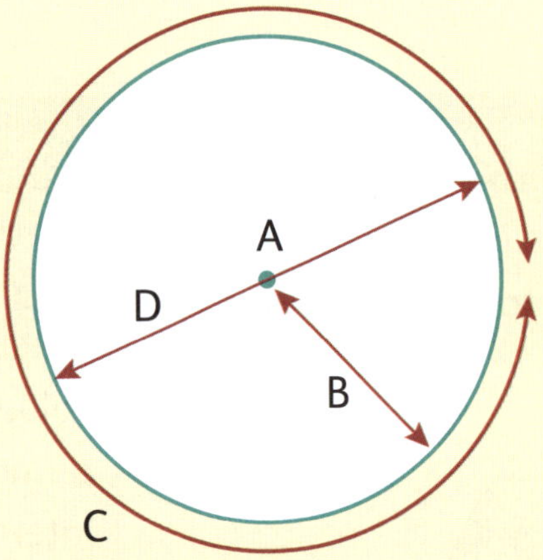

2 Draw each circle accurately. The measurements given are diameters.

a

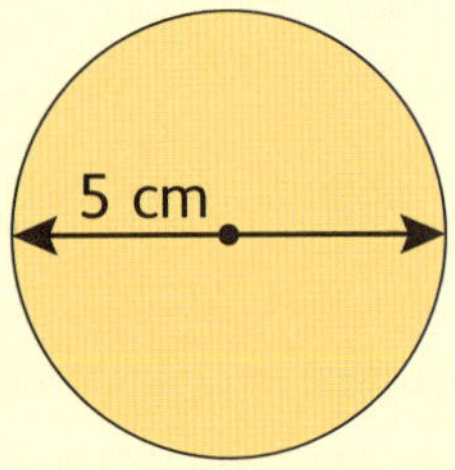

b

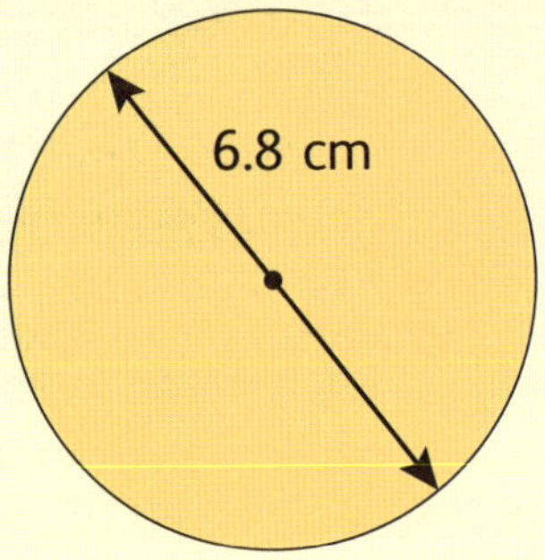

3 Calculate the radius of a blue and a yellow circle.

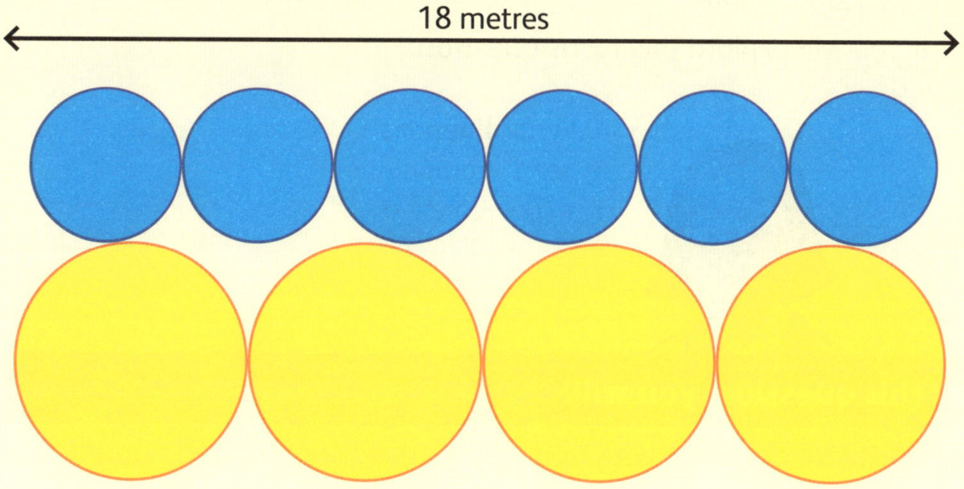

20 Self check

Check you can do this:	If you have difficulty, try this:
1 Draw circles of various sizes with radius given.	Look back at page 280. Remind yourself how to use a compass, or to use a pencil attached to a push pin with some string.
2 Identify and label radius, diameter and circumference of a circle.	Look back at the diagram on pages 280–281 and remind yourself of the parts of a circle.
3 Identify chords, sectors, segments and arcs.	Look back at pages 282–283 and remind yourself of these terms.
4 Recognise curves in a diagram.	Look back at pages 282–283 and remind yourself what we mean by curves.
5 Create designs and patterns.	Look back at the work you did on page 286. Practise drawing different patterns using circles.

Chapter 21 Polygons

In this chapter you will explore this focus question:

To answer this question, you will:

- make and explore polygons with up to eight sides
- use properties of polygons to solve problems
- use triangles and quadrilaterals to make regular and irregular polygons.

Starting point

Look at this diagram of a football field.

What shapes can you see in this picture of a football field?

Which are polygons? How can you tell?

Explore the shapes found in other sports fields.

Remember: the term **polygon** comes from 'poly' meaning many and 'gon' meaning angle. Polygons have three or more straight sides. **Triangles** and **quadrilaterals** are some polygons you have learned about before.

Key words
polygon
triangle
quadrilateral

Using grids to form polygons

Key maths idea

You can use a grid to help you draw line segments and polygons. Polygons must be closed shapes enclosed by straight line segments.

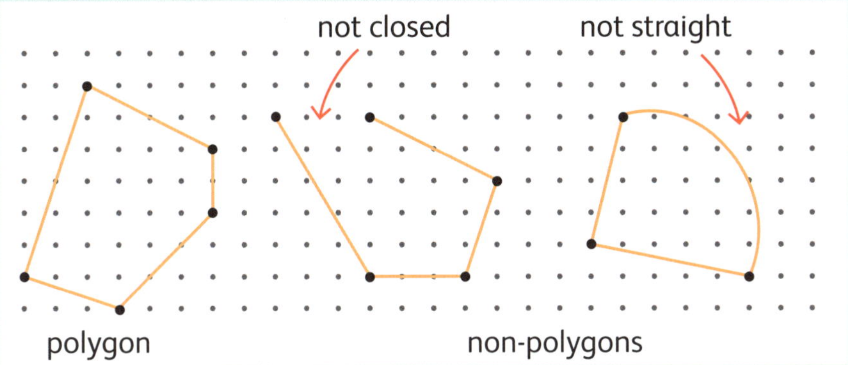

We classify polygons according to how many sides they have:

pentagon	5 sides
hexagon	6 sides
septagon or **heptagon**	7 sides
octagon	8 sides.

Convex polygons have inside angles that are all less than 180 degrees.
These shapes are all convex:

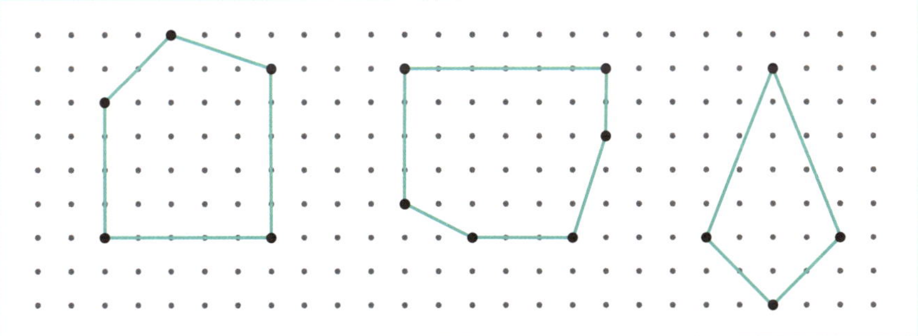

Concave polygons have one or more interior angles that are reflex (between 180 and 360 degrees).

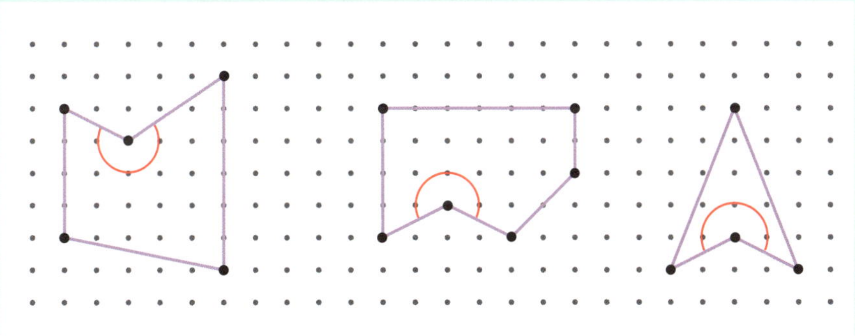

Key words

pentagon
hexagon
septagon/ heptagon
octagon
convex
concave

1 Classify each shape with the correct shape name.

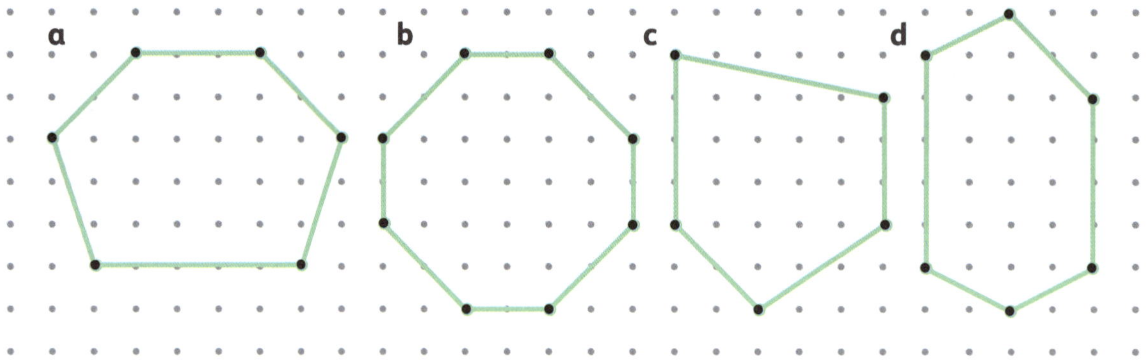

2 Look carefully at these shapes. Identify each as convex or concave and classify it with the correct shape name.

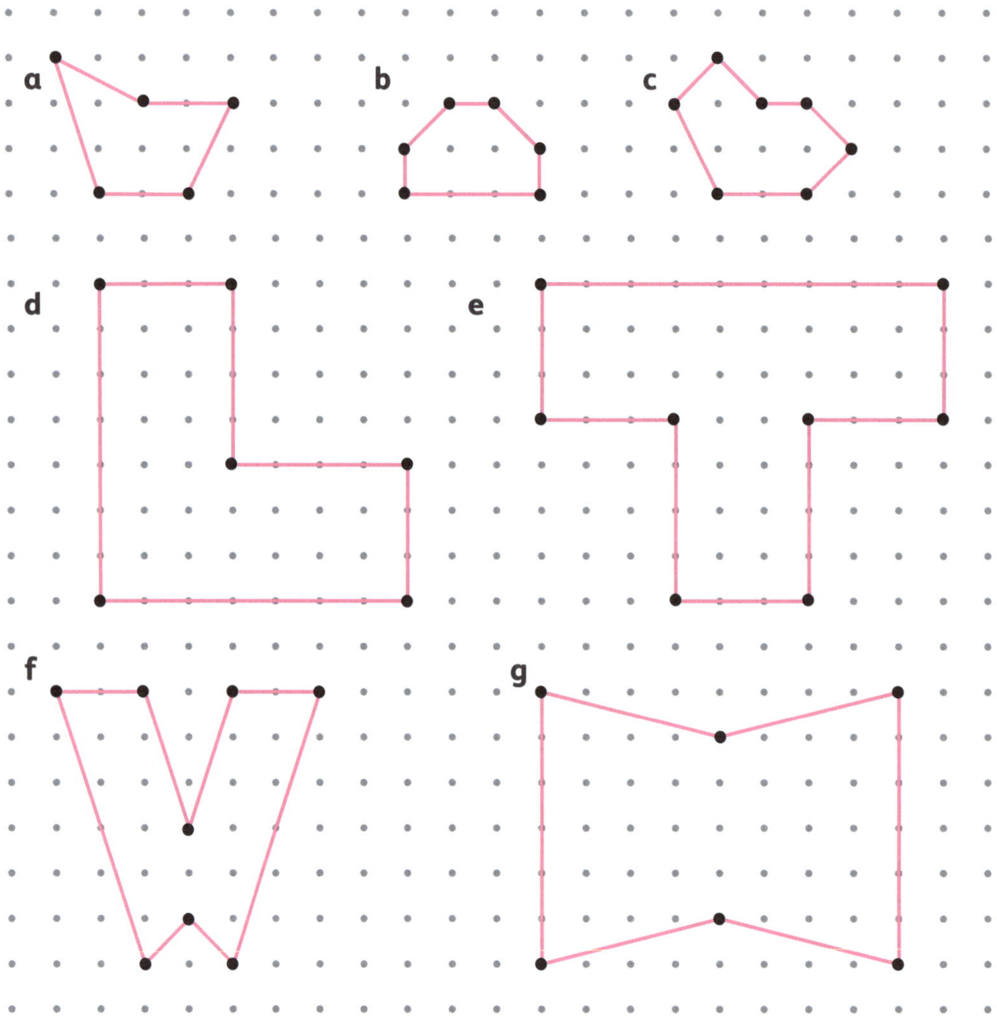

3 Draw three different examples of a concave quadrilateral.

Drawing parallel and perpendicular lines

Key maths idea

You can also use a grid to help you identify parallel and perpendicular line segments.

These sets of lines are **parallel**.

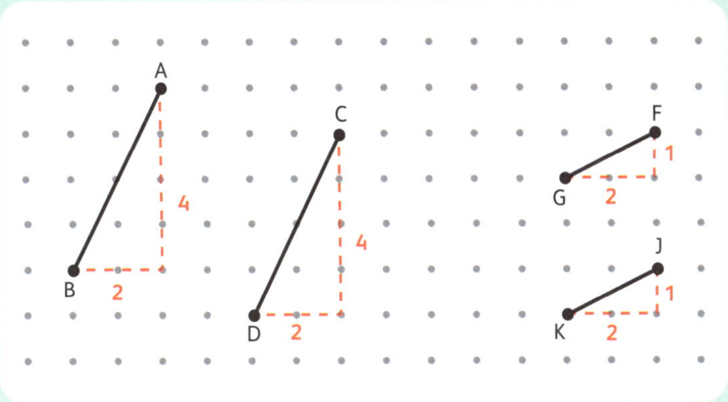

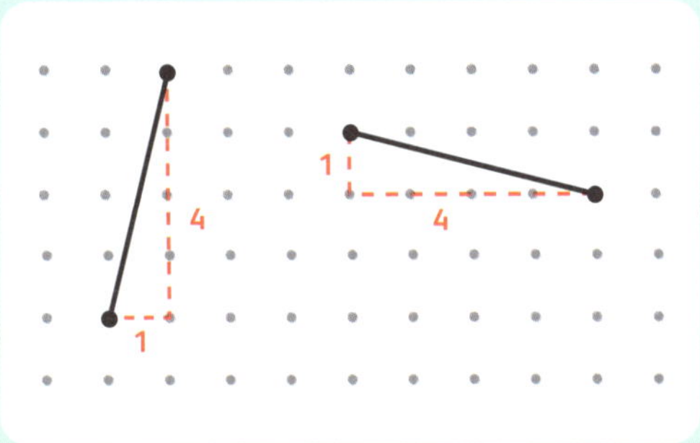

These lines are **perpendicular**. If you join them together, they form a right angle.

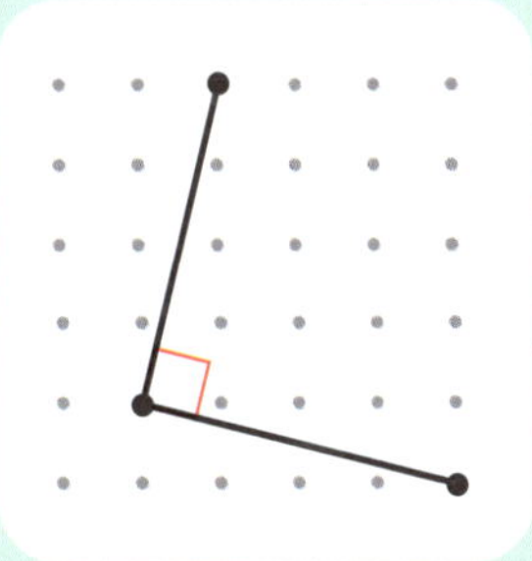

You can use this skill to draw polygons with parallel lines, such as squares, rectangles, parallelograms and trapezia in different orientations.

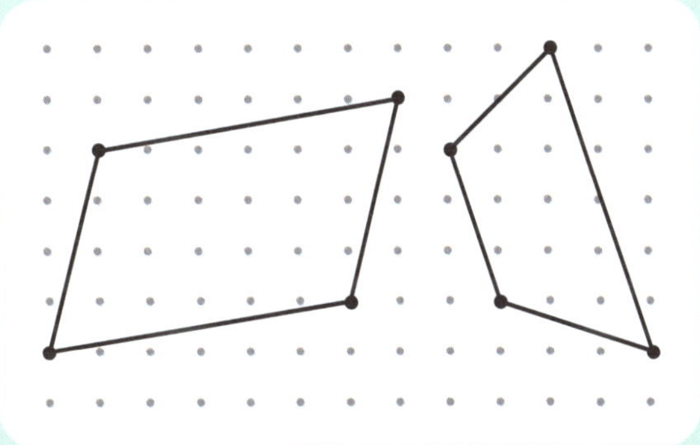

Key words

parallel

perpendicular

Look at line AB. Identify how to move from point A to point B using horizontal and vertical lines. Go 2 units right and 4 up.

Draw another point, C. Go 2 units right and 4 units up. Call this point D. Join CD.

AB is parallel to CD.

Can you see how I did the same with FG and JK?

1 Copy these sets of parallel and perpendicular lines onto a grid.

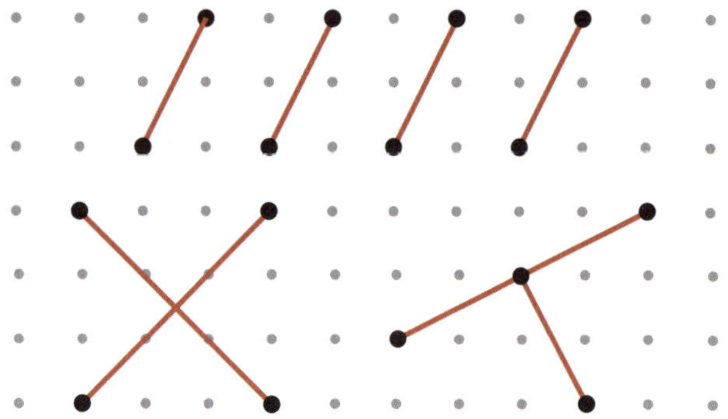

2 Copy each line on a grid or a geoboard. Draw a parallel line segment for each.

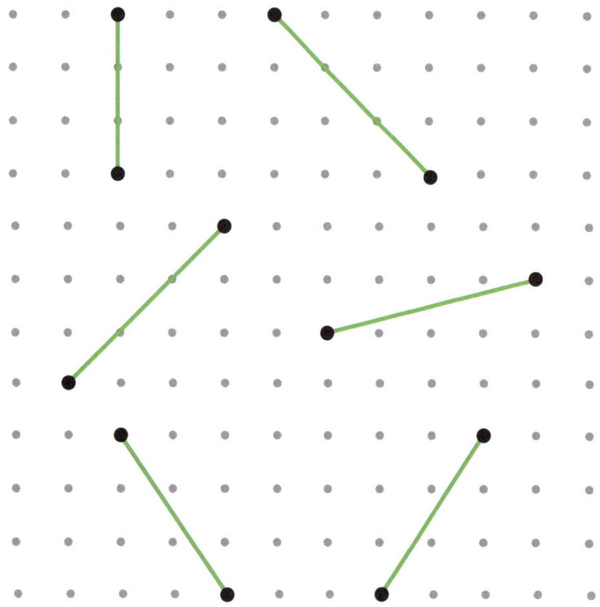

3 Copy each line onto a grid. Draw a line perpendicular to each line segment.

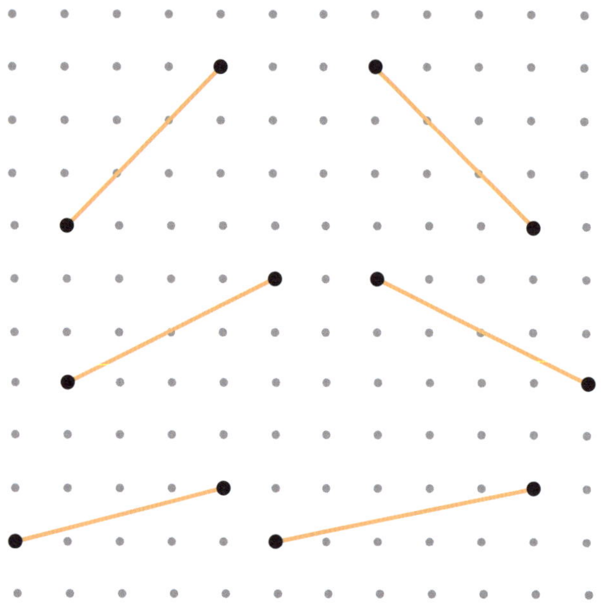

4 Use your new skills to copy and complete each tilted square.

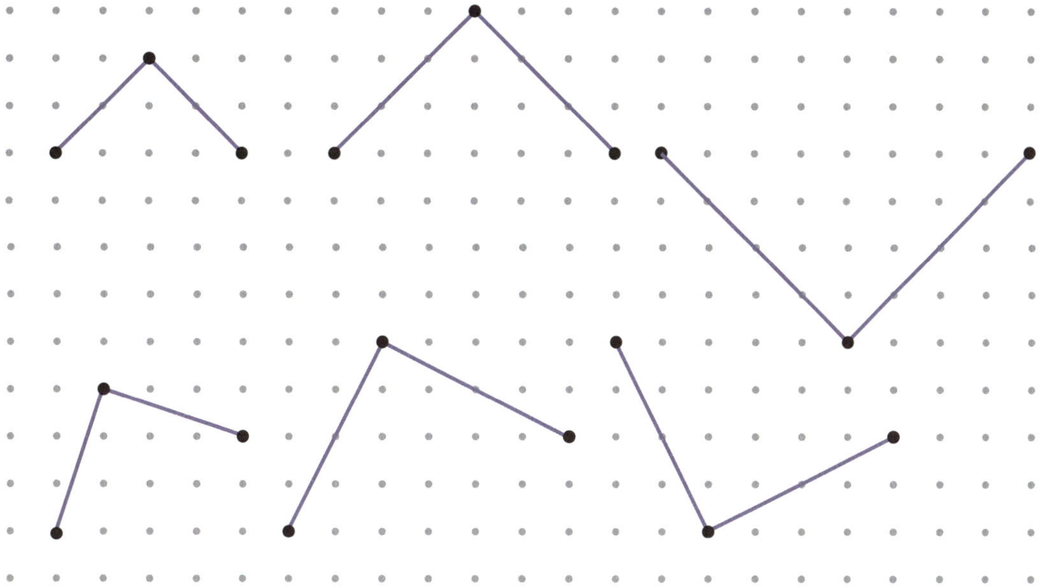

Composite polygons

Key maths idea

We can join shapes to make **composite** shapes. Here are some examples of pentagons made up of other polygons.

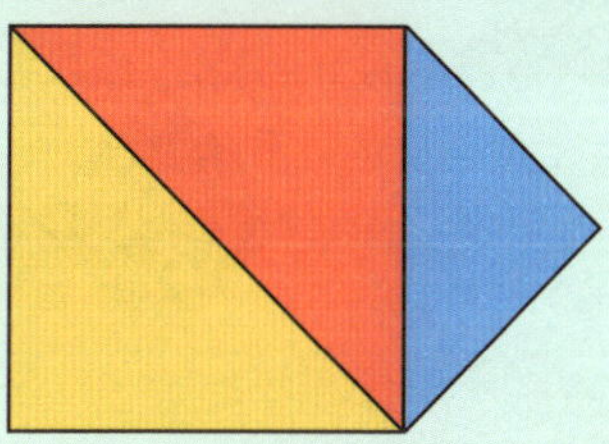

three triangles

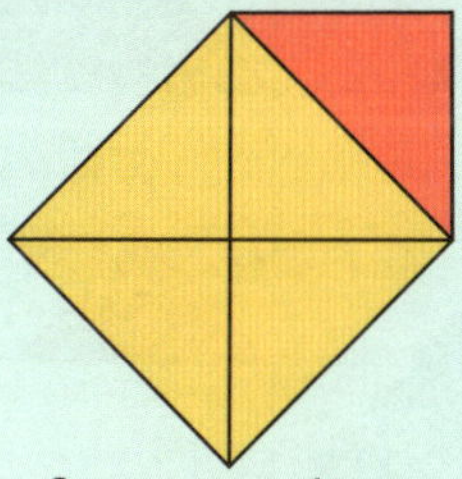

five triangles

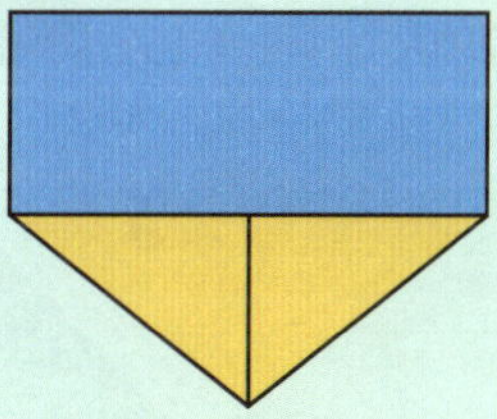

two triangles

You have worked with **tangrams** before in maths. A tangram is a Chinese puzzle game made of seven polygons that can be used to make a variety of different shapes and pictures. When you work with tangrams, the rule is that you must use all seven shapes to make the image or **design**.

You can make a tangram by folding and cutting up a square or tracing this tangram example.

Key words

composite
tangram
design

1 This picture shows the letters E and R made from a tangram. Work in groups. Explore shapes you can make using tangram pieces:

a Choose five other capital letters and make them from tangram pieces.

b Use the tangram pieces to design two different birds or animals.

2 Make three different composite hexagons from triangles and quadrilaterals. Sketch and label each shape.

3 Form three different composite octagons from triangles and quadrilaterals.

4 Investigate the fewest number of triangles you need to use to form each shape. Sketch your examples in a table. Explain any patterns you notice.

a kite	a trapezium	a square	a parallelogram
a pentagon	a hexagon	a septagon	an octagon

Look back at Chapter 13 if you need to remind yourself about the different types of polygons.

Maths detective: Shape shenanigans

Use the clues to draw these seven shapes on squared paper. Work accurately – use the grid lines to help you!

Case 1

Draw these 7 shapes on squared paper.

Work accurately – use the grid lines to help.

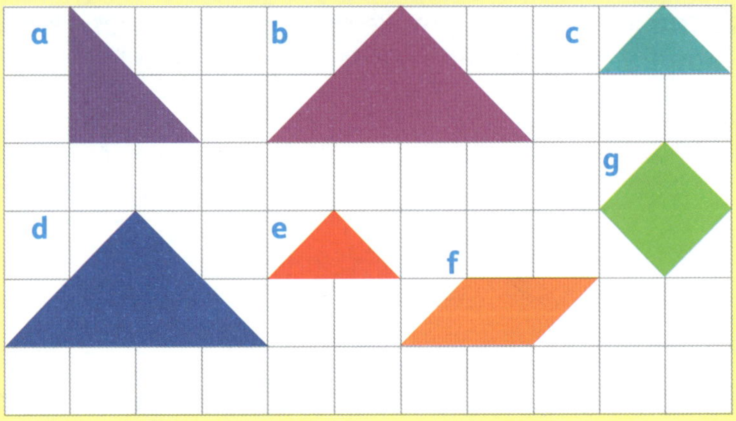

Case 2

Now cut out the shapes.

1 Join two shapes to make a trapezium.

2 Join three shapes to form a parallelogram.

3 Join four shapes to form a rectangle.

Case 3

What different triangles can you form by joining the cut-out shapes? Can you use all seven shapes?

Real-world maths: Geometry in design

Geometry is an important part of **design** – the process of planning and creating anything that we make or produce. Furniture, buildings, fabrics, bridges, roads, **art** and even packaging … everything that humans make involves a design process.

Design **elements** include shape, colour, **symmetry**, space, line and texture. You can see these elements in many objects around you. Sometimes you may notice **motifs**. These are images or shapes that are repeated to make a pattern.

Most designers use computer **applications** (commonly known as **apps**) to draw polygons and other shapes.

Key words
design
art
element
symmetry
motif
application (app)

These apps are used for:

- graphic design
- marketing
- engineering
- architecture.

1 A designer made these drawings of some teacup designs.

a What do they all have in common?

b Choose three of the designs and describe to a partner how the designer has used line, colour and shape. Your partner should try to guess which designs you are describing.

c Draw your own teacup design. Choose one of the polygons you have learned about to create your design.

2 Look around at the patterns and designs you see on curtains, rugs and clothing. Find examples of motifs.

a Draw a sketch of some designs you found and identify the motif in each design.

b Choose an area of design (for example, cars, clothing, crockery, carpets or curtaining). Put together a booklet showing how different types of polygons are used in the designs of some of the objects in this area of design.

3 Use a drawing program to create your own pattern that you could use to decorate a saucer or teacup. Investigate how to make the shapes different sizes and how to move and rotate them in different positions.

4 Use an app to create a class logo. Use some of the polygons you have learned about.

What I have learned

1 Explain how to draw a segment parallel to this line.

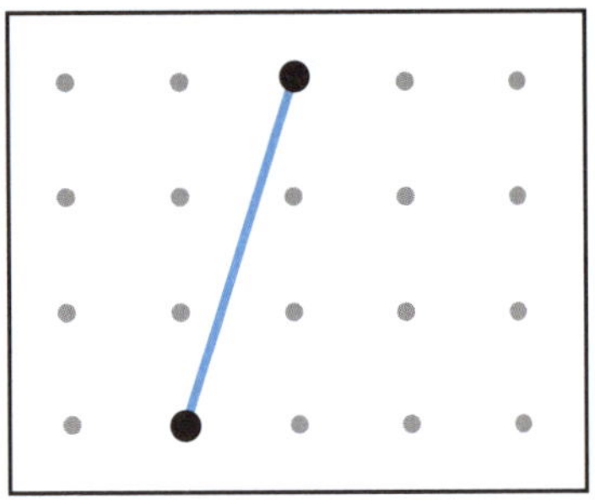

2 Explain how to draw a line perpendicular to this line.

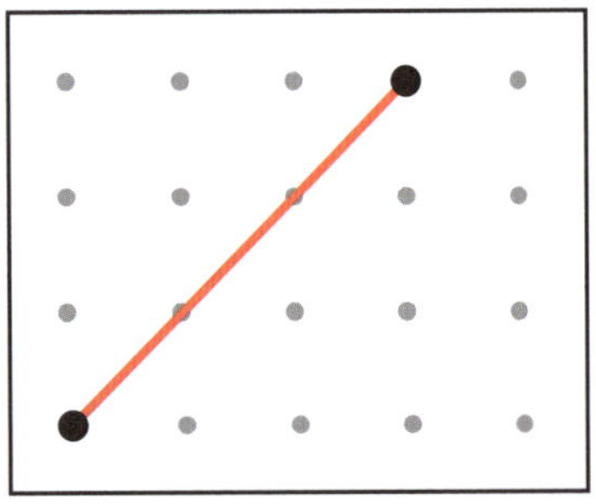

3 What name is used for polygons that have:

a five sides b six sides c seven sides d eight sides?

4 Explain why it is impossible for a triangle to be concave.

5 Draw an example of a concave quadrilateral.

Practice questions

1 Draw two squares on a large sheet of paper. Then cut each square to form four identical triangles.

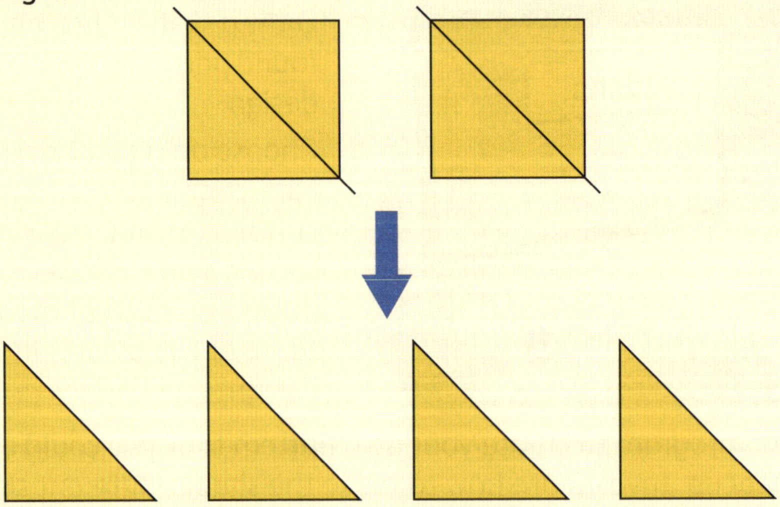

Form composite polygons using all four triangles. Sketch and name the different shapes you can make.

2 Draw three identical rectangles on a large sheet of paper.
Cut each rectangle to form six identical triangles.

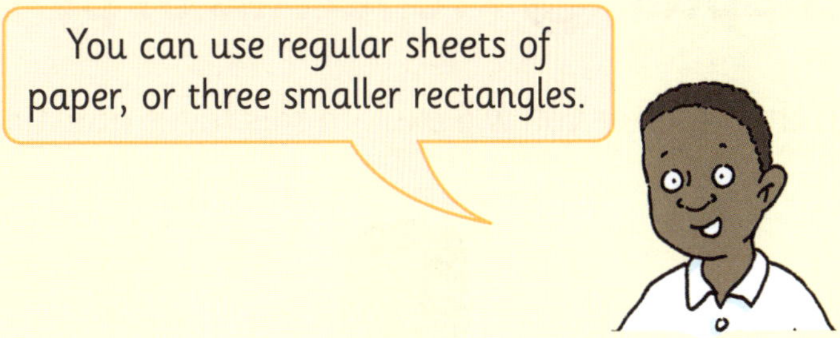

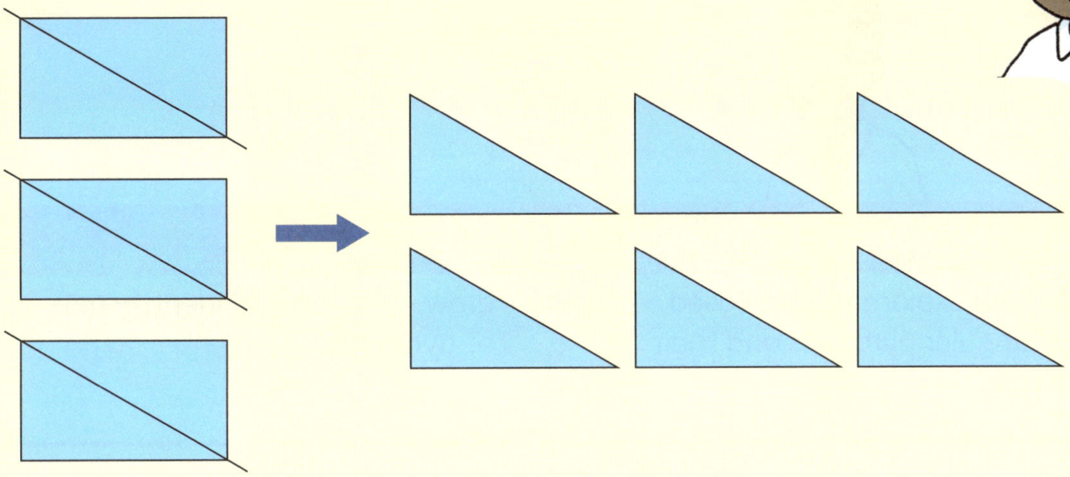

Investigate the composite polygons you can form from these triangles.

21 Self check

Check you can do this:	If you have difficulty, try this:
1 Sketch polygons on a grid.	Go back to page 289 and remind yourself how to create the vertices of a polygon using points of the grid. You join them using a ruler to make straight lines.
2 Sketch parallel and perpendicular lines using a grid.	Go back to page 291 and remind yourself how to make parallel and perpendicular lines using a grid.
3 Form composite polygons.	Look back at the work you did on page 293. Remember that a composite shape is made up of other shapes put together.
4 Make patterns with polygons and use an app to explore and draw polygons.	Go back to the work you did on page 295. Remember, a pattern is made by repeating lines and shapes.

Chapter 22 Properties of pyramids

In this chapter you will explore this focus question:

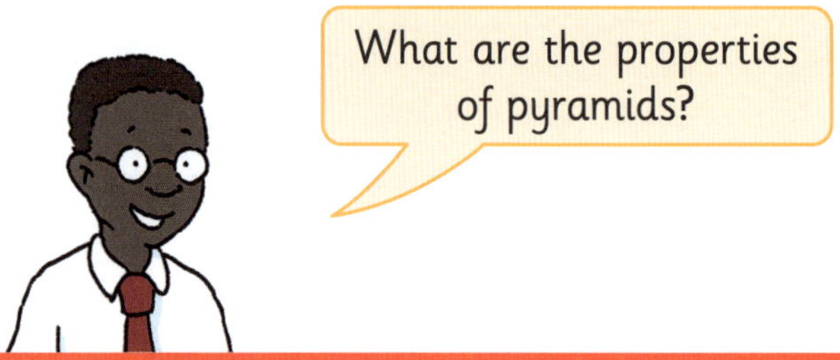

To answer this question, you will:

- model and explore square-based and triangular pyramids and their properties
- draw, describe and identify nets of pyramids.

Starting point

Look at these two nets. What is the same and what is different?

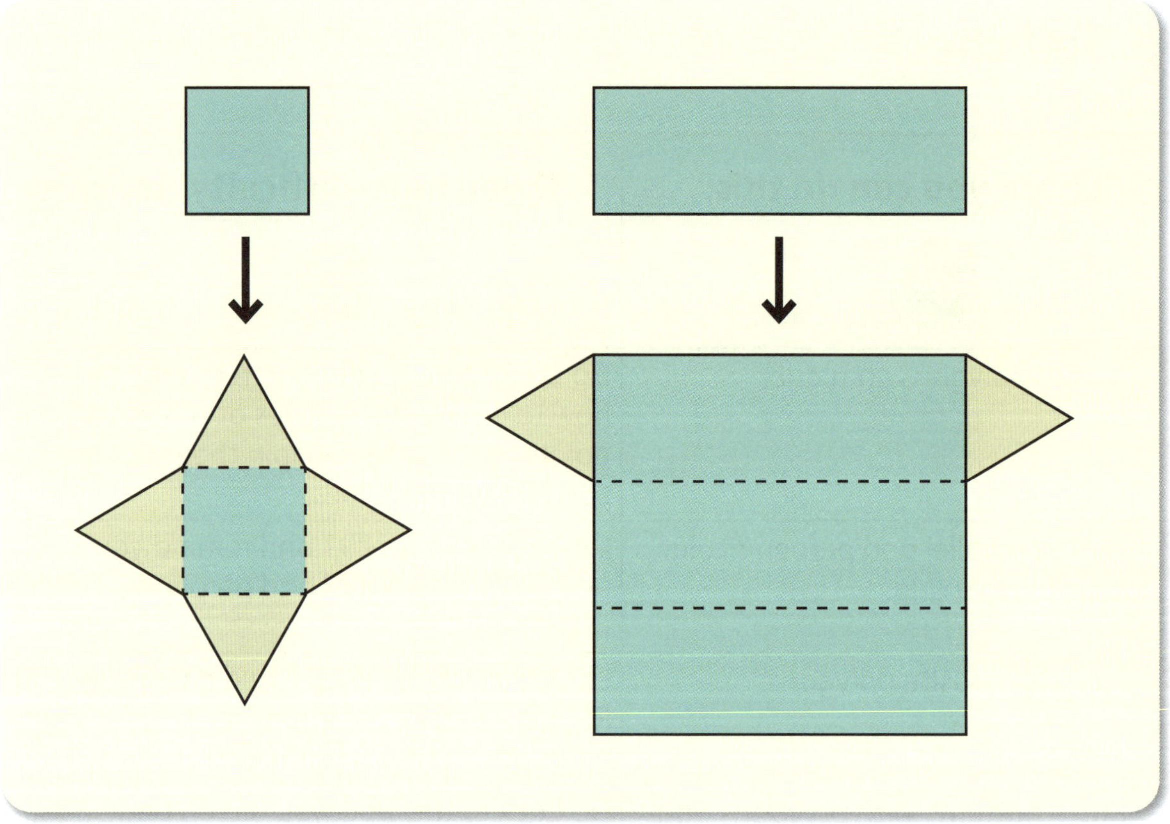

Can you visualise the solid shapes they will form?

Properties of pyramids

Key maths idea

A **pyramid** is a solid shape. It has a polygon as its **base** and its remaining **faces** are all triangles, which meet at the top **vertex**, called the **apex**. A triangular-based pyramid has a triangle as its base and four **vertices**. It has six **edges** where the faces meet.

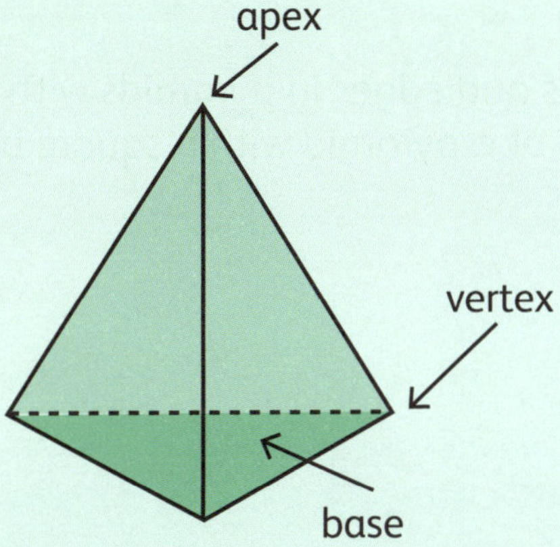

If all four triangles are equilateral, the pyramid has a special name: a **tetrahedron**.
A pyramid may also have a square base. Here is one **net** for a square-based pyramid:

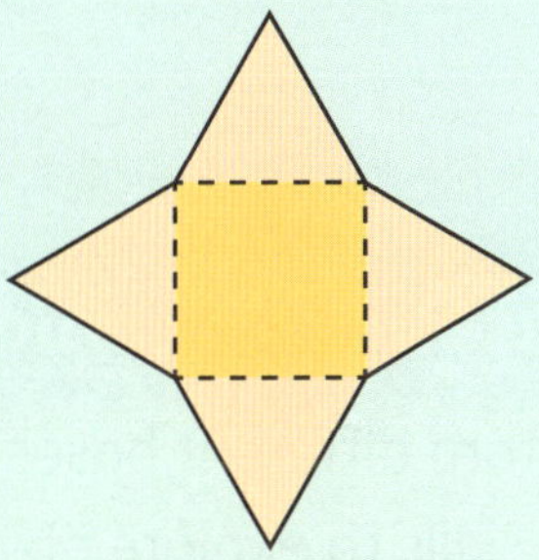

Key words

pyramid
base
face
vertex
apex
edge
tetrahedron
net

1 Use sticks and clay to make a pyramid frame by following these instructions.

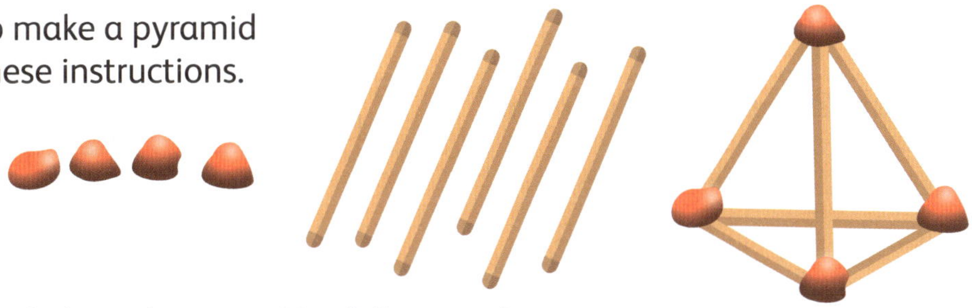

2 Make a net for a triangle-based pyramid by following these steps.

Draw and cut out one triangle for the base and three more for the remaining faces.

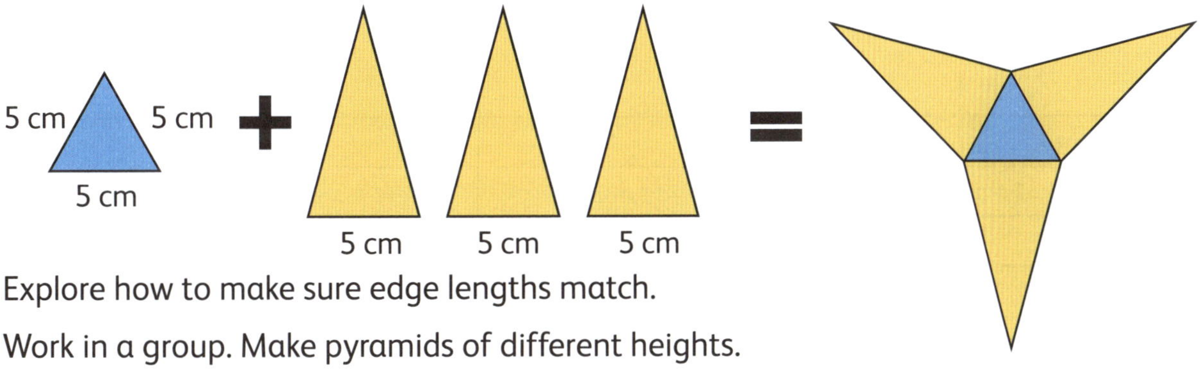

Explore how to make sure edge lengths match.

Work in a group. Make pyramids of different heights.

3 Now design your own net for a square-based pyramid. Make sure the edge lengths match.

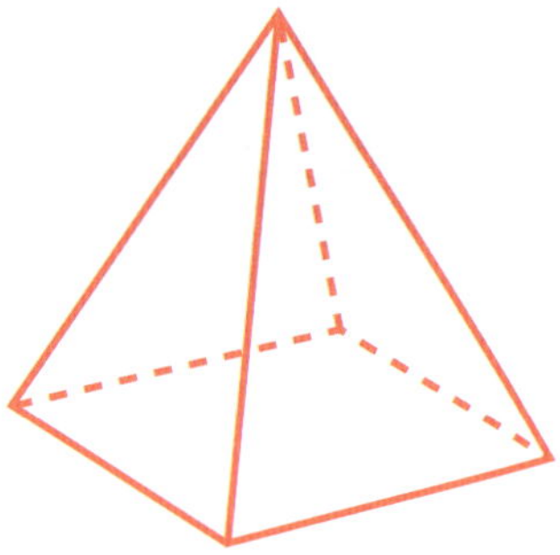

4 Determine the number of vertices, faces and edges in pyramids with a triangle base. Repeat for the properties of a pyramid with a square base.

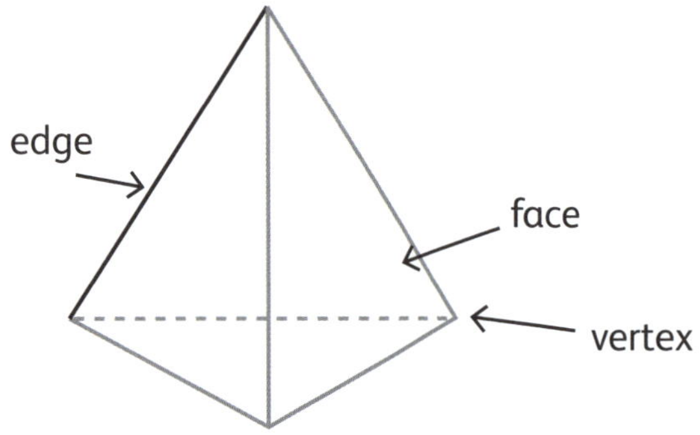

Maths detective: Bases of pyramids

Pyramids can be formed from different bases.

Use your maths detective skills to explore how to make or form pyramids with different shaped-bases. The clues will help you copy and compete the table for each pyramid with different shapes as a base.

Base shape	**Sketch of net**	**Number of faces**	**Number of vertices**	**Number of edges**
triangle	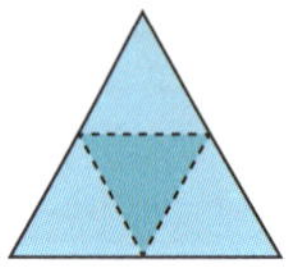	4	4	6
square	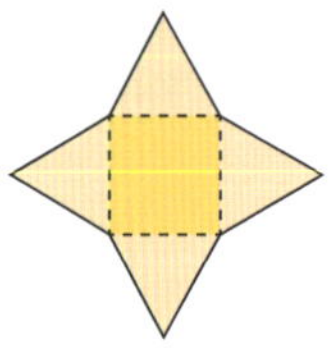			
pentagon				
hexagon				
septagon				
octagon				

Real-world maths: Pyramids in architecture

1 Pyramid shapes have been used in architecture over many centuries.

The pyramid-shaped entrance to the Louvre art gallery in Paris, France.

The pyramids of Giza in Egypt.

- **a** Research and sketch these or other pyramids from around the world. Notice how they are constructed in different ways.
- **b** Write a short paragraph to compare and contrast the pyramids.

2 Pyramid shapes are also popular for packaging and gift wrapping, for example for individual tea bags or single milk portions.
Find and research at least two types of pyramid-shaped packaging and their templates. What do the developers of pyramid shaped templates need to consider when they design their templates?

- **a** Draw the two templates.
- **b** Then think about what each package contains, how it is packed for transport from the factory, and how it is displayed in a store. Can you think of ways to improve the shape or materials of the packaging?
- **c** Present your findings to your partner or to the class.

What I have learned

Use these words to describe the similarities and differences between a pyramid with a triangular base or a square base.

faces   vertices   edges   base

apex   triangular   square

Practice questions

1 Which of these are the nets of a tetrahedron?

a

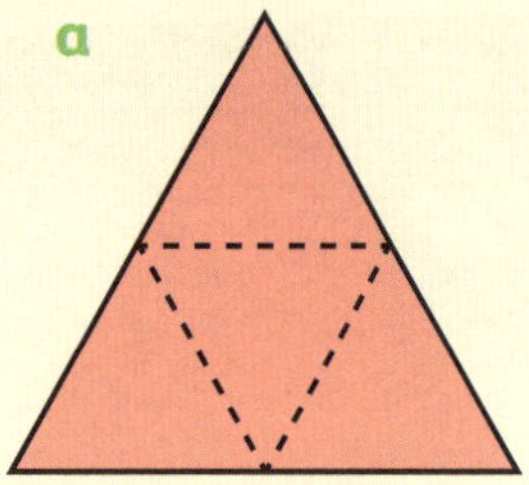

b

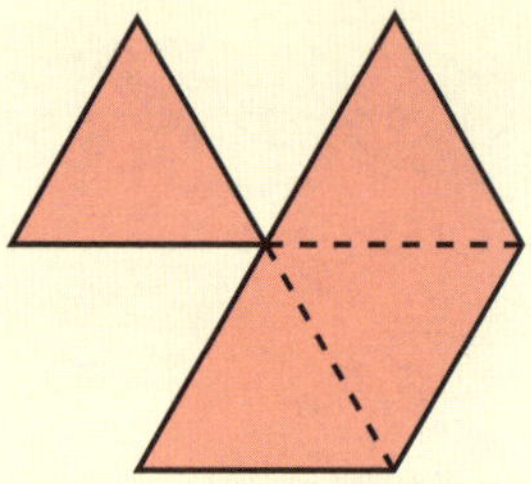

c

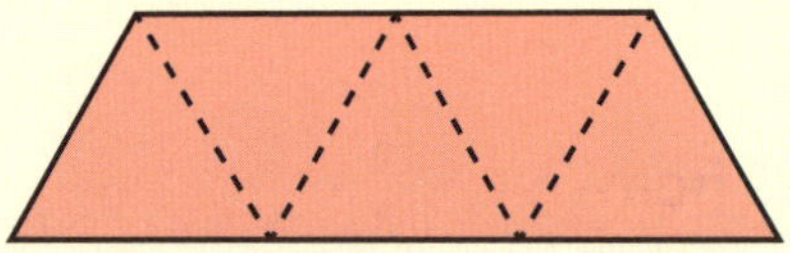

d

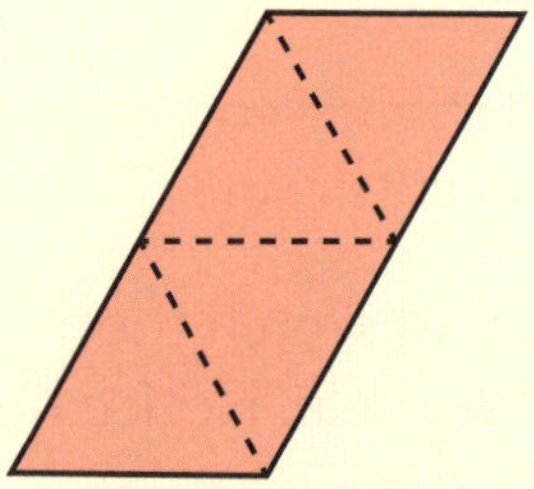

2 Which of these fold correctly to form a pyramid?

a

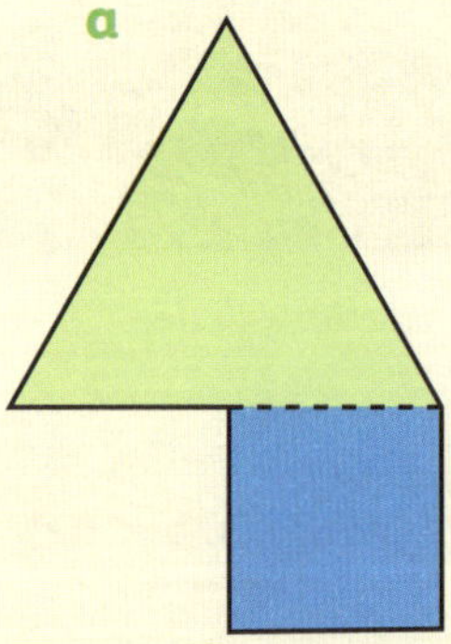

b

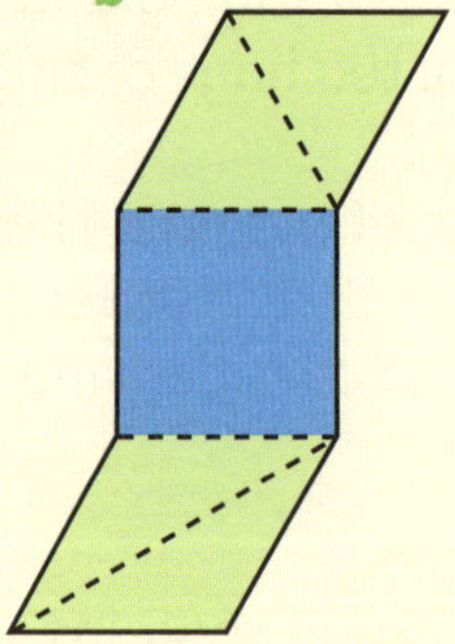

c

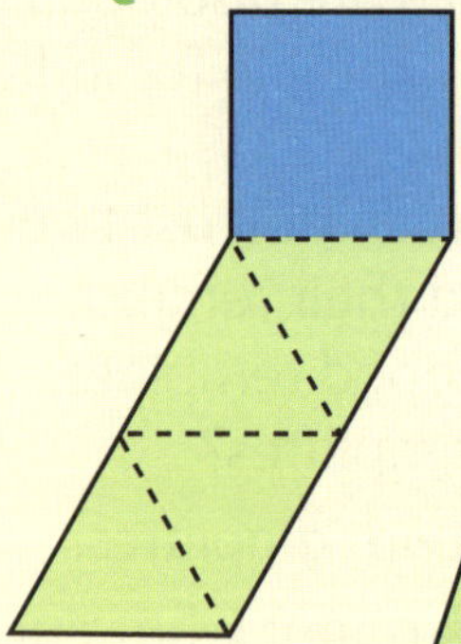

d

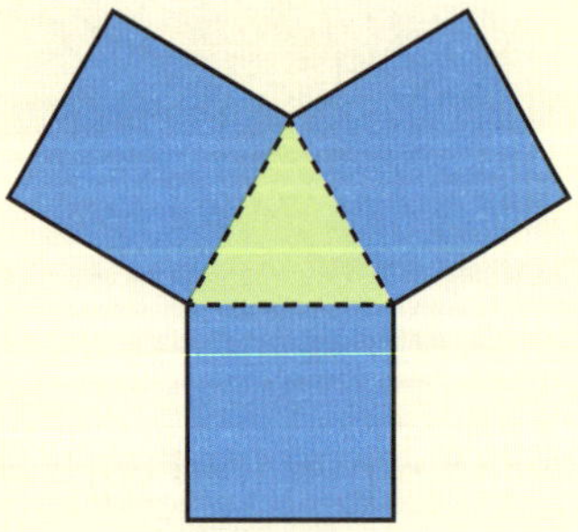

e

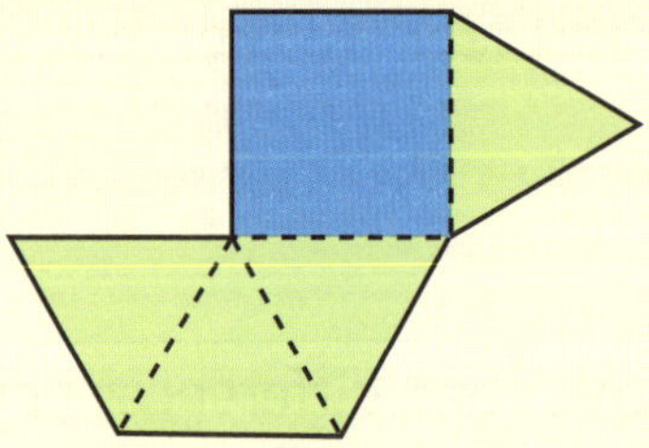

f

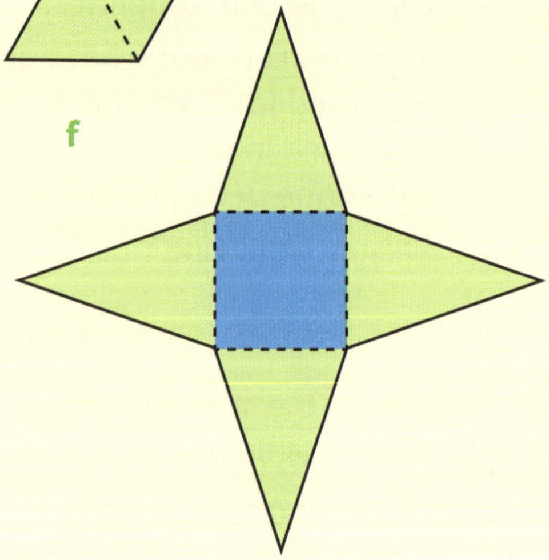

22 Self check

Check you can do this:	If you have difficulty, try this:
1 Identify the net of a square-based and a triangle-based pyramid.	Look back at the nets on page 299. Identify which has a square base and which has a triangular base.
2 Describe the properties of square-based and a triangle-based pyramid.	Go back to page 299 and remind yourself of the definitions of square-based and triangular pyramids.
3 Construct a pyramid frame from construction materials.	Look at the pyramids you made from sticks and clay. Other materials you can use are paper and tape, or fudge sticks and putty.
4 Explain similarities and differences between different pyramids.	Remind yourself what is the same and what is different (think about the bases and other faces of each type of pyramid).
5 Identify the edges, vertices and faces of different pyramids.	Look at any of the diagrams of pyramids in this chapter. Make sure you can identify the vertices (corner points), faces (flat sides) and edges (where the faces meet).
6 Sketch a design and create a model of a pyramid shape.	Practise drawing and making them from clay or paper.

Unit 3: Algebra

Chapter 23 Variables and equations

In this chapter you will explore this focus question:

How do I write and interpret expressions and equations involving variables?

To answer this question, you will:

- understand that a variable is a changing quantity and an unknown quantity
- work with variables in algebraic expressions and equations
- write algebraic expressions to represent situations where a quantity increases or decreases by an unknown amount.

Starting point

1 Use words, calculations and symbols to describe these balance situations.

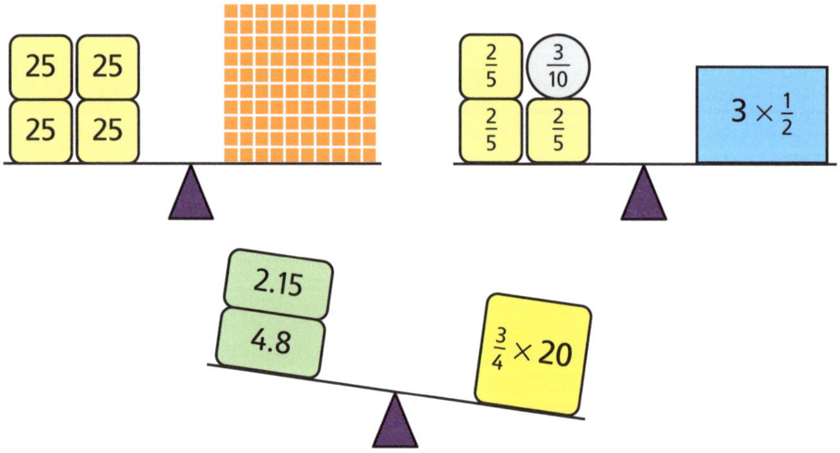

Now create your own balance diagrams using different operations.

2 How can you describe this balance without knowing the values of *a*, *b* or *c*?

Variables for changing quantities

Key maths idea

You can use a **variable** to represent a quantity that changes.
When a quantity changes, it **increases** or **decreases**.
For example, think of a job where the wages are \$12 per hour.
The total pay depends on the number of hours worked, which may vary from week to week.

You can use **algebra** to represent the variables.
Let a stand for the number of hours worked.
Then the total pay is $12 \times a$. We can write this **expression** as $12a$.

You can use a table to show how the total changes as the value of the variable a changes:

a	1	2	3	5	20
$12a$	12×1	12×2	12×3	12×5	12×20
Total pay	\$12	\$24	\$36	\$60	\$240

An expression is a mathematical statement which includes operations but no equality:
$3 \times 4 + 5$ $5a - 4$ $10y - 4z$

Key words

variable
increase
decrease
algebra
expression

1 Write an expression to represent the total pay for each job shown below.

Choose a variable to represent the amount of time worked.

Complete a table of values to show how the total pay changes depending on the changing variable.

a

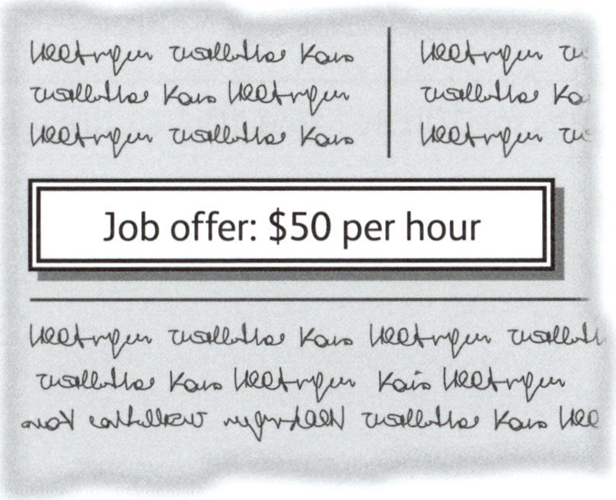

b

2 A tray holds 18 buns. A crate holds 24 buns.

Write an expression for the total number of buns if there are *x* trays and *y* crates.

3 A restaurant has 50 kg of potatoes and buys *n* more sacks of potatoes. Each sack contains 15 kg of potatoes.

a Write an expression for the total mass of potatoes.

b Then complete a table of values to show how the total mass of potatoes changes (increases or decreases) as *n* changes.

n	1	2	3	5	20
Total mass	50 + 15 = 65 kg				

4 A ticket for a boat ride costs $25 per person, plus a $5 booking fee.

a Write an expression for the total cost of a boat trip for *n* people.

b Complete a table of values to show how the cost varies for different values of the variable *n*.

n	1	2	3	5	10	15	20
Total cost							

5 Complete each row of the table to show how the value of each expression changes as the variable *n* changes.

n	1	2	3	5	20
$10n$					
$10n + 1$					
$10n - 1$					
$10n + 10$					
$10n + 11$					

Write a short sentence to explain how the values for $10n + 1$ and $10n - 1$ are related.

Equations and unknown values

Key maths idea

An **equation** consists of two expressions that are equal to each other.

Here is a numerical equation:

$5 \times 20 = 60 + 40$

The expression 5×20 is equal to 100.

The expression $60 + 40$ is equal to 100.

The equation shows that both expressions are equal:

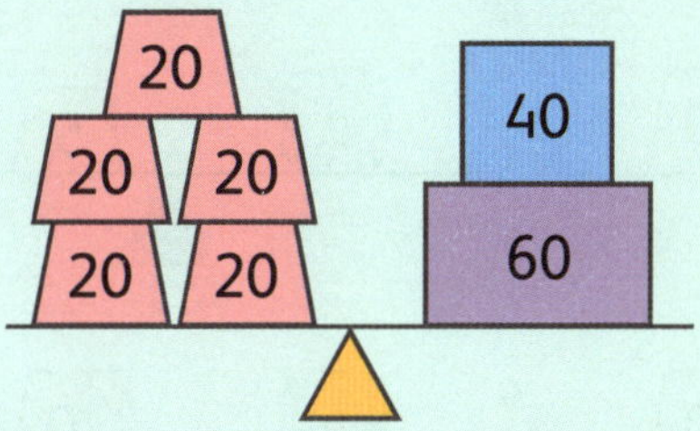

$5 \times 20 = 60 + 40$

Some unknown value problems may use shapes to represent the unknowns.

$100 - \blacktriangle = 80$

You **solve** this problem by finding the value for the symbol that balances the expressions.

If $\blacktriangle = 20$, then $100 - \blacktriangle = 80$

Therefore, the solution is that $\blacktriangle = 20$

We use algebra to represent unknowns in an equation.

You can use a variable to represent a **quantity** that does not change but is **unknown**.

Look at this missing number equation:

$10 + a = 17$

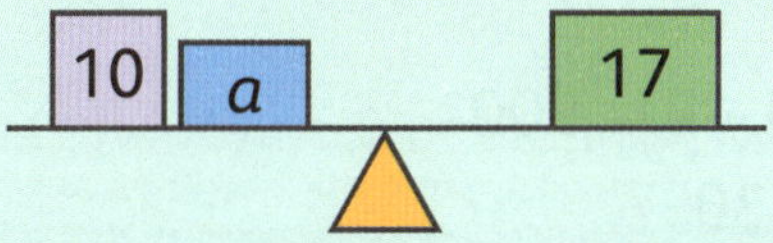

The expression $10 + a$ should balance the expression 17.

You can see that a must be equal to 7, because

$10 + 7 = 17$

Calculating the value of the unknown variable is called solving the equation.

When you solve an equation without needing to do formal calculations, this is called solving by inspection.

Key words

equation
solve
quantity
unknown

1 Solve these shape puzzles by calculating the value of each symbol.

♥ + 10 = 90

99 – ■ = 95

1 000 ÷ ▲ = 125

★ + ★ + ★ = 60

2 How many solutions can you find for this?

25 – ♥ = ■

3 Calculate the value of a in each balance.

a
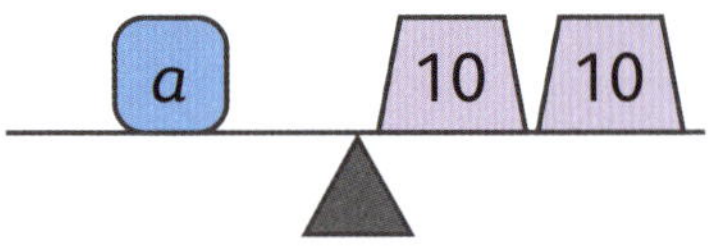

b
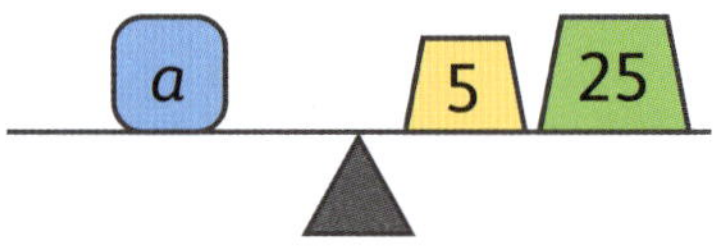

c
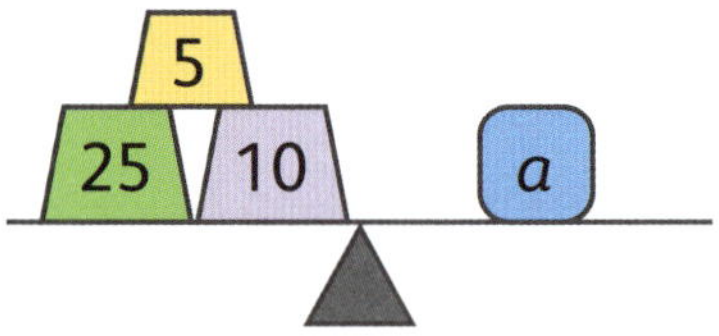

d
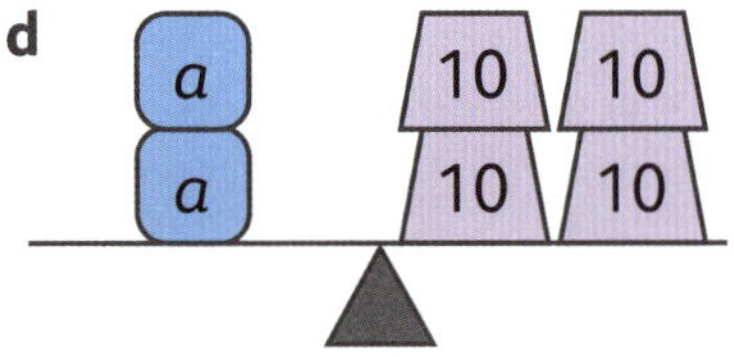

e
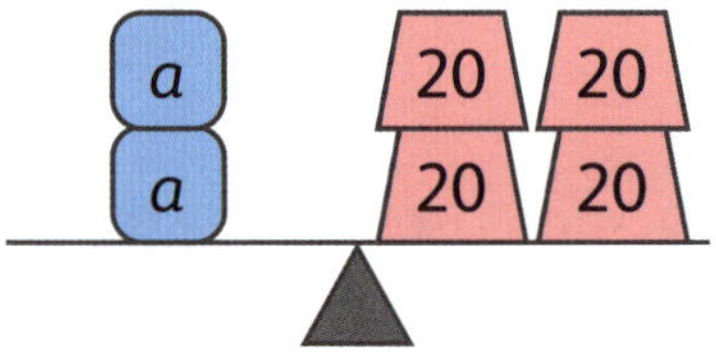

f
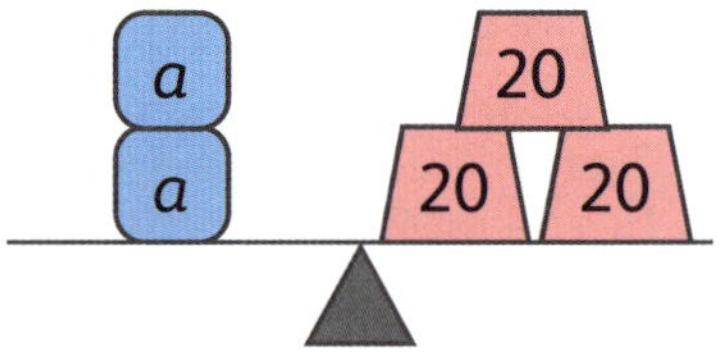

g
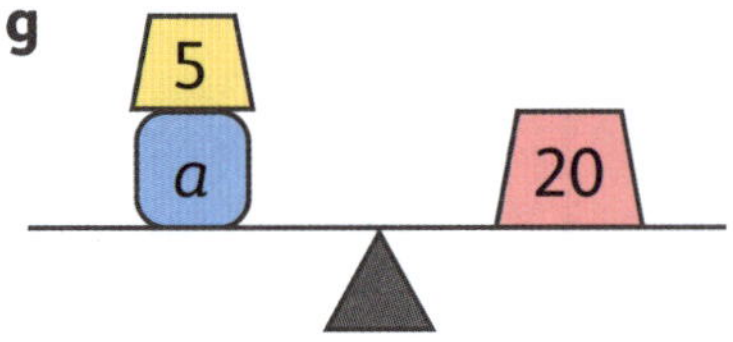

h
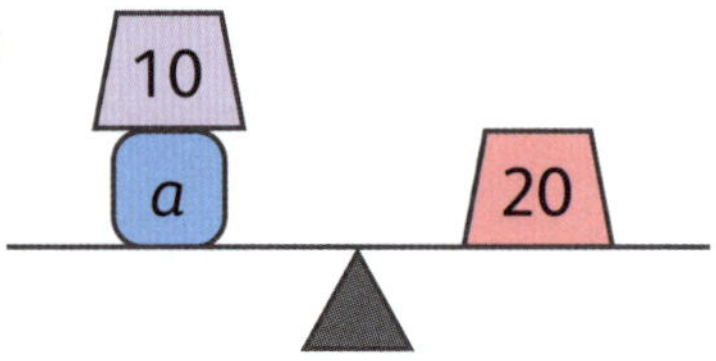

i
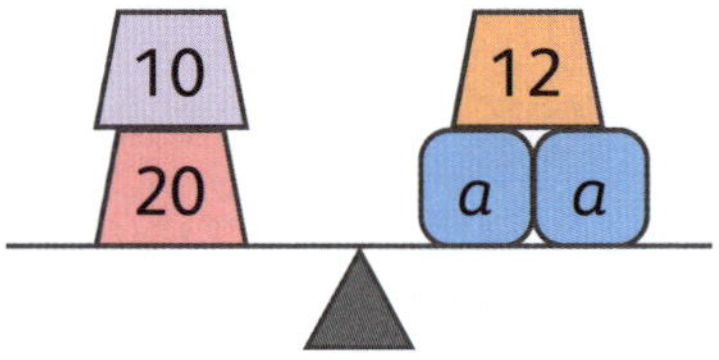

4 Solve each equation.

a $x + 2 = 10$ $x =$ ______

b $x - 2 = 10$ $x =$ ______

c $2x = 10$ $x =$ ______

d $10 = \frac{x}{2}$ $x =$ ______

e $10 = 4x + 2$ $x =$ ______

f $10 = 4x - 14$ $x =$ ______

5 Now solve each of these equations.

a $100 = 20 + j$

b $1\,000 - k = 900$

c $2.5 + m = 5.2$

d $2n = 30$

e $3q + 1 = 22$

f $50 = \frac{r}{4}$

g $99 + 3s = 249$

h $700 - 10t = 0$

Maths detective: Calculating unknown values

Use the clues to find the solutions to these challenges.

Case 1

Complete each equation for each unknown angle situation. Solve the equations to find the missing angles.

1 $x + 50 =$ ______

2 $90 +$ ______ $=$ ______

3 $y +$ ______ $=$ ______

a

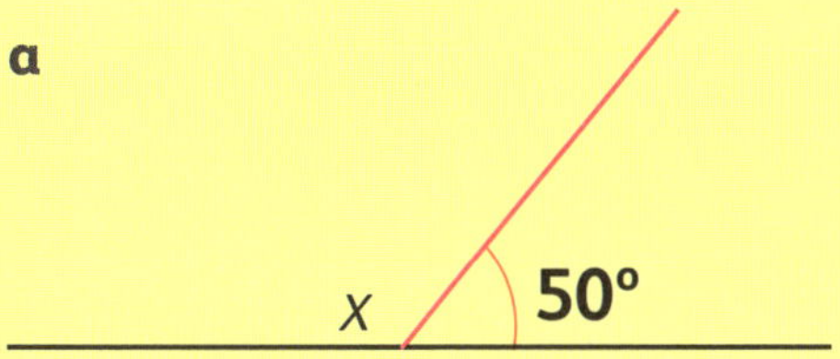

b

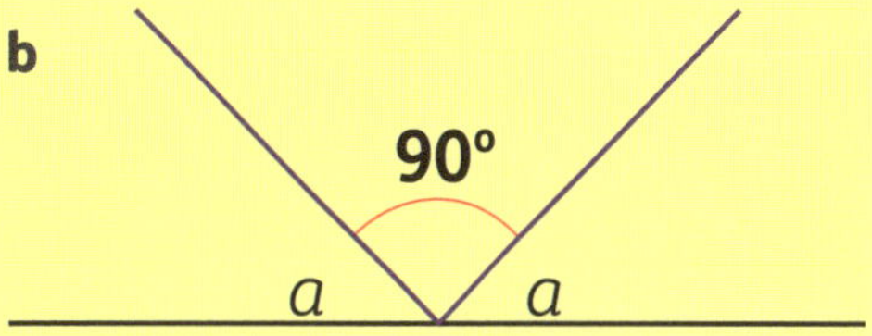

c

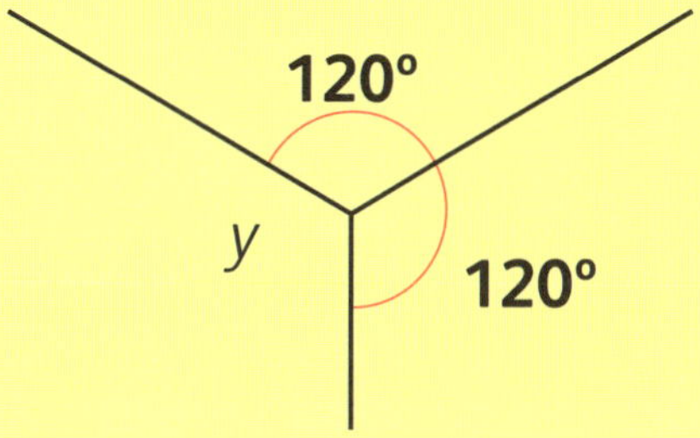

Case 2

Write and solve an equation to calculate the value of p.

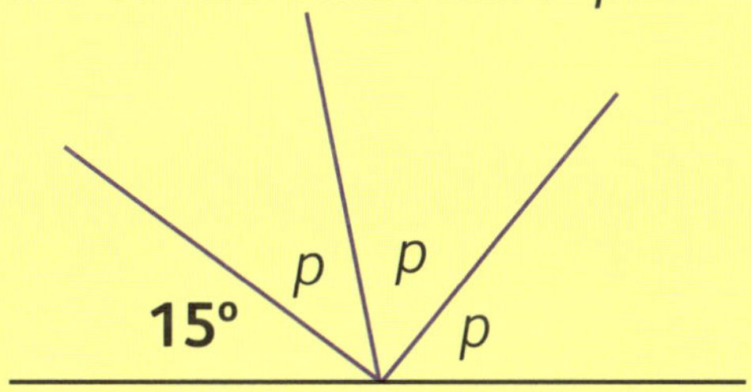

Case 3

1 How many solutions can you find for this equation?

♥ × ■ = 360

2 Now find different pairs of values that solve this equation:

$\frac{1}{e} \times 360 = f$

> You will need to find pairs of values.
> For example, one solution is:
> ♥ = 3 ■ = 120

Real-world maths: Patterns and storage

1 You can use algebra to represent real-life problems involving patterns and storage. Here is a tile.

This square tile has sides of length 12 cm.

What is the width and height of a rectangular display made from *m* rows of *n* tiles?

Write an expression for the:

a width **b** height **c** area.

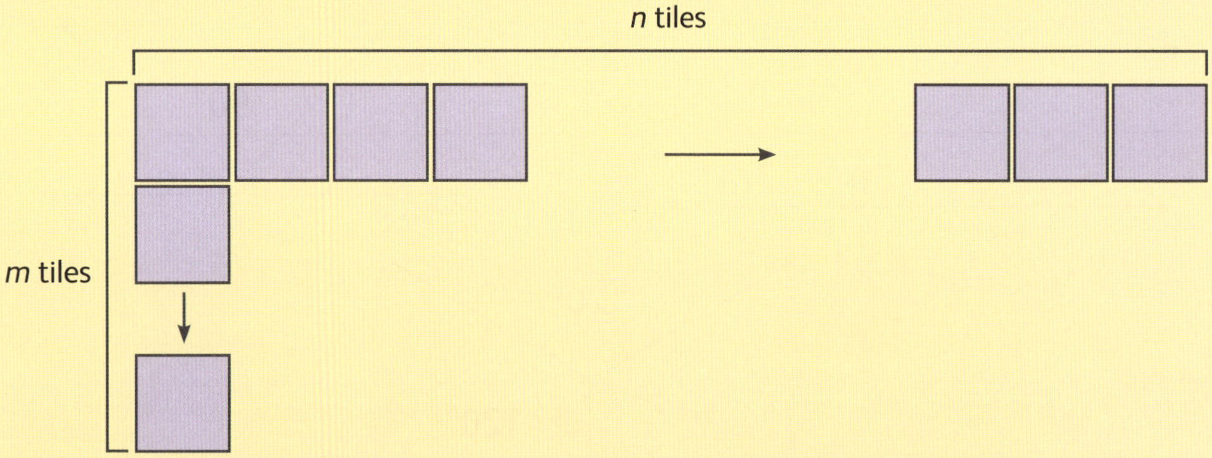

2 Mr Jones plans to tile an area which covers a width of 2.4 m and a height of 1.44 m. A pack of 20 tiles costs $60.

a Use your expressions to calculate the number of tiles needed and the total cost of buying all the tiles required.

b Will there be any tiles left spare?

What I have learned

Define each term and use examples to support your descriptions.

1 variable

2 expression

3 equation

4 solve

Practice questions

1 Each bucket holds 5 litres. There is one water barrel holding 85 litres and Flo fills *n* buckets.

a Write an expression for the total volume of water in this picture.

The total volume will be ____________ .

b Copy and complete this table of values for the total amount of water collected in one barrel and *n* buckets.

n	**1**	**2**	**3**	**10**	**25**	**50**
Total volume						

c If Flo has exactly 200 litres water, how many full buckets does she have?

d How many buckets of water will Flo need for a total of 1 000 litres of water?

2 Calculate the value of each unknown symbol:

a ▲ + 2 = 18 **b** 40 – ★ = 25 **c** ■ ÷ 3 = 12

3 Calculate the value of *x* in each balance equation.

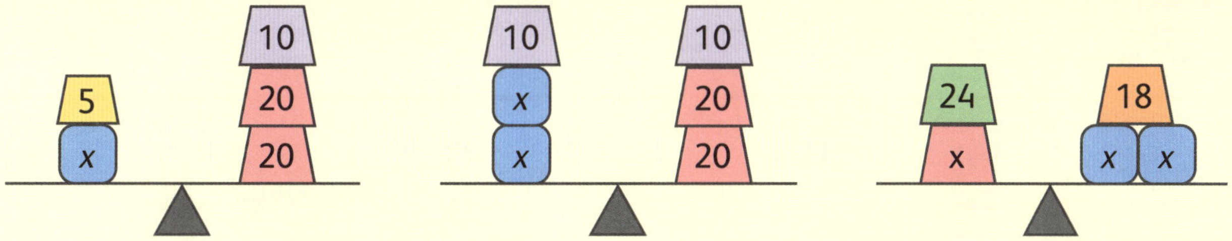

4 Solve each equation:

a $5a = 60$ **b** $9a - 3 = 60$ **c** $100 - 6a = 46$ **d** $100 - 5a = 0$

5 Solve each equation:

a $\frac{10}{20} = a$ **b** $\frac{b}{4} = 20$ **c** $\frac{100}{c} = 20$ **d** $90 = \frac{2d}{4}$

23 Self check

Check you can do this:	If you have difficulty, try this:
1 Write algebraic expressions with variables to represent problems.	Look back at page 305, and make sure you know how to use a letter to write an expression.
2 Substitute values to evaluate an expression.	Look back at page 305 and make sure you know how to replace the letter with the given value to find the value of the expression.
3 Write equations using unknown values.	Go back to page 310 and remind yourself how to work with both sides of the equation to solve it.

Chapter 24 Solving problems with equations

In this chapter you will explore this focus question:

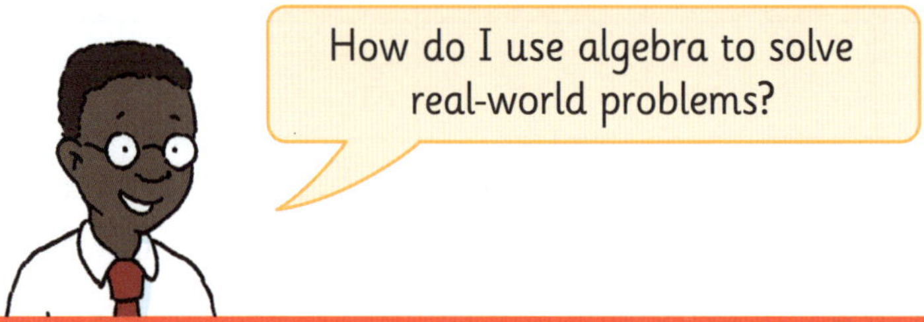

To answer this question, you will:

- use algebraic thinking in problem situations
- solve equations
- write algebraic sentences to solve worded problems.

Starting point

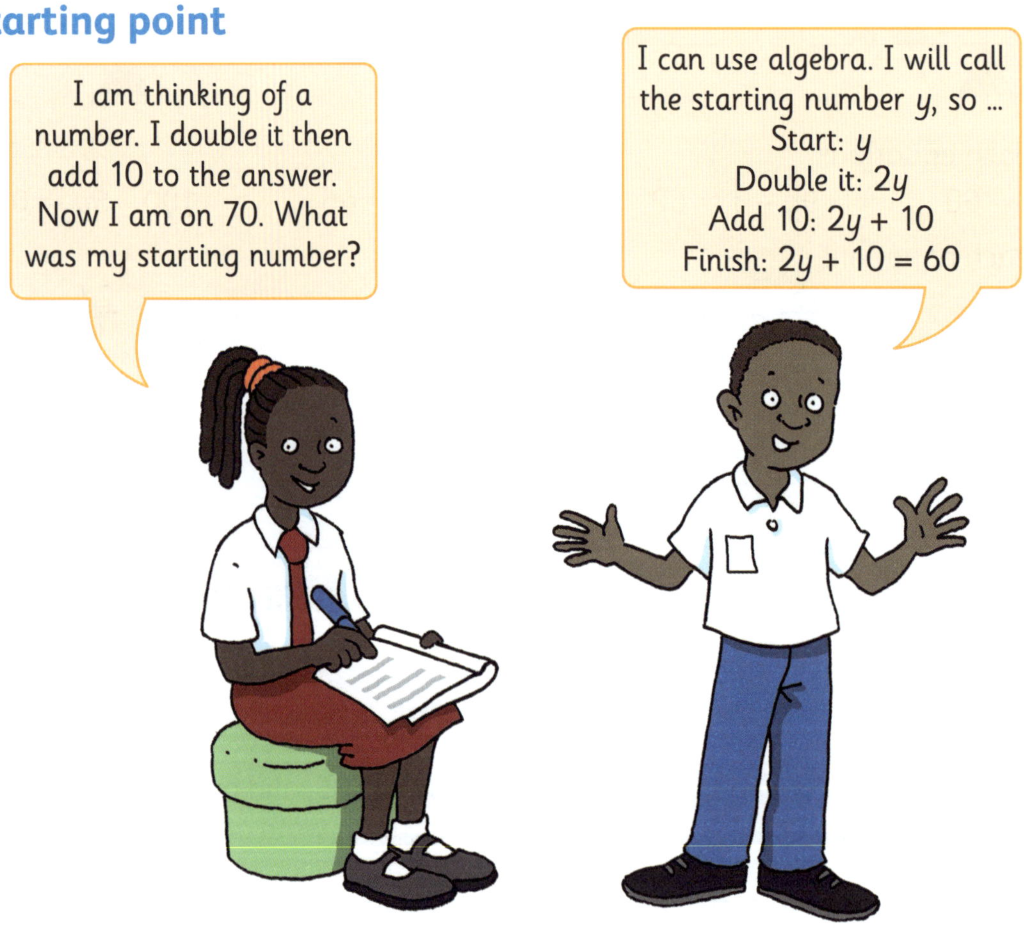

Solve the equation to calculate the starting number.

Invent your own 'Thinking of a number…' puzzles involving one or two steps. Show how to solve them using algebra. Use a variable for the starting value.

Using a formula

Key maths idea

A **formula** uses **variables** to show how quantities are related. It can express a **general rule** about the relationship between variables.

For example, the formula below gives Area (A) of a rectangle.

The formula for the area of a rectangle is:

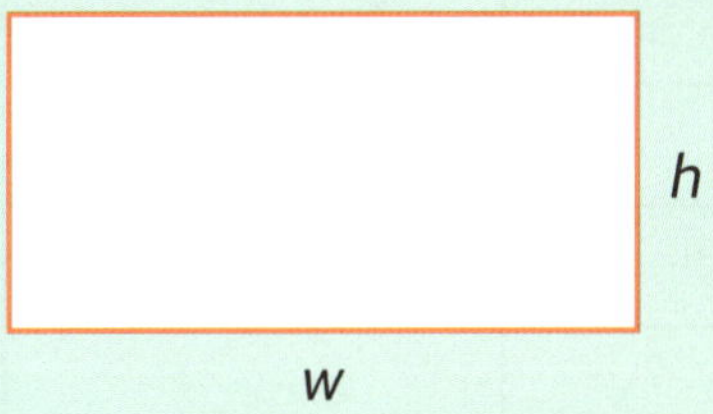

Area = width (w) × height (h)

$A = wh$

When $w = 5$ cm and $h = 12$ cm,

$A = 5 \times 12 = 60$

$A = 60 \text{ cm}^2$

A formula may require **whole numbers** or **fractional numbers**. For example, you may substitute a fractional number into a formula, or the formula itself may include a fraction or decimal.

Key words

formula
variables
general rule
whole numbers
fractional number

1 Use the formula to find the area of each rectangle.

a $w = 8$ cm, $h = 5$ cm

b $w = 4$ m, $h = 5.5$ m

c $w = 1\frac{1}{2}$ m, $h = 2\frac{3}{4}$ m

d $w = 99$ cm, $h = 5$ cm

2 Here is a formula that a supermarket gives for cooking a roast.

Cooking time (t) = 20 mins + 30 mins per 500g

Warning: This is a guide only. Check safety guidelines first.

Work out the cooking times for the following roasts:

a a chicken of 1.5 kg

b a turkey of 5 kg

c a goose of 2.75 kg

Calculate t for each roast.

3 Write a formula for the perimeter of each regular shape.

a

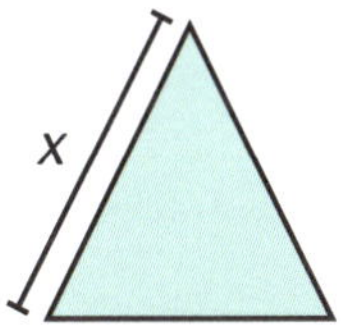

b

c

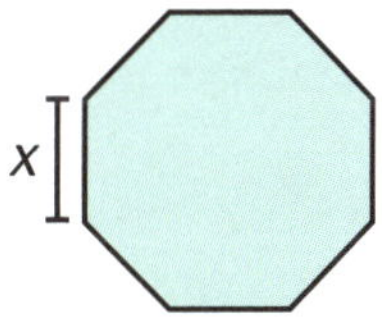

d

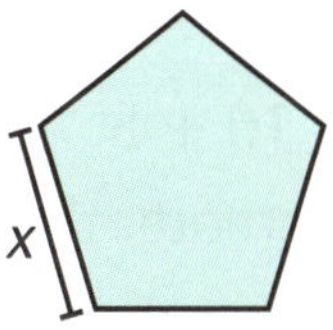

Calculate the perimeter of each shape when:

$x = 10$ cm $\quad x = 12$ cm $\quad x = 2.5$ cm

Algebra

4 Mr Jones builds a rectangular pen for sheep.

The area of the pen is 48 m^2.

Fencing costs $50 per metre.

Calculate the cost of the fencing required for the perimeter.

Complete a table such as this.

Length	Width	Area (m^2)	Perimeter (m)	Cost
1	48	48	98	$4 900
2	24			

Equations and inequalities

Key maths idea

An **equation** balances two **expressions**.

$5 \times 3 = 10 + 5$

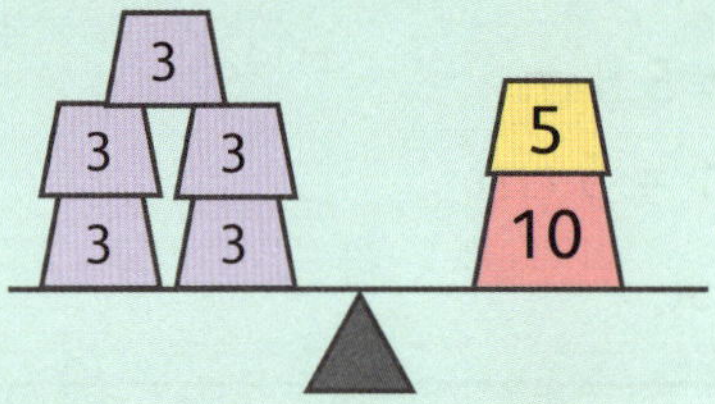

$4a = 3b$

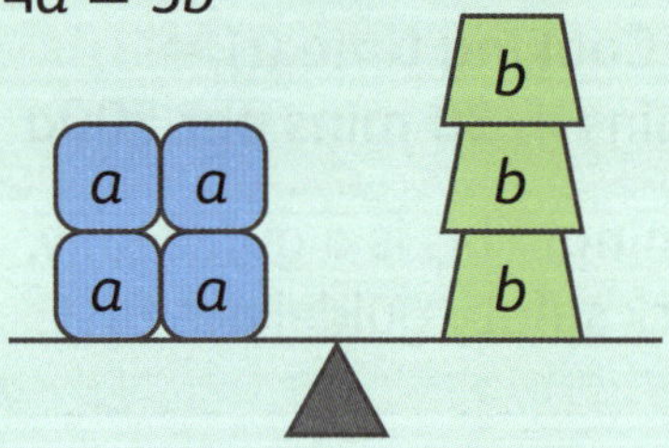

I can tell that one b must be greater than one a, for this to balance. I will write $b > a$.

An **inequality** shows how two expressions are related, but not equal.

$10 + 10 > 10 + 5$

$10 + 10$ is greater than $10 + 5$

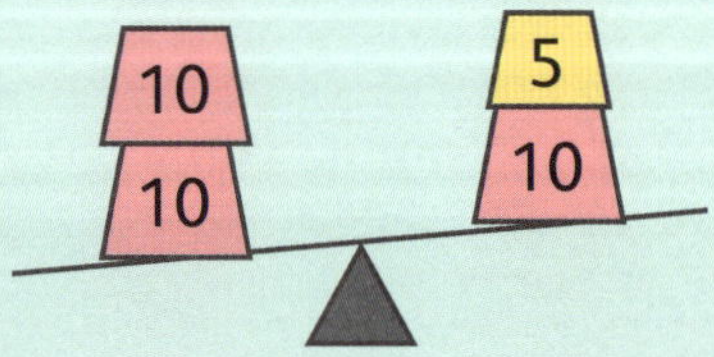

Key words

equation

expression

inequality

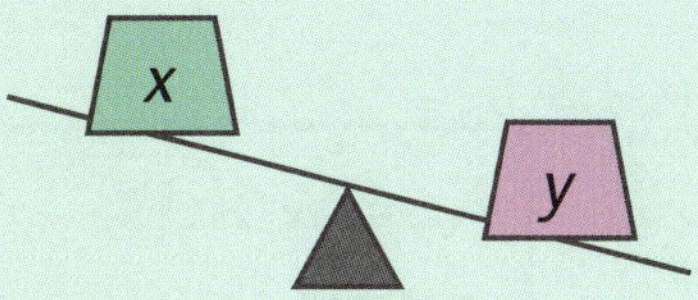

$x < y$

x is less than y

The $\neq$ symbol can be used to show that two expressions are not equal, without saying which expression is greater.

$2a \neq 20$ means $2a$ is not equal to 20.

1 Copy and complete each inequality.

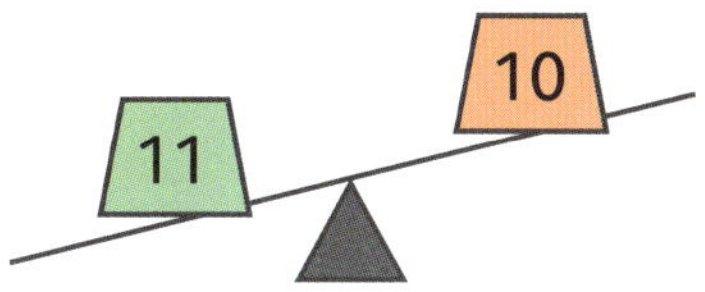

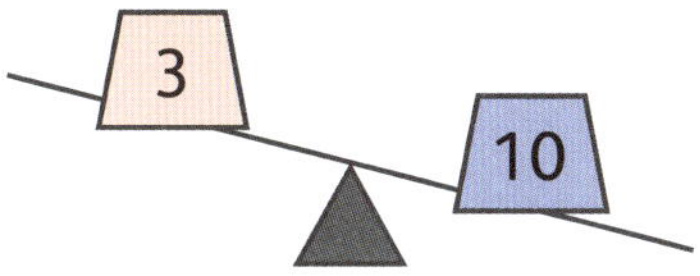

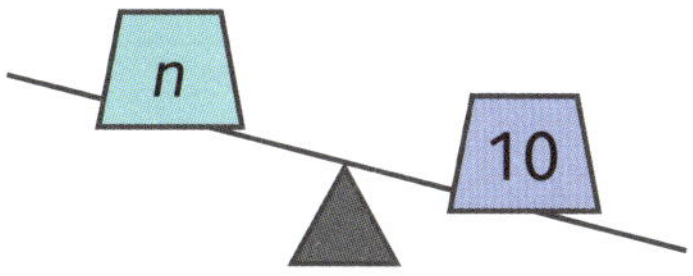

a 10 ☐ 11 **b** 3 ☐ 10 **c** n ☐ 10

2 Write an inequality for each balance.

a

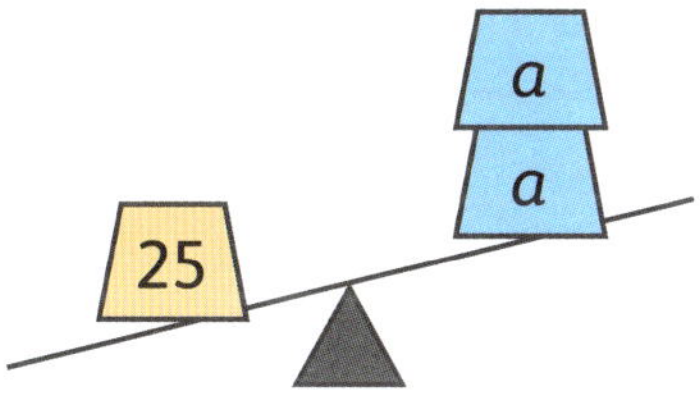

b

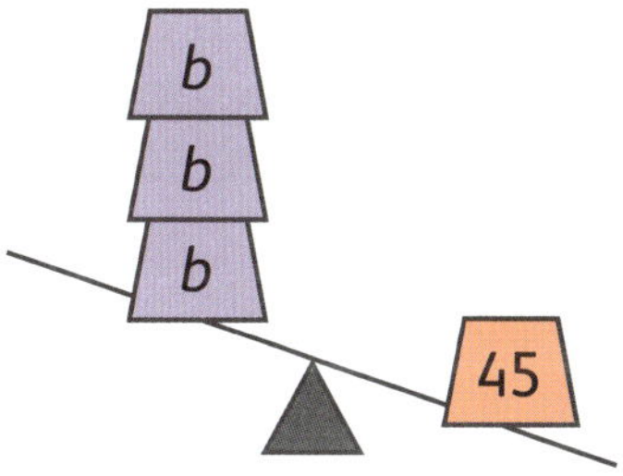

c

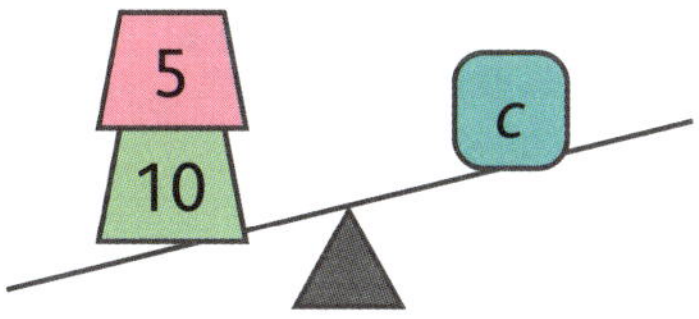

d

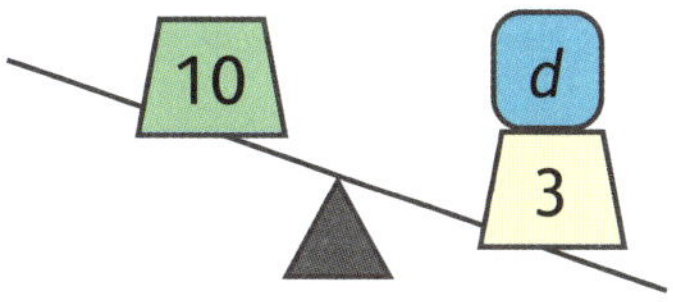

3 Choose the correct symbol to complete each numerical statement.

a 24 ☐ $\frac{1}{2}$ of 50 **b** 9×6 ☐ 8×7

c 25×3 ☐ 15×5 **d** $8 + 8 + 8 + 9$ ☐ 8×4

4 Substitute the values shown. Choose the correct symbol to complete each algebraic statement.

a $4x$ ◯ 14, when $x = 3$

b $5x$ ◯ 15, when $x = \frac{1}{2}$

c $100 - 5x$ ◯ $20 + y$, when $x = 4$ and $y = 5$

d $100 + 3x$ ◯ $100 + 4y$, when $x = 4$ and $y = 3$

Patterns and rules

Key maths idea

Variables and algebra can be used to explore and generate **patterns**.

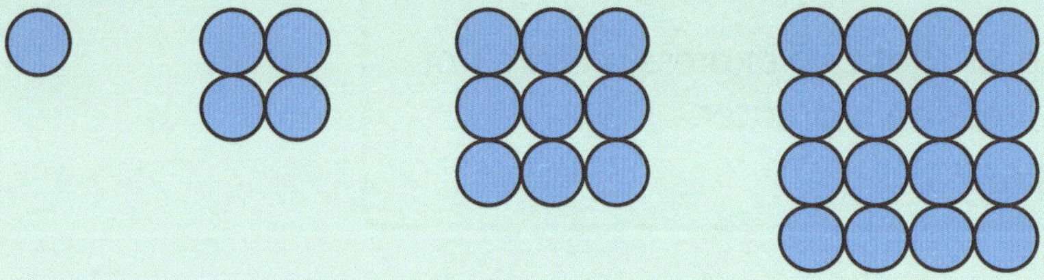

Here are the first four patterns in a sequence.

Key words

pattern
square number
relationship

Maths detective: Explore more patterns

Use algebra to explore the patterns and find the solutions to these challenges.

Case 1

Look at the four patterns in the dot sequence above.

This table of values shows the number of circles required for each pattern.

Pattern number	**1**	**2**	**3**	**4**	**5**	**10**	***n***
Circles	1	4	9				

Write a formula using words or algebra for the number of circles required to make the *n*th pattern.

Case 2

Examine this version of the **square number** pattern.

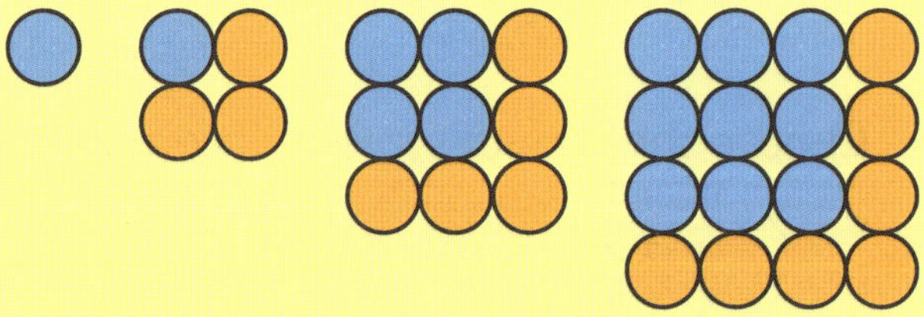

Explain how the 4th pattern would be related to the 5th pattern.

Explain how the 10th pattern would be transformed into the 11th pattern.

Explain how the *n*th pattern would be transformed into the next pattern.

Case 3

Make or draw the next stage in this pattern.

Case 4

Complete the table of values to show the number of blocks used for each pattern.

Height	1	2	3	4	5	9	10
Blue blocks used	1	3					
Yellow blocks used	0						
Total blocks	1						

Write a clear explanation of the **relationship** between the number of yellow and the number of blue tiles in each pattern.

Case 5

1 What would the pattern look like that used 144 blocks in total?

2 How many blocks would you use for the 15th stage of this pattern?

3 How many blocks would you use for the 99th stage of this pattern?

4 How many blocks would you use for the pattern that was *n* blocks high?

5 What would be the tallest pattern you could make with 1 000 blocks?

Algebra

Real-world maths: Working out prices, sizes and lengths

You can use algebra to solve word problems.
Choose a variable to represent the unknown value and solve the equations.

1 Mrs Gordon charges \$3 for a glass of lemonade and spends \$10 on ingredients. After costs, she made \$29.
How many glasses did she sell?
Use x to represent the number of glasses sold:
She makes $3x - 10$ dollars.

$3x - 10 = 29$

Solve the equation to find out the number of glasses she sold.

2 Mr Gordon builds a fence. It costs \$25 for each panel and \$75 for a gate.
He makes a fence with some panels and 1 gate.
In total it costs \$500. How many panels does he use?
Write n for the number of panels.

$25n + 75 =$ ____________

3 A garden is in the shape of a rectangle. Its length is 10 m greater than its width. The perimeter is 70 m. What is the length and width? Use w to stand for the width.

Write an equation for the total perimeter, then solve the equation to calculate the width.

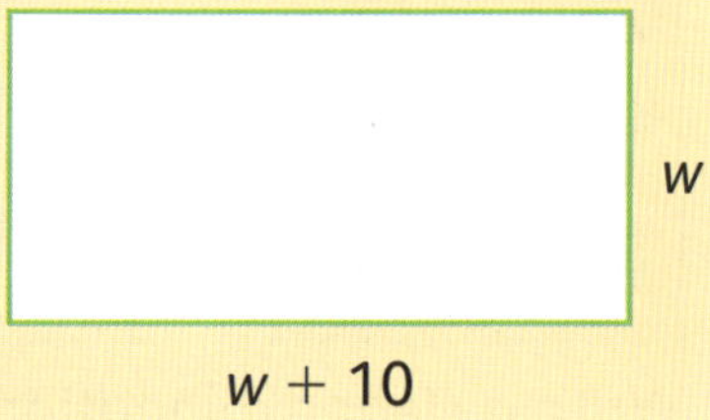

What I have learned

Match each word or symbol to the correct definition.

formula | ≠ | solve | substitute

Definitions
A – a symbol used to link two equations in an inequality
B – an equation used to link variables
C – to use particular values in the place of variables in an expression
D – to find the specific value of an unknown variable in an equation

Practice questions

1 What whole number values of n make each inequality true?
a $2n < 17$ **b** $100 - 10n > 31$ **c** $5n < 20 - 2n$

2 Use x to represent the starting number. Write and solve equations to calculate the final number.

a

I am thinking of a number. I multiply it by 7 then subtract 11. I finish on 66.

Start: x
Multiply by 7: ______
Subtract 11: ______
Finish: ______ = 66

b

I am thinking of a number. I add 50 then subtract 48. I finish on 52.

3 The formula for the perimeter (P) of a rectangle is $2w + 2h = P$. If P = 20 m, what different values could there be for w and h?

4 Use the formula to complete the table of values for different values of the variable t.

Formula: $s = 5t + \frac{1}{4}$

Value of *t*	$\frac{1}{2}$	1	2	10	20
Value of *s*					

24 Self check

Check you can do this:	If you have difficulty, try this:
1 Substitute values into formulae.	Look back at page 313 and remind yourself how to replace the variable (letter) with the known value (number or quantity) in order to work out the value of the expression.
2 Use inequality symbols correctly.	Go back to pages 314–315 and remind yourself how to use the $<$, $>$ and $\neq$ symbols.
3 Interpret and solve word problems using algebraic expressions and equations.	Look back at the problems you solved on page 318 and remind yourself how to make algebraic sentences to help you solve a story problem.
4 Deduce general rules from patterns.	Look back at page 316 and remind yourself how to look at a pattern and work out the rule.

Unit 4: Statistics and probability

Chapter 25 Averages

In this chapter you will explore this focus question:

To answer this question, you will:

- estimate and calculate the mean of a set of data
- find the modal value of a set of data
- find the median value of a set of data
- find the range of a set of data.

Starting point

The children have built towers using all the blocks they have. Discuss how you could complete these sentences about the towers:

1 Most of the towers are about ________ tall.

2 The towers are between ________ and ________ tall.

3 The total number of blocks was ________ .

4 To make all the blocks equal in height, they would need to ________ .

Line plots, mode and range

Key maths idea

This **line plot** shows the data for the number of pets from a questionnaire given to 13 people.

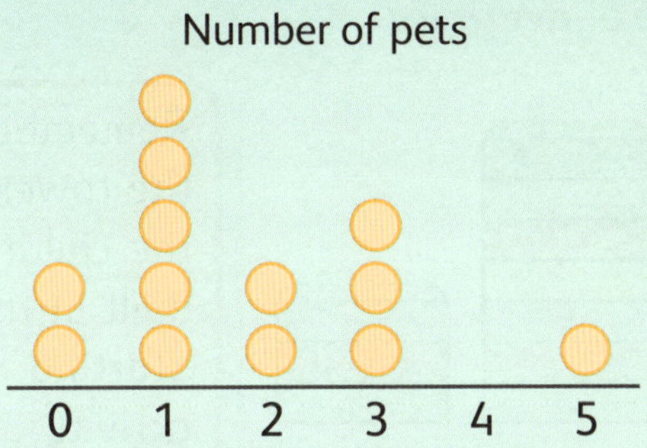

2 people have 0 pets.

5 people have 1 pet.

2 people have 2 pets.

The **range** is the difference between the lowest and highest number. The lowest number of pets is 0. The highest is 5. 5 – 0 = 5, so the range is equal to 5.

The **mode** means the most common response.

The line plot shows clearly that 1 is the mode for the number of pets in this group.

Key words

line plot

range

mode

1 Collect information for the number of pets owned by people in your class.

Plot the results as a line plot.

What is the range?

What is the mode?

2 This line plot shows the results of the question: How many people live in your house?

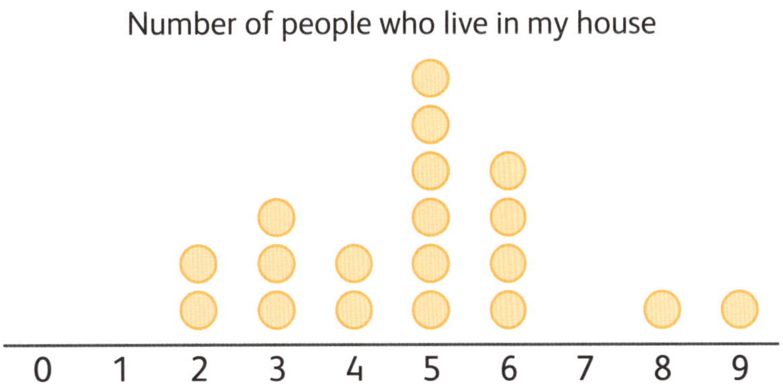

a How many people were surveyed?

b What is the mode response?

c What is the range?

d Collect the same information for the students in your class. Display the results as a line plot and compare the mode and range of this set of data with the set of data given in Exercise 2.

3 Collect data about the number of students in each class of your school.

a Predict the range and mode for your set of data.

b Present the data as a line plot.

c Now compare the data to your predictions.

Mean

Key maths idea

The **mean** is a measure of **average**.

Remember the towers the children built in the Starting Point activity.

There were 4 towers. What is the mean average height?
First, add all the items:
4 + 2 + 7 + 3 = 16

Then share the total between the number of towers.
There are 4 towers, so:
16 ÷ 4 = 4

Calculating the mean is a way to find the average.

The average height is 4 blocks.

Here are the test scores for 5 students: 20, 24, 15, 16, 25.
The mean is calculated in two steps:
Step 1: Add all the items 20 + 24 + 15 + 16 + 25 = 100
Step 2: Divide by the number of items 100 ÷ 5 = 20
The mean score is 20.
You can **estimate** the mean of a set by trying to **predict** the number that would be an appropriate average.

Key words

mean
average
estimate
predict

1 This is the data for the distance that 10 different students travel to school.

3.5 km	2.1 km	1.8 km	4 km	6.2 km
8 km	0.4 km	0.1 km	2.9 km	2.9 km

- **a** What is the shortest distance?
- **b** What is the greatest distance?
- **c** What is the range?
- **d** What is the mode of this set of data?
- **e** Now think about the mean. Look at the data. Predict the mean average for this set of data.
- **f** Now calculate the mean using the two-step process.

Step 1: Add all the distances together.

Step 2: Divide by the number of items.

2 Here are the scores from tests in class A and class B.

Class A: 12, 19, 15, 13, 21, 16

Class B: 18, 10, 22, 11, 14

- **a** Which class had the highest score?
- **b** Which class had the lowest score?
- **c** What is the range of scores for each class?
- **d** Now think about the average score in each class:
  Predict the class that scored the highest mean.
- **e** Now calculate the mean for each class.

3 Collect some data for a sports event in your class. You could measure:

- the distance of a thrown beanbag
- the time taken to run 400 m
- the number of catches between two people in 1 minute.

- **a** Collect the data.
- **b** Predict the mean score.
- **c** Calculate:
  - range
  - mode
  - mean.

Median

Key maths idea

The **median** is a different way to measure a kind of average.

To find the median, arrange all of the data items in order. It does not matter whether you arrange them in **ascending order** (from smallest to greatest) or **descending order** (from greatest to smallest). The middle value is the median. If there is an even number of data values, add together the two middle values and divide by two.

For example, here are the ages of 5 people from a household:

Step 1: Arrange the items in order:

4 12 34 36 63

Step 2: Find the middle value:

4 12 34 36 63

The median age in this household is 34 years.

Here is an example of a data set with an even number of values. The median of this set is the value halfway between the two middle items:

3 7 8 9 11 21

So the median is 8.5.

Key words
median
ascending order
descending order

1 Here are data for the sizes of 20 pairs of shoes sold.

Kind of running shoes	Sizes sold
Men's	40, 47, 40, 46, 43, 40, 40, 44, 40, 38
Women's	35, 39, 30, 33, 37, 35, 36, 35, 34, 35

Calculate the mean and median for the men's shoes.

Calculate the range and mode for the men's shoes.

2 Now predict the values for the mode, range, median and mean of the women's shoe sizes. Calculate to check your predictions.

3 a Collect five different pens or pencils. Measure the length of each item.
Now calculate the mean and median length.

b Collect one more pen or pencil. Now measure the length and recalculate the mean and median for your six items.

c How have the mean and median changed?

Key words
pulse
heartbeat

4 Your teacher will show you how to find your **pulse**. Your pulse is your **heartbeat**. You can feel it at your wrist or your neck.

a Count your heartbeat for a few seconds. Estimate how many times you think it beats each minute.

b Now use a stopwatch, or your teacher can time you, for 15 seconds. Count the heartbeats for 15 seconds. Multiply it by 4 to get your heart rate per minute.

c Everyone in the class writes their heart rate on the board.

d Work out the mean, median and mode, as well as the range of your class's heartbeats.

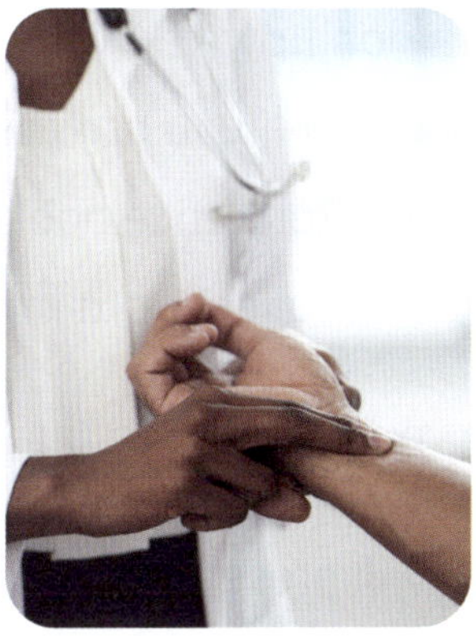

Maths detective: Find the missing values

Use the clues to calculate the missing data values for each set.

Case 1

3	14	8	10	15	?

The median of this set is 11.

Case 2

2	2	2	3	5	?	20

The mean is 7.

Case 3

The range is 9.
The mean is 10.
The mode is 11.
The median is 9.5.

Statistics and probability

Real-world maths: Scientific data

Scientific data uses averages to make sense of experiments and observations. Investigate these questions:

1 Choose an animal native to Jamaica. Investigate the average lifespan, weight, size and number of offspring of the creature.

2 Explore data about environmental issues that affect Jamaica:

How has the mean global temperature changed over the last 20 years?

Plan how you will investigate the topic you choose:

- Will you use the internet?
- Reference books from the library?
- Collect data from different sources?

Present your findings to your class on a poster or in an information booklet. Think about how you will present your data. Will you use tables, graphs or other diagrams?

What I have learned

1 Write a definition and example for each term.

a range b mode c median d mean

2 Here is a set of data items.

5 kg 20 kg 9 kg 10 kg 6 kg 10 kg

Show very clearly how to find the range, mode, mean and median of this set of data. Write step-by-step instructions for each different measure.

Practice questions

1 Identify the range, mode and median of each range of values.

a 10 kg, 12 kg, 25 kg, 9 kg, 10 kg, 13 kg, 10 kg

b 20 minutes, 22 minutes, 19 minutes, 18 $\frac{1}{2}$ minutes, 29 minutes, 20 minutes, 16 minutes

c 6 months, 8 months, 11 months, 4 months, 2 months.

2 Work out the mean (average) of each set of values from Exercise 1.

3 Here are the results from a long-jump practice session.

Practice session	Best jump (metres)
1	2.76
2	2.86
3	2.80
4	2.93
5	2.95

a Calculate the range, median and mean.

b There is one more jump of 291 cm. Predict how this will affect the mean and median.

c Calculate the new median and mean including this data item.

4 Here are some data from a science experiment collecting weather reports for two weeks.

Week 1

	Rainfall (mm)	Hours of sunshine	Maximum temperature (°C)	Wind speed (km/h)
Sun	2	11	18	15
Mon	8	7	22	8
Tues	11	2	14	6
Wed	5	15	21	19
Thurs	9	8	20	21
Fri	0	15	25	20
Sat	5	10	19	32

Week 2

	Rainfall (mm)	Hours of sunshine	Maximum temperature (°C)	Wind speed (km/h)
Sun	0	16	24	30
Mon	0	16.5	26	12
Tues	0	16	24	2
Wed	3	16.5	25	1
Thurs	0	17	25	3
Fri	0	16	26	2
Sat	0	16.5	25	5

a Calculate the range, mode, mean and median for each week.

b Use this data set to compare the conditions for each week.

25 Self check

	Check you can do this:	If you have difficulty, try this:
1	Find the range of a set of data.	Go back to page 321 and make sure you recall how to identify the range of a set of data values.
2	Find the modal value of a set of data.	The 'modal value' is the same as the mode. Go back to page 321 and make sure you recall how to identify the mode.
3	Estimate and calculate the mean of a set of data.	Revise page 322. Remember, to find the mean, you add together all the values and divide by the number of addends.
4	Calculate the median value of a set of data.	Look back at page 324. Remember, median means 'middle value'. Remind yourself how to find the median of both odd and even numbers of data values.
5	Explain what the statistical averages mean that we use to describe a set of data.	Look back at the work you did on each of the types of average (range, mean, median, mode) and discuss when it is useful to use each of these types of averages.

Chapter 26 Probability

In this chapter you will explore this focus question:

To answer this question, you will:

- list the possible outcomes of a simple event
- determine the probability of given outcomes
- carry out probability experiments and draw conclusions about them.

Starting point

Discuss these kinds of statements. What do you understand by the words 'probably', 'certain' and 'impossible'?

1 Roll a die.

Do you agree with this statement?

Roll one die a few times. Is a 6 a difficult score to roll?

2 Roll two dice.

Add the numbers on the dice to give you a score.

3 Try to predict the score before you roll. Do you notice any totals more often than others? What is the lowest score you rolled?

Language of probability

Key maths idea

An **event** is something that may happen. We can describe events using the language of **probability** to tell how **likely** they are.

- An **impossible** event can **never** happen.
- A **certain** event will **always** happen.
- A **possible** event may be likely or **unlikely** but not impossible.

You can arrange these words on a **probability scale**:

impossible	unlikely	even chance	likely	certain
0		$\frac{1}{2}$		1

Where would you place these events on the probability scale?

- It will rain tomorrow.
- The 2-D shape you are thinking of has more than two sides.
- I will guess the exact number you are thinking of.
- You will fly to the moon on a green table.
- The number you are thinking of is odd or even.

How do you understand the numbers under the probability scale? Explain why we can express any event as having a **chance** between 0 and 1.

Key words

event
probability
likely
impossible
never
certain
always
possible
unlikely
probability scale
chance
uncertain

1 Draw a probability scale from impossible to certain. Show where you would place each of these events on the scale.

2 Add three more events to different places on your probability scale.

3 Complete each sentence with a different event:

a It is impossible that …

b It is **very likely** that …

c It is **uncertain** whether …

d There is an even chance that …

Possible outcomes

Key maths idea

If you roll one die, there are six possible **outcomes**.

You could score

1	2	3	4	5	6

Each outcome is equally likely.

The chance of scoring a 6 is exactly the same as the chance of scoring a 4.

There are 6 balls in this bag, but there are only two possible outcomes: red or yellow. Both of the possible outcomes have an **even chance**. That means there are two possible outcomes that are **equally likely**.

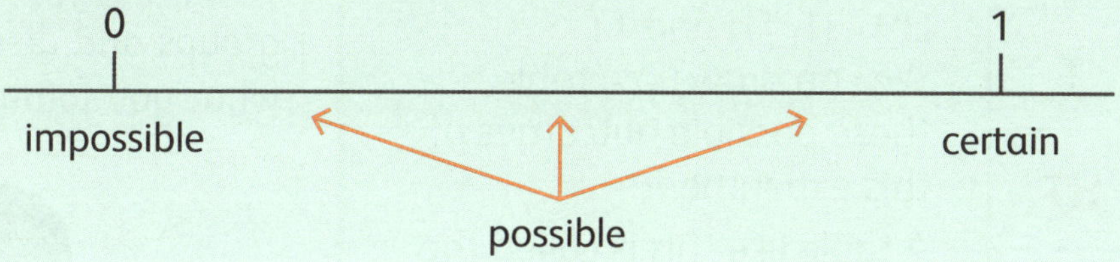

Key words

outcome

even chance

equally likely

1 There are 4 possible outcomes from spinning this spinner

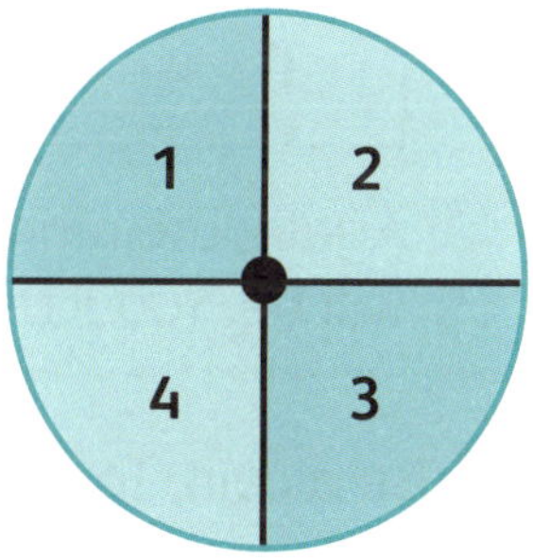

Describe the probability of spinning:

a a number greater than 0

b a number less than 2

c an odd number

d a score of 6.

2 There are 9 possible outcomes from spinning this spinner.

Use the language of probability to describe the chances of:

a Spinning an odd number

b Spinning a number greater than 2

c Spinning a number greater than 9

d Spinning a number less than 9

e Spinning a number that is not on a yellow section.

3 List all the possible outcomes from spinning this spinner. Describe the probability of each outcome.

Statistics and probability

Probability experiments

Key maths idea

You can conduct probability experiments to make predictions.

The more trials you conduct, the more accurate your predictions will be.

Examples

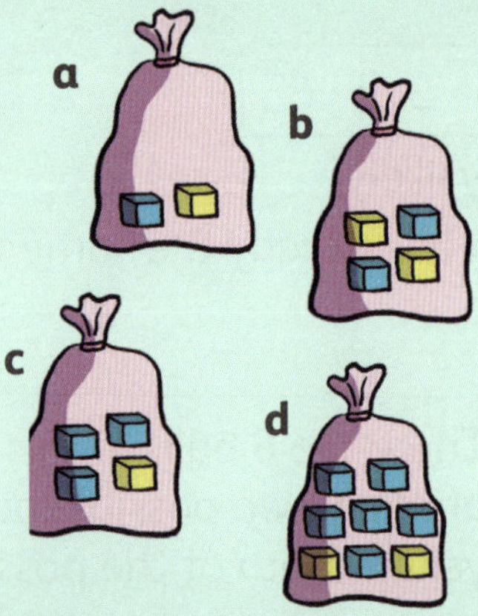

1 You can use probability to predict the contents of each bag without looking inside. If you used bag A, and completed 10 trials, you would expect the number of yellow and blue picks to be closely matched.

2 You can also predict what happens when you flip two coins simultaneously.
If H = heads and T = tails, there are four possible ways the coins can land:

	H	T
H	HH	HT
T	TH	TT

HH, HT, TH and TT.
We can draw up a table of these possible outcomes of this experiment:
A table like this is called the **sample space**. It shows all the possible outcomes.

Again, the more flips, the more accurate your predictions will be. You'll try this in the next activity.

Key word
sample space

Talk about

What would happen if you completed 10 trials with bag B? And bag C and D? Discuss this as a class. Assign a trial to a few different groups and discuss what you found.

1 Use a coin. Toss the coin 10 times and record whether it lands on heads or tails.

	Tally
Heads	
Tails	

a Predict the number of heads and tails you will get from 25 flips.

b Now complete 25 flips and compare the results with your predictions.

c Predict the number of heads and tails you will get from 50 flips.

d Now complete 50 flips in total and compare the results with your predictions.

2 Look at the outcomes from 8 trials of this probability experiment. Do you agree with Ayla and Adiel's interpretations?

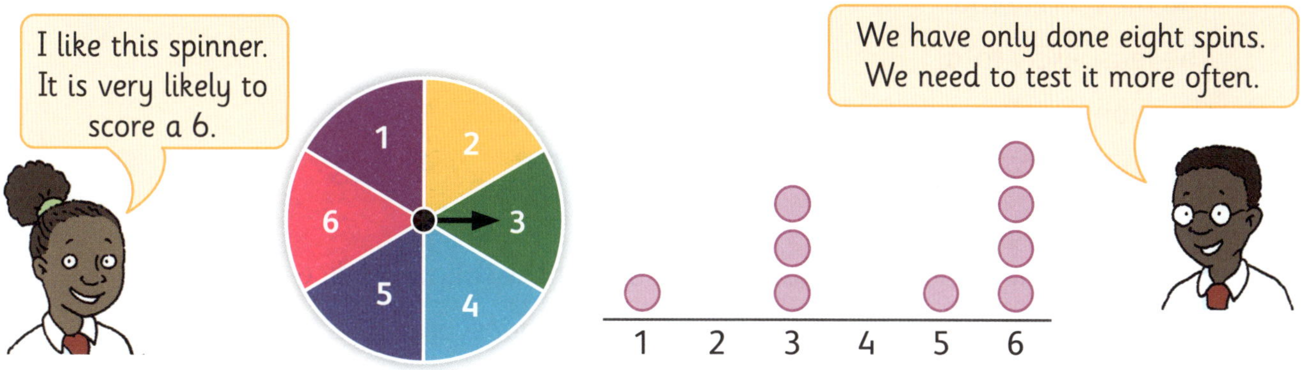

a Use a spinner with six sections. Record your results using a line plot.

b Predict what your results will be and discuss the probability of each outcome.

3 Predict the results of picking a ball at random and then replacing it from each bag below. Make a prediction for the results after:

a 1 trial **b** 2 trials **c** 10 trials **d** 1 000 trials.

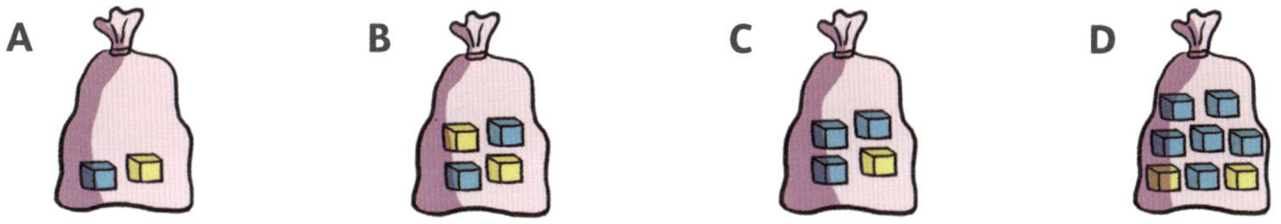

Maths detective: Dice games

Use the clues to predict the outcomes of these experiments.

Case 1

Roll two dice. Add your scores.

What are all the possible outcomes?

Complete a table like this to collect all the possible outcomes.

		Green die score					
		1	2	3	4	5	6
Red die score	1	2	3				
	2						
	3			6			
	4						
	5						
	6		8				

Use your table to predict:

1 The most likely total.

2 The least likely total.

3 Describe the chances of rolling a total of 6 or more.

Case 2

Now create a table of possible outcomes for one of the following dice games:

1 Roll two dice. Multiply the two numbers to give your score.

2 Roll two dice. Find the difference between the two numbers to give your score.

Case 3

Invent your own game using this spinner.
Spin twice and add or multiply your scores.
What are all the possible outcomes?
Describe the probability of each outcome.

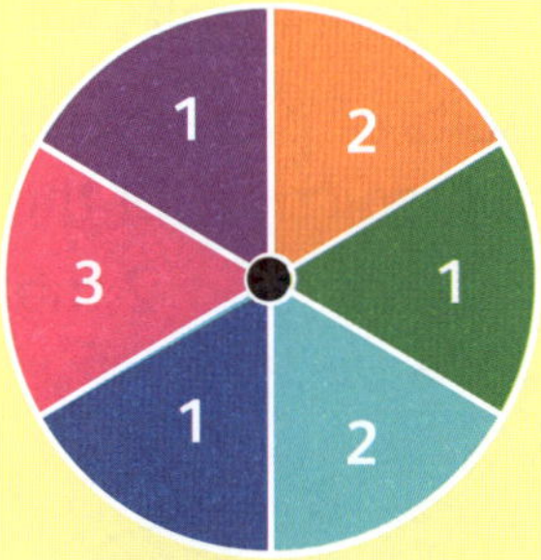

Real-world maths: Spin and win

1 Probability is used in gambling and on gameshows.

Look at this spinning wheel. What are the possible outcomes?

Describe the probability of winning a big prize.

2 Investigate other situations which use probability for gambling.

Write a short explanation of how gambling can have negative outcomes for people.

What I have learned

Use the language of probability to describe the chances of picking a yellow cube from each bag.

Practice questions

1 Sam uses one of the bags for 10 experiments.

The results are

Blue	~~||||~~ |||
Green	||

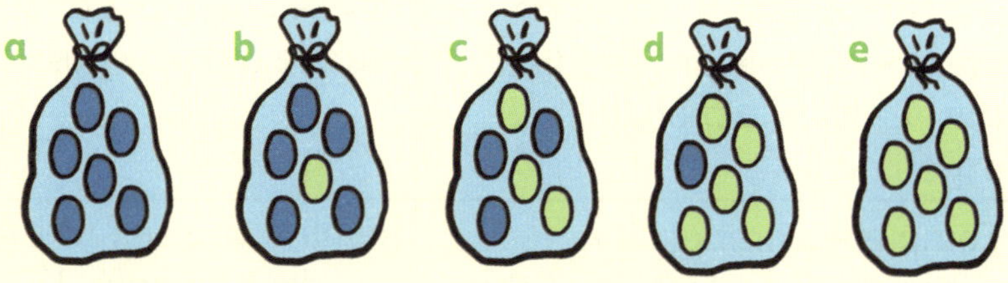

Which bag was she using? Can you be certain? Justify your answer.

2 Here is an 'I am thinking of a number' game.

1	2	3	4	5
6	7	8	9	10
11	12	13	14	15
16	17	18	19	20
21	22	23	24	25

a Read each guess. Sort them in order, from least chance to best chance of being correct.

b Write your own guesses. Try to think of a guess that is:

- impossible
- certain
- even chance.

26 Self check

	Check you can do this: 🙂	If you have difficulty, try this: 😐
1	Use the language of probability to describe the chance of events.	Go back to page 330 and remind yourself of the terms on the probability scale.
2	Find all the possible outcomes of an experiment.	Go back to page 331 and remind yourself how to list the possible results of a given experiment.
3	List the sample space of an event in a table and predict the results of a probability experiment.	Revise page 332 and remind yourself how to use a table to show possible outcomes.
4	Complete a probability experiment, and interpret the results.	Remind yourself of the coin flipping experiments you did on pages 332–333.
5	Express the probability of an event as 0, 1 or between 0 and 1.	Look back at the probability scale on page 331 and express probabilities using numbers between 0 and 1.
6	Make an informed decision about whether an event is probable.	Look back at the work you did on pages 333–334. How did you predict and decide the probability of each outcome?

Glossary and Index

D

fractional number: any number that has parts that are a fraction of a whole (proper fractions, improper fractions, mixed numbers, decimals)........................44, 313

fraction: a number usually expressed in the form $\frac{a}{b}$; a number that results from dividing one whole number by another 44

frequency: the rate of repetition or occurrence.. 115

friendly society: an organisation formed voluntarily by individuals, who pay regularly small amounts of money in return for old-age pension or life and health insurance 274

G

gallon: a unit of measurement of volume, especially for liquids, equal to about 4.5 litres 82

general rule: the rule that allows us to deduce the value of any number in a pattern..223, 313

geometric shape: a closed figure formed by joining lines together in a particular way........ 89

gram: a unit of mass equal to 0.001 kilograms 80

greater than or equal to: the symbol ≥, used to show that one quantity is bigger than or equal to the other quantity 30

greater than: larger; bigger in amount 29, 225

group(ing): collecting elements according to their specific characteristics..........................13

H

heartbeat: a single pulsation of the heart..325

hectare: a square unit used to measure large areas, equal to 10 000 m^2......................184

height: the distance upwards, from base to top 75

hexagon: a polygon of six sides......................289

highest common factor (HCF): the largest factor that divides two or more numbers..46, 159

honesty: truthfulness, sincerity 274

horizontal (lines): parallel to the ground or to the bottom or top edge of something.....91, 215

hundredth: one of a hundred equal parts of something.. 33

I

imperial: a system of measurement that uses units such as inches and feet for length and pints and gallons for volume......... 75

impossible: not able to happen 330

improper fraction: a fraction that has a numerator that is greater or equal to the denominator 162

inch: unit used for measuring length, equal to about 2.5 centimetres.......................... 75

increase: to become or make larger 32, 305

inequality: a mathematical statement that shows that two quantities are unequal..........314

infinite set: a set with an endless number of elements 10

insurance company: a company whose business is creating, providing and selling insurance products 274

interest: money paid on a loan from a financial institution or money earned from keeping money in a bank or other financial organisation 273

intermediate colour: colour created by mixing a primary with a secondary colour....285

intersect: to cut or divide by going through 92

intersecting (lines): lines that cross each other at a common point........................... 92

intersection: area at which two or more things intersect 14

inverse: opposite in relation to something else..250

invest: to put up money in order to make a financial return 273

irregular polygon: a polygon that does not have all its sides and angles equal...........178

irregular: not even in shape or form..............205

isosceles: a triangle with two sides of equal length...196

J

jointed sets: sets that share common elements...14

K

kilo-: one thousand.. 73

kilogram: a unit of mass equal to 1 000 grams..80

O

oblique (lines): lines that run at a slant or an angle 90

obtuse angle: an angle greater than 90° but less than 180° 98

obtuse-angled: a triangle that contains one obtuse angle 196

octagon: a flat shape with nine straight sides 289

odd (numbers): the numbers that cannot be divided exactly by 2 and leave a remainder 10, 45

opinion poll: the process of asking people the same questions on a particular subject to find out what most of them think 110

opposite: across from 198

order of operations: the sequence in which we solve an algebraic expression with multiple operations 231

ordering: the arrangement or sequence of things 141, 161

orientation: a geometric notation which says about the position of the object based on the specific direction or with relation to the points of the compass 215

outcome: one of the possible results in a probability experiment 331

P

parallel (lines): lines that stay the same distance apart and never cross 92, 291

parallelogram: a flat shape that has four sides – the two sets of opposite sides are parallel and of equal length to each other 198

partition: division into smaller parts 27, 56

partner scheme: a business strategy that connects companies to multiple types of partners 274

pattern: any regularly repeated arrangement 19, 49, 94, 223, 246, 264, 316

pentagon: a shape with five sides and five angles 289

percentage: an amount of something, expressed as a rate per hundred 273

perimeter: the distance around a shape or the length of its boundary 77, 176

period: a specific length of time 62

perpendicular (lines): lines that meet at a right angle (90°) 92, 291

pictograph: the pictorial representation of data using images, icons or symbols 114

pie chart: a way of showing information about how a total amount is divided up, consisting of a circle that is divided from its centre into several parts 115

pint: a unit of measurement of volume, especially for liquids, equal to about half a litre 82

place value: the value of a number decided by its position; for example, whether it is a one, a ten, a hundred, etc. 19, 25

plane of symmetry: an imagined flat surface that cuts a shape into identical parts 210

plane symmetry: a solid has plane symmetry if you can slice it into identical parts along a plane 210

point (dot): a position in space 90

polygon: a closed path made from line segments that meet at vertices (corners) 77, 176, 205, 288

possible: that might or might not happen 330

predict: to estimate that something will happen in the future, especially as a result of knowledge, observation, experience or scientific reason 322

prediction: a statement about what you think will happen in the future 223

primary colour: any of the three spectral colours (red, green and blue) that can be mixed together to make any other colour 285

prime number: any number that can only be divided exactly by itself or by 1 without any remainders 11, 50

principal: the original amount of money borrowed in a loan 273

probability: a possibility or chance of anything happening; likelihood 330

probability scale: a way to express how likely it is that an event will happen by using numbers 330

problem-solving approach: the process of understanding a problem and finding the optimal solution 266

product: the result you get when two or more numbers are multiplied together 149, 244

proper fraction: a fraction in which the numerator is less than the denominator 162

properties: the basic or special qualities of something 196

protractor: a tool for measuring and drawing angles 97

pulse: the regular beating of the arteries, caused by the movement of the blood through the body, that can be felt at the wrist or neck 325

pyramid: a solid shape with a square base and four triangular faces 299

Q

quadrilateral: a flat shape that has four sides and four inside angles 198, 288

quantity: the total amount or number of something 307

questionnaire: a list of questions that several people are asked so that information can be collected about something 110

quotient: the result you get when you divide one number by another 244

R

radius: a straight line extending from the centre of a circle to the circumference 280

random: occurring or done by chance 108

range: the difference between the lowest and highest value 321

rectangle: a flat shape with four right angles and two pairs of sides that are equal in length 176

reflection: when the source image is reflected in a mirror line; flip 217

reflex angle: an angle greater than 180° and less than 360° 98

regular polygon: a polygon that has all its sides equal in length and all its inside angles equal in size 180, 205

relationship: the way in which two things are connected 73, 317

remainder: the amount that is left when one number cannot be exactly divided by another 46

representative: typical of, or the same as others in a larger group of people or things 108

research: the detailed study of a particular subject 107

rhombus: a flat shape that has four equal sides 198

right angle: a 90° turn; quarter turn 92, 98

right-angled (triangle): a triangle that contains one angle of 90° 196

rotation: turn 217

rounding: a mathematical technique of adjusting the number's digits to make the number easier to use during calculations 142, 162

S

sampler space: describes the total number of possible outcomes for any event 332

sampling: the way a researcher chooses the sample group 107

savings account: a bank account on which interest is usually paid 273

savings: the money saved and put in a bank or other safe place 268

scalene: a triangle with three sides of different length 196

schedule: an ordered list of tasks to be done indicating the times and dates when they are planned to occur 58

sector: the area enclosed between two radii and an arc 282

segment: section, part cut off from a geometric figure 282

septagon/heptagon: a polygon of seven sides 289

sequence: the arrangement of two or more elements in a successive order 48

set: a group of objects with same characteristics 8

similar (shapes): shapes that have the same number of sides and angles, but their sizes are different 215

simplest form: the smallest possible equivalent fraction of the number 159

simplify: to make simpler 159

slanted: inclined at an oblique or sloping direction 90

slide: to shift, change position 215

solve: to find the solution, explanation or answer to a problem 307

spectrum: the set of colours produced when a beam of light strikes a prism 285

spreadsheet: a software for the entry, organisation and manipulation of data 122

square centimetre (cm^2): an area equal to a square that is 1 centimetre on each side 182